From Computer to Brain

Springer
New York
Berlin
Heidelberg
Hong Kong
London
Milan
Paris
Tokyo

William W. Lytton

From Computer to Brain

Foundations of Computational Neuroscience

With 88 Illustrations

William W. Lytton, M.D.
Associate Professor, State University of New York, Downstate, Brooklyn, NY
Visiting Associate Professor, University of Wisconsin, Madison
Visiting Associate Professor, Polytechnic University, Brooklyn, NY
Staff Neurologist, Kings County Hospital, Brooklyn, NY

SUNY
450 Clarkson Ave., Box 31
Brooklyn, NY 11203
USA

Cover illustration: Roy Wiemann, (2002).

Library of Congress Cataloging-in-Publication Data
Lytton, William W.
From computer to brain : foundations of computational neuroscience / William W. Lytton.
p. cm.
Includes bibliographical references and index.
ISBN 0-387-95528-3 (alk. paper) — ISBN 0-387-95526-7 (pbk. : alk. paper)
1. Computational neuroscience. I. Title.
QP357.5 .L98 2002
006.3—dc21 2002070819

ISBN 0-387-95528-3 (Hard cover)
ISBN 0-387-95526-7 (Soft cover)

Printed on acid-free paper.

Printed in the United States of America.

9 8 7 6 5 4 3 2 1 SPIN 10883094 (Hard cover)
SPIN 10883036 (Soft cover)

Typesetting: Pages created by the author using a Springer LaTeX macro package.

www.springer-ny.com

Springer-Verlag New York Berlin Heidelberg
A member of BertelsmannSpringer Science+Business Media GmbH

for Jeeyune and Barry

Foreword

In *From Computer to Brain: Foundations of Computational Neuroscience*, William Lytton provides a gentle but rigorous introduction to the art of modeling neurons and neural systems. It is an accessible entry to the methods and approaches used to model the brain at many different levels, ranging from synapses and dendrites to neurons and neural circuits. Different types of questions are asked at each level that require different types of models.

Why learn this?

One of the reasons why someone might want to learn about computational neuroscience is to better predict the outcomes of experiments. The process of designing an experiment to test a hypothesis involves making predictions about what the possible outcomes of the experiment might be and to work out the implications of each possible result. This is a difficult task in most biological systems, especially ones like the brain that involve many interacting parts, some of which are not even known. A model may reveal assumptions about the system that were not fully appreciated.

One of the earliest and most successful models is the Hodgkin-Huxley model of the action potential (Chap. 12). For their classic papers on the giant squid axon, they integrated the differential equations on a hand-powered mechanical calculator. Computers today are millions of times faster than those in the 1950s and it is now possible to simulate cortical neurons having thousands of dendritic compartments and dozens of different types of ions channels, and networks with thousands of interacting neurons. The complex dynamics of these networks is exceptionally difficult to predict without computational tools from computer science, and mathematical tools from dynamical systems theory.

But there is another reason to delve into this book. Computational neuroscience also provides a framework for thinking about how brain mechanisms give rise to behavior. The links between facts and consequences are often subtle. We are told, for example, that lateral inhibition enhances contrast discontinuities, but without a quantitative model, such as that for the Limulus compound eye (Chap. 8), it is not at all obvious how this occurs, especially when the signals are varying in time as well as space. The jump from mechanism to behavior becomes even more difficult to understand for the cellular basis of learning and memory, where the memory of a single item can be distributed over millions of synapses in distant parts of the brain.

Why read this book?

Computational neuroscience is such a young field that ten years ago there were no good books for someone who was getting started. That has now changed and there are now several excellent textbooks available, but most of them focus on one type of model, such as Hodgkin-Huxley style models or abstract neural networks, or presume a high level of mathematical sophistication. This book gives a balanced view of the wide range of modeling techniques that are available, in a way that is accessible to a wide audience.

Another reason for reading this book is to enjoy the playfulness that the author brings to a subject that can be dry and technical. His imaginative use of examples brings mathematical ideas to life. This is a book that will bring a smile to your face as well as inspire your imagination.

Terrence J. Sejnowski
Howard Hughes Medical Institute
Salk Institute for Biological Studies
University of California at San Diego

Preface

As a college student, I got interested in the big questions of brain science: the origin of language and thought, the nature of memory, the integration of sensation and action, the source of consciousness, and the mind-body problem. Although I worked on the mind-body problem all through sophomore year, I didn't arrive at a solution. After studying some more psychology, some biology and some physics, I took a wonderful computer science course. I decided that the way to understand the mind's brain was to analyze it in detail as a computational device.

When I graduated from college and looked at graduate programs. I explained to interviewers that I wanted to apply computers to brain science. No one I spoke to in biomedicine or in computer science knew of anyone doing such things. In retrospect, such people were around, mostly located in departments of mathematics and physics. Because I couldn't find what I was looking for, I put off my research training for almost a decade. When I came back, a new field had emerged and was gaining recognition. Shortly afterwards, this field, whose various branches and tributaries had gone by a variety of names, was dubbed *Computational Neuroscience.*

As I got involved in research, I began to appreciate that massive theoretical and informational frameworks had to be, and were being, built in order to approach these grandly challenging problems that interested me. So, like many of my colleagues, I became a bricklayer, laboriously adding brick upon brick to the wall, without always looking up or down. Later, given the opportunity to teach an undergraduate course, I was happy to have a chance to reflect more expansively on the problems, techniques, aspirations and goals of my field. I found that the students asked basic questions that required me to peak out from under the residue of brick-dust and think about the architecture.

When I started to teach my course, I reviewed the various textbooks in computational neuroscience, most of them brand new and many excellent. I realized that many of these new texts were committed to one particular angle or theoretical bias, and did not reveal the broad scope and wide interplay of ideas that I found so exciting. Additionally, they generally required too much math for most of the students in the class.

In fact, most students came into the course either without strong math background, or without strong biology background, or without either. At first, I was concerned that I would be unable to stuff in enough "remedial" material to bring everyone up to a point where they could understand what I was talking about. However, I found that I could cover a wide swath of topics by only teaching the little pieces that I needed out of each one. Additionally, for students who were not particularly comfortable with math, I made an effort to explain things a lot: first in metaphor, then in pictures, then directly in words, then in equations, and finally with computer simulations that brought it to life interactively. I have included the former set of approaches directly in this book, and have made available programs to allow a student to directly interact with, and alter, all of the figures, at least those that aren't just pictures of things.

As I wrote this book, I was sort of hoping that by the end of it I might finally figure out the answer to the mind-body problem. Alas, no such luck. But I've tried to convey some of the information that will be required for a solution. The rest is left as an exercise for the reader.

Bill Lytton
East Flatbush, NY
June, 2002

Acknowledgments

I first want to thank the folks at University of Wisconsin who helped me organize, launch, and teach the course on which this book is based: *Zoology 400, Introduction to Computational Neuroscience.* In particular a number of students helped with the development of the course, the lecture notes, and the software: Kevin Hellman, Stephen Cowan, William Annis (Fig. 10.4), Ben Zhao, Winston Chang, Kim Wallace (Fig. 14.3), Adam Briska, Luke Greeley, Sam Sober, and Sam Williams.

Version 1.0 of the software for this book, written in Perl CGI, was supported by the University of Wisconsin Office of Medical Education, Research, and Development. I particularly want to thank Mark Albanese and John Harting for their encouragement, and thank Jeff Hamblin, who wrote the software the first time around.

A number of longtime collaborators engaged me in many helpful discussions that found their way into this book: Dan Uhlrich, Peter Lipton, Josh Chover, Jay Rosenbek, Lew Haberly, Sonny Yamasaki, and Peter Van Kan.

Many thanks to the enablers of Springer-Verlag: Paula Callaghan, Jenny Wolkowicki, and Frank Ganz. They were unfailingly helpful and friendly and gave me plenty of leeway.

The book was written in Latex using Emacs running under Redhat Linux. As a fan of software, I particularly wish to acknowledge Neuron, a wonderful program that was used to produce all of the figures in the book. The software available for this book is written in the Neuron simulator. I thank Mike Hines and Ted Carnevale for developing and supporting Neuron, and responding to my requests for language augmentations.

I need to acknowledge my biological and intellectual patrimony. The hybridization of Dad, a psychoanalyst, and Mom, an applied mathematician, produced a computational neuroscientist. Intellectually, I am most strongly indebted to Terry Sejnowski, who trained me.

I want to thank my wife, Jeeyune, who put up with a lot — my working on this book during our most intimate moments, and my son, Barry, who put up with a fair amount – my working on it while I was supposed to be watching "SpongeBob SquarePants." Oh, and Barry drew Fig. 9.6.

I especially must thank two colleagues who helped out enormously by reading the manuscript in detail and making numerous suggestions that improved the book greatly: Mark Stewart and Ning Qian.

I'd like to place the blame for any remaining errors on someone else, but I can't find a fall guy. So for whatever errors remain, I'm the guy.

Contents

1
Introduction

1.1 For whom is this book?

This book is for undergraduates and beginning graduate students. Of course, other graduate students, postdoctoral students, and anyone concerned with the fate of Western Civilization should read this book as well. Why focus on undergraduates? Because the malleable mind of the undifferentiated stem-person of college age is likely to be capable of making conceptual leaps that more ossified brains cannot. A theme that will recur throughout the book is the difficulty of unifying the compute and the neuro.

There are two major barriers to grand unification. First, computer science (like math, physics, and engineering) is made up of grand unifiers and their all-encompassing schemes, while neuroscience (like the rest of biology) is boatloads of facts. The differentiated mind of the engineer cannot swallow so many facts without a unifying framework. The differentiated mind of the biologist knows too much and is distracted by the many facts that contradict any preliminary framework someone tries to build. The fabled undergraduate mind, however, is notoriously unburdened by facts and yet willing to accommodate them. It (that mind, whether he or she) also typically seeks big pictures and is willing to take the leaps of faith required to acquire them.

The second barrier: these are early days. The field is a newly emerging hybrid and is itself still undifferentiated (an undifferentiated field needs an undifferentiated mind). It is a field still driven more by passion and fashion than by cool reason, emerging like a star from a gaseous cloud (OK that's a bit fanciful). Anyway, some of my colleagues will read this book and say that I've missed the whole point of the field. I've said the same of their books. Specifically, many researchers in the field come from the aforementioned engineering, physics, and math tradition. Their efforts are directed at developing a theoretical framework and their work tends to be colored by the framework they have chosen. I, on the other hand, come from the bio side. As a consequence, this book is a largely atheoretical approach to a theoretical science. I present stuff that is either fun or interesting or important and maybe sometimes all three.

In this computational neuroscience funhouse, I have included biological facts, computer science facts, equations, and theories both true and false. In some cases, I have included false theories because I don't know that they're false, though I may suspect it. In other cases, I introduce unlikely hypotheses just to roll them around and play with them. This interplay of facts and ideas makes up much of the work (or play) of modeling. Through experiencing it, the student will get a better idea of how and why modeling is done. In the last chapter, in particular, I bring the reader to my own particular circle of purgatory — the hall of perpetual mystification. I've tried to illustrate the complexities and contradictions of the field without unduly confusing the reader.

1.2 What is in the book?

Although my target undergraduate students are pluripotential, they are not yet omniscient. Specifically, they are majoring in philosophy, physics, math, engineering, biology, zoology, psychology, physical education, or business administration. As a result they know a lot about some things and little or nothing at all about others. For this reason, I have tried to cover all the basic bases. I have relegated much of this to a final chapter (Chap. 16), which can be read piecemeal as needed. In addition, many of the fundamentals have seeped into the main text as well. I have included a lot of basic computer science since an understanding of computational neuroscience requires a fairly sophisticated working knowledge of computers. This broad, blanket coverage means that certain chapters may seem trivial and others overly demanding for the particular student.

For the nonmathematical reader, the biggest challenge will be the advanced math topics, notably matrix algebra and numerical calculus. Though these topics are hard and can easily fill up a year of classroom instruction, I have tried to extract just the parts needed for present purposes and to make this accessible to anyone with mastery of high-school algebra. I have written out most equations in words so as to make them more accessible to anyone who is allergic to math. Additionally, the computer is a great leveler in this regard. Many once abstruse concepts in mathematics can now be quickly illustrated graphically. Computer in hand, this book can be enjoyed as a lighthearted romp through calculus, electrical engineering, matrix algebra, and other sometimes-intimidating topics.

For the nonbiological reader, the biggest challenge will be the profusion of facts and jargon words. This onslaught of information can be intimidating and discouraging. There is so much to know that it can be hard to know where to start. Often it is impossible to connect one set of facts to another set of facts. That is the goal of computational neuroscience. When you first encounter these facts, it will be without the benefit of such a model.

Although I have tried to write clearly and comprehensibly, I have also tried to use a lot of jargon. This can be annoying. I try to use jargon kindly and responsibly, to help the reader learn the words needed to read and converse knowledgeably in the many subfields that make up computational neuroscience. I have tried to always define jargon words immediately upon use in the text. As further assistance, I've provided a glossary. In addition to introducing jargon words, I also introduce some jargon concepts — touchstone ideas that are frequently referenced by people in a particular field. Having so much to introduce, concepts and phrases are sometimes mentioned, but not followed up on. They are presented to provide the reader with vocabulary and mental reference points for further reading or just plain thinking.

1.3 Do I need a computer for this book?

This book is meant to be read independently of any computer work. I have not put explicit exercises in the book but have made them available online (see below). One of the neat things about computational neuroscience is that it is so readily accessible. It is hard to get hold of the particle accelerators, centrifuges, and chimpanzees needed for most scientific study. But computers are everywhere, making computer-based research accessible to undergraduates and even to nonacademic folks. This is more true of computational neuroscience than it is of other computer modeling fields. To do weather prediction you need a supercomputer. A simple desktop PC will do for most of the material in this book. If the first stage of learning a field is to talk the talk by learning vocabulary, running the computer exercises will enable you to walk the walk as well.

Software

All of the figures in this book were put together using Neuron, a computer simulation program written by Mike Hines at Yale University. This program is freely available at *http://www.neuron.yale.edu*. Although Neuron is primarily designed for running the type of realistic simulations highlighted in the latter part of the book, it is flexible enough that I was able to use it for all the other simulations as well.

Software to produce all of the simulations and to run the emulator of Chap. 5 is available at these sites:

http://www.springer-ny.com/computer2brain
http://www.cnl.salk.edu/fctb
http://www.neuron.yale.edu/fctb

I will be pleased to consider additions or augmentations to this software, particularly if the contributor has already coded them.

Examples in the software are primarily presented through a graphical user interface. The reader or teacher who is interested in pursuing or presenting the subject in depth will want to become familiar with the Neuron program and with HOC, the Neuron programming language. This will allow the programs to be manipulated more flexibly in order to look at different aspects of a particular modeling problem.

Many of the examples presented in this book could also be readily programmed up in Matlab or Mathematica, or in other simulation programs such as Genesis or PDP++.

1.4 Why learn this now?

The genetic code was cracked in the mid-20th century. The neural code will be cracked in the mid-21st. Genetic science has given way to its applications in biotechnology and bioengineering. Neuroscience is still up-and-coming, the next big thing. Furthermore, as genetic manipulations and basic neuroscience add more raw facts to the broth, the need for meaning, structure, and theory becomes greater and greater. Enough information is coming together that the next generation of computational neuroscientists will make the leap into understanding. That means grant money, prizes, and fancy dinners! (If the movies are a guide, it also means evil robots, mind control, and dystopia, but let's not ruin the moment.)

There's such a variety of things to learn about in computational neuroscience that the student is in the position of the proverbial kid in the confectionery: so many problems to work on; so many amazing facts and theories from so many interrelated fields. Of course, this profusion of riches can also be frustrating. One doesn't know which gaudy bauble to pick up first, and, having picked one and discovered that it is not quite gold, strong is the temptation to drop it and pick up one that seems gaudier still.

1.5 What is the subtext?

In an age of ubiquitous computers, any topic can be discussed in their context, as attested to by the recent publication of *From Computer to Stain: Dry Cleaning in a Digital Age,* the inspiration for the title of this book. However, computational dry-cleaning is still just dry-cleaning done with computers. In fact, most of computational biology is just biology done with computers. Computational neuroscience is a little different. The computer itself represents the state of our knowledge about how complex information processing devices like the brain might work. For this reason, I have covered more computer science than would usually show up in a neuroscience book.

Present computer science curricula generally emphasize sophisticated abstractions that pull one away from the machine. Similarly, a branch of computational neuroscience has concerned itself with finding general principles of neural computation and has shied away from the messy meat of the brain. My contention is that the meat is the message for the brain and for the computer. Lovely abstract theories must grow out of an understanding of the machine. The most useful theories will be different for different machines. If there is a grand unified theory, it will stand abstract and austere away from the daily marketplace of synapses or transistors.

For this reason I have gone into some detail about the design and operation of an ancient computer, the PDP-8, a machine with the power of a modern pocket calculator. Such a simple machine can be readily described in a chapter. It also is small enough that one quickly runs into its limits and has to overcome them with programming tricks, commonly called "hacks." Hacking is now frowned upon in computer science, since it is mostly used to break into other people's machines. Biological evolution is one long history of hacking — using a piece of machinery for a new purpose and gradually working it into shape so that it seems to have been engineered from scratch for that purpose.

Understanding the process and products of evolution means understanding the problems of engineering with limited resources (and unlimited time). Programming a PDP-8 or rebuilding a diesel engine with pieces of scrap in a Third-World country requires ingenuity, ability to compromise, and willingness to make mistakes and start again. This process may leave us with a program or an engine with vestigial organs, tangled distribution routes, and inefficient procedures. Just as building the machine was a study in frustration, so examination of the machine will be a frustrating study that will also lead to dead-ends and back-tracking.

In this book, I repeatedly contrast my emphasis on the brain with the tendency of others in the field to focus on theory rather than detail. Perhaps I occasionally disparage these poor theoretical guys as cyborgs and hedgehogs. This is all in fun. Integration of theory and fact is a necessary goal in computational neuroscience. I try to give both their due in this book but I have not been successful in integrating them. The section titles — Computers, Cybernetics, and Brains — demonstrate this.

Our brains are full of contradictions but we learn to live with them. If we want to study the brain, we must be prepared for the kinds of ambiguities and occasional false leads that characterize life with our own brains.

1.6 How is the book organized?

Computational neuroscience is a new field whose essential paradigms are still the subject of debate. For this reason, it is not possible to present the basic material with the conceptual coherence of an introduction to well-established fields like chemistry or physics. The field remains a hodge-podge of exciting ideas and remarkable facts, some of which cannot be neatly conjoined. Instead of progress in a neat sequence from one idea to another, this book will at times seem to jump back and forth from one thing to another. This is an inherent difficulty of trying to teach both the computational and biological approaches in a single text. In general, I try to fill in the gaps where they can be filled in and point them out where they remain unbridged.

The organization of the book is as follows. I start with a brief introduction to neuroscience, touching on many but not all of the subfields that may need to be considered. I then go top-down with a description of how computers work. I start by noting how computers represent information. I then go into still more detail about the bits-and-bytes level of computer function. From there, I switch from transistors to neural network units and explore the concepts of artificial neural networks. This will entail a comparison between transistors and neurons and an explanation of how the artificial neural network units represent a compromise position between the two. From there, I show how these units can be connected together into artificial neural networks. I further expand on the artificial neural network paradigm, showing the use of these networks to explain the retina of a simple sea creature, the horseshoe crab. I look at another simple brain system, but this time one found in humans, the brainstem reflex that stabilizes the eyes in the head when the head is moved. Then, in a more speculative vein, I go still higher in the brain, looking at how artificial neural networks can be used to emulate aspects of human memory. This will involve an explicit compare-and-contrast with computer memory design.

Following this, I turn bottom-up, more seriously exploring the biological concepts of nervous system function. I start with a detailed description of the neuron with some ideas of how the different parts of the neuron can be modeled. I then explore in greater detail the two major techniques of realistic neuronal modeling: compartment modeling and the Hodgkin-Huxley equations. I then look at an example of how artificial neural network models of learning can inform our understanding of the brain and how study of the brain leads us to reconsider the details of these artificial neural networks.

The final chapter covers some details of the mathematical and scientific approaches and techniques used in this book. It includes unit analysis, binary arithmetic, linear algebra, calculus, and electronics. Comfort with handling units and scientific notation is needed for finding your way around science. Knowledge of binary is important for finding your way around a computer. Linear algebra is useful for finding your way around a network. Calculus is important for assessing movement and change. Electronics is needed for understanding electrical signaling in neurons. In each case, the material has been presented graphically and algebraically to make the subject accessible to those who do not feel comfortable with mathematical notation. It is expected that many readers will be unfamiliar with some or all of these areas and will want to read these sections as the technique comes up in the main text.

Since there are many topics touched on that do not always relate cleanly to one another, I tried to provide additional guidance. Each chapter begins with a brief introduction entitled "Why learn this?" Similarly, each chapter ends with "Summary and thoughts," meant to synthesize concepts and remind the reader of what was learned. This section is "... and thoughts," rather than "... and conclusions" because in many cases the conclusions await.

Part I

Perspectives

2

Computational Neuroscience and You

2.1 Why learn this?

A major goal of computational neuroscience is to provide theories as to how the brain works. Such mind–body theorizing has been a subject of philosophical, theological, and scientific debate for centuries. The new theories and taxonomies for organizing information about the brain will be built upon this historical foundation. It is valuable to see where we are starting.

2.2 Brain metaphors

Mechanical models or metaphors for the brain date back to the time when the brain first beat out the heart as leading candidate for siting the soul. Plato likened memory to the technique of imprinting a solid image onto a block of wax. Over the centuries, the nervous system has been compared to a hydraulic system, with pressurized signals coursing in and out; a post office, with information packets being exchanged; or a telephone switchboard, with multiple connecting wires to be variously assorted. Today, the digital computer, or sometimes the Internet, is cited as a model for brain function. Do these modern mechanisms hold greater promise than prior metaphors for helping us understand our most intimate organ?

In many ways, the brain is not much like the standard digital computer. Yet, both as a direct model of certain aspects of brain functioning and as a tool for exploring brain function, the computer enjoys many advantages over previous models. Take, for example, the post office. The difficulties of actually utilizing the postal service to test out the feasibility of a particular brain model must give one pause. (However, below I discuss a similar human-based system that was proposed more than a century ago as a calculating technique for weather prediction.) The telephone switchboard, on the other hand, is a considerably more manipulable organizational and technological artifact. In fact, the early analog computers of the 1930s and 1940s were, in appearance and in some functional aspects, aggrandized telephone switchboards. Although simple neural models were run on such machines, the technical difficulties of programming them made them far less useful than digital computers as a tool. However, the basic concepts of analog computing may be useful for understanding brain function.

2.3 Compare and contrast computer and brain

When we liken the brain to a computer, we mean several things. First, we mean that several definable computer actions are analogues of things that the brain appears to do. Such computer actions include memory, input/output, and representation. Second, we mean that computers have been used to do a variety of tasks that were previously believed to be exclusively the province of human intelligence: playing chess, reading books aloud, recognizing simple objects, performing logical and mathematical symbol manipulations. Finally, although no machine has yet passed the Turing test (a machine passes if it fools a conversation partner into thinking that it is a person), those who work intensively with computers develop a distinct sense of communicating or even communing with the machine.

Modeling is the work and play of computational neuroscience, as it is for much of physics, engineering, business, and applied mathematics. It's a tricky thing. To learn something about the thing being modeled, we need to reduce the model to the essentials. If we reduce too far, however, we may miss a critical component that is responsible for interesting properties. For the Wright brothers and other early aviators, the process of building a heavier-than-air flying machine was a task of bird emulation. To those who said that heavier-than-air flight was impossible, they could point to birds as a counterexample. As they evaluated the basic bird, it would have seemed clear that many aspects of bird design were not required for flight. For example, the beak seems quite clearly designed more for eating than for flying. However, the beak's aerodynamic design might still tell us something valuable about fuselage design. The importance of other bird features for flight would not have been as apparent. For example, it might a priori seem likely that wing beating was critical for flight. It is critical for small-creature flight but not for the flight of large birds or airplanes. Many early, misguided attempts were made to design a full-sized ornithopter (e.g., that aircraft with flapping wings that beats itself to death). On the other hand, the Wright brothers had a key insight when they noticed that birds steered by tilting their body to the side (rolling) rather than by using a rudder like a boat.

When we model birds we know what we want to model. The function of interest is flying. We can focus on flying and ignore feeding, foraging, fleeing, etc. The brain, however, is doing many things simultaneously and using hidden processes to do them. Therefore, we can model a brain function, such as chess playing, and yet gain little or no insight into how the brain plays chess. The brain is utilizing unconscious properties that we are not aware of when we play chess. In this example, I would guess that an important underlying ability used in chess is the capacity of the brain to complete partial patterns. This ability is seen in the normal unawareness of the blind spot. It is also seen in the abnormal confabulatory tendency of demented or psychotic individuals to forge links between false perceptions so as to build an internally consistent, although irrational, story.

In this book, as we look in detail at how a computer works, and compare and contrast its functioning with that of the brain, a variety of differences will become apparent. We consider various brain features and wonder whether or not these are critical features for information process, for memory, or for thought. Certainly, many aspects of brain design are not critical for brain information processing but are there for other purposes: metabolism, growth and differentiation, cell repair, and general maintenance.

If we wanted to use a modern jet aircraft as a model to help us better understand birds and the phenomenon of flight, we would want to take note of similarities and differences that might clarify essential concepts. Both have wings; it seems reasonable to expect that wings are essential for heavier-than-air flight (note that helicopters are considered rotating wing aircraft). However, the wings are made of very different materials so there is apparently nothing critical in the design of feathers. Closer analysis would reveal that airplanes and large birds like albatrosses have similarly shaped wings (smaller birds and insects use different-style wings suited to their small size).

Using the computer as a model to understand the brain raises questions about similarities both in detail and in function. Airplanes fly like albatrosses but computers don't think like brains. Both brains and computers process information, but information processing may not be central to the process of thinking. Therefore, we will wish to explore not only differences from the bottom, differences in materials and design principles, but also differences from the top, differences in capacity and capability.

Starting with the manufacturing side, there are already a variety of differences that can be explored. Computers are made of sand and metal, while brains are made of water, salt, protein, and fat. The computer chip is built onto a two-dimensional matrix, while the brain fills three dimensions with its wiring. The time and size scales involved are also believed to be vastly different. Of course, this depends on exactly what is being compared to what. As we will see, typically a transistor in the computer is compared to a neuron in the brain. With this comparison, the time scales are about 1 ms for the neuron vs. 1 ns for the transistor (see Chap. 16, Section 16.2 for discussion of units). The spatial scale is about 1 mm for the largest neuron vs. less than 1 μm for a modern CMOS transistor. Thus the neuron is much bigger and much slower. However, if it eventually turns out that the proper analogue for the transistor is the synapse, or a particular type of ion channel or a microtubule, then we would have to reevaluate this comparison.

Additional differences arise when one considers functional issues. Brains take hints; computers are remarkably stupid if given a slightly misspelled command or incomplete information. The digital computer has a general-purpose architecture that is designed to run many different programs. The brain, on the other hand, has dedicated, special-purpose circuits that provide great efficiency at solving particular problems quickly. Calculations on a digital computer are done serially, calculating step by step in a cookbook fashion from the beginning to the end of the calculation. The brain, on the other hand, performs many calculations simultaneously, using parallel processing. Digital computers use binary; transistors can take on only two values: 0 or 1. In this book, we utilize binary extensively, and consider its applicability to the brain. This may not be a fair approximation since the brain uses a variety of elements that take on a continuum of analog values.

2.4 Origins of computer science and neuroscience

Neuroscience and computer science came into being at about the same time and influenced each other heavily in their formative stages. Over time, the fields have diverged widely and have developed very different notions of seemingly shared concepts such as memory, cognition, and intelligence.

D.O. Hebb proposed over 40 years ago that a particular type of use-dependent modification of the connection strength of synapses might underlie learning in the nervous system. The Hebb rule predicts that synaptic strength increases when both the presynaptic and postsynaptic neurons are active simultaneously. Recent explorations of the physiological properties of neuronal connections have revealed the existence of long-term potentiation, a sustained state of increased synaptic efficacy consequent to intense synaptic activity. The conditions that Hebb predicted would lead to changes in synaptic strength have now been found to cause long-term potentiation in some neurons of the hippocampus and other brain areas. As we will see, similar conditions for changing synaptic strength are used in many neural models of learning and memory. These models indicate the great computational potential of this type of learning rule.

One difference between the neuroscience and computer science viewpoints has to do with the necessary adoption of a big-picture approach by computer scientists and a reductionist approach by many neuroscientists. These two approaches are typically called top-down and bottom-up, respectively. The top-down approach arises from an engineering perspective: design a machine to perform a particular task. If you're interested in intelligence, then design an artificial intelligence machine. The bottom-up perspective is the province of the phenomenologist or the taxonomist: collect data and organize it. Even granting that most U.S. science today is federally mandated to be hypothesis-driven, an essential element of biology is the discovery of facts. Hypotheses are then designed to fit these facts together. As outlined here, these positions are caricatures. Most biologists want to consider how the brain thinks, and many computer and cognitive scientists are interested in what goes on inside the skull.

In this book, we concern ourselves with many ideas that have been promulgated for understanding higher levels of nervous system function such as memory. However, it is important to note that much of the data on real nervous systems has been gathered from either the peripheral nervous systems of higher animals or from the nervous systems of invertebrates such as worms, leeches, and horseshoe crabs. The genesis of the action potential or neuron spike, one of the most important ideas to come out of computational study of the brain (Chap. 12), was the result of studying the peripheral nervous system of the squid. The low-level source of much of our knowledge of the nervous system contrasts sharply with the ambition to understand the highest levels of mental functioning, and helps explain why some of the topics to be discussed may seem quite remote from human neural function, while other subjects will be very relevant but highly speculative.

2.5 Levels

The notions of bottom-up and top-down approaches to the problem of nervous system function, and the corresponding contrast between acknowledged facts at the lower level and uncertain hypotheses at the higher, lead naturally to hierarchical divisions. Two such divisions that are commonly used are called the levels of organization and levels of investigation. Each of these divisions into levels creates a hierarchy for brain research that leads between the reductionist bottom and the speculative top.

The levels-of-investigation analysis was historically a product of top-down thinking. This approach, pioneered by computationalists, starts at the top with the big-picture problem of brain function and drips down to the implementation in neurons or silicon. The levels-of-organization analysis was in part a reaction to this. By putting all of its levels on an equal footing, the levels-of-organization approach invited the investigator to start anywhere and either build up or hypothesize down.

Levels of organization

Levels of organization is fundamentally a bottom-up perspective. The basic observation that leads to this division of the knowledge comes from the "grand synthesis" that connected the physical with the vital world. Modern biology explains genetics and physiology in terms of the interactions of molecules. This allows connections to be made all the way from physics to physiology. Physics is the more fundamental science. The basic concepts of biology can be understood from the concepts of physics, while the converse is not the case. However, understanding biology directly from physics would be a hopeless task for two reasons. First, there is no way one can predict what would occur in a biological system using knowledge of atoms and electron orbits. Second, the conceptual leap from physics to biology is simply too great to be made without interposed models from other fields. Specifically, much of biology can be understood from cell biology, which can be understood from molecular biology, which can be understood from biochemistry, which can be understood from organic chemistry, which can be understood from physical chemistry, which can be understood from physics. In comparison with this known hierarchy of knowledge, the levels of organization of the nervous system remain tentative. Any hierarchy will likely embody a fundamental, trivial law: big things are built out of smaller things.

Following the scheme of others, we can build a hierarchy of levels of organization and levels of study. From smallest to largest:

study method	object of study
physics	ions
chemistry	transmitters and receptors
cell biology	neurons
computer science	networks
neurology	systems
psychology	behavior or thought

Although the general order of dependencies in the nervous system can be assumed to be based on size and the simple inclusion of one structure within another, the exact structures that are of functional importance are not clear. The ambiguity starts when one considers the appropriate items for anchoring the two ends. At the top, one can choose to regard either behavior or internal mental representation as the highest level suitable for scientific investigation. There is a long history of debate in the psychology literature between proponents of these two positions. Behavioralists believe that since physical movement is the only measurable evidence of nervous system function, this is the only appropriate area of high-level functional study. Other psychologists believe that putative internal representations of the external world are also suitable subjects of investigation, even though these cannot be measured directly. Computational approaches generally make the latter assumption, not only postulating internal representations but often making them the central question for further study.

At the small end of the organizational scale, most investigators would consider the concentrations of ions and neurotransmitters and their channels and receptors to be the smallest pieces of nervous system that are worth paying any attention to. A dissenter from this view is the physicist Roger Penrose, who believes that the underlying basis of neural function will lie in quantum mechanics and that it's necessary to study the subatomic realm.

In between quantum mechanics and behavior, there is still more room for debate both as to which levels are relevant, and as to which levels can be adequately built on a previous level without further investigation at an intermediate level. To go back to the physics-to-biology spectrum described above, the conceptual jump from the concepts of physics to the concepts of organic chemistry would not be possible without the intermediate concepts developed by physical chemistry. This is because the representations of electron orbitals and chemical bonds used in physical chemistry provide conceptual links between the detailed equations describing electron orbitals used in physics, and the schematic stick diagrams used for bonds in organic chemistry. Similarly, the neuroscience levels of organization suggests that neurons can be adequately described by taking account of properties at the level of transmitters and receptors. That's probably not going to turn out to be true. It's likely that intermediate-sized ultrastructural components of the neuron such as spines, dendrites, and synapses may have their own critical properties that cannot be understood without independent study of these structures in themselves.

As we move up the scale, higher levels of neural organization are less well understood and can be farmed out somewhat arbitrarily to various interested specialty areas. Much study of networks has come out of computer science, but the organization of networks is also studied in mathematics by geometry and topology. The level of cortical columns is not shown in this diagram. It is unclear whether this level would go below or above the level of the network. I gave *systems* to neurology, a clinical field that subdivides brain function into motor, sensory, and various cognitive systems based on changes seen with brain damage. Engineers mean something different when they study *systems* in the field called "signals and systems." *Systems* neuroscience has yet another connotation, referring to neurophysiological techniques related to investigating the origins of perception and behavior.

Levels of investigation

The levels-of-investigation approach comes from David Marr, a computationalist who produced some very influential early models of different brain areas. This viewpoint is from the top down. The top level is the level of *problem definition* (this was called the computational-theoretic level by Marr). Marr suggested that understanding any particular brain function requires that we first understand what problem the brain is solving. Problem in hand, we can deduce what additional information the brain would need to solve it. The next level is that of *algorithm definition.* An algorithm is like a cookbook recipe, defining a step-by-step approach to accomplish some task. The third and final level is the level of *implementation*, where the algorithm is finally translated into machinery, whether neural or silicon, that can actually perform the task.

Marr's three levels of problem, algorithm, and implementation are the current approach a software engineer would take in designing a big program (e.g., a word processor or a Net browser) using a modern computer language. If writing a Web browser like Netscape or Explorer, for example, we would first define the problem — delivering information from remote sites to a user in a user-palatable form. We would then write algorithms that would simply assume that we have or can develop the underlying tools needed for the subsidiary processes. For example, a basic algorithm for processing a Web page would be 1) request the page, 2) wait and accept the data, 3) confirm that a full data set was received, 4) parse the data to determine content type, and 5) parse fully to present in a graphical form on the screen. Individual steps would then be implemented. It is important to avoid considering details of implementation in working out the algorithm since we are interested in readily porting our browser between machines that use different low-level implementations.

This Marr trinity of problem, algorithm, and implementation can be collapsed into the familiar concepts of software and hardware. A problem is provided. Algorithms are written into the software. The software is compiled so as to run on a computer — the physical implementation level. A software engineer using modern computing machinery doesn't routinely run into the limits of what the machine can do. The Marr top-down approach is ideal in this engineering environment. However, as will be discussed in the next chapter, when the limitation of the machine becomes part of the problem, another engineering approach is needed.

2.6 New engineering vs. old engineering

Over time, science and technology have advanced from being based on everyday commonplace observations to being based on sophisticated theories. Similarly, engineering has moved away from the tinkerer or hacker mentality toward reasoned conceptual approaches to technical problems. Working from theory, rather than empirically, the modern engineering approach is close to David Marr's notions of levels of investigation: from problem to method to implementation.

Modern building design is predicated on principles of tension and stress. By contrast, the great cathedrals of Europe were largely built using rules of thumb and intuition born of experience. Sometimes they fell down. Similarly, computer science has given up ad hoc hacking and developed tools and theories to allow software design problems to be addressed from basic principles.

From one perspective, Marr's insistence on first defining the problem is unavoidable. Until we know that the brain can do a certain thing, we cannot study it. A blind man who has never had sight, and had not spoken with someone who has, would have an impossible task trying to study vision based simply on being told that it represented an alternative to hearing. On the other hand, insistence on an initial problem definition can lead to what has been called premature definition. Fondly held hypotheses can be blinders that preclude appreciation of new facts that could shed light on the problem. This risk is particularly great in the general area of brain/mind studies, where the appeal to intuition is hard to resist.

The complexity of the unconscious workings of the many subdivisions of brain function makes them resistant to an introspective, intuitive understanding of what is going on behind the scene. This can make it impossible to frame the problem correctly. For example, Marr believed that vision primarily performed the task of taking the two-dimensional retinal representation of the world and re-creating a three-dimensional internal model of the world, just as one might look at a family photograph and determine that your cousin was standing to the side and in front of your aunt. This definition of the *problem* of vision seems intuitively reasonable to a sighted person. This clearly describes something that the brain can do and that needs to be done under some circumstances. However, this turns out to be a bad start for studying vision in the brain. As it turns out, any single *problem definition* for vision will turn out to be a bad choice, since the brain does not simply have one type of vision but instead utilizes many different types of vision that are processed simultaneously.

Marr's is a sophisticated engineering approach to vision. Given the complexity of brain and our limited intuition, a naive engineering approach is more reasonable. A congenitally blind man starts with no clue as to where to begin studying this mysterious phenomenon called "vision." He might therefore start asking questions about different things that vision can do. "Can it detect objects behind other objects? Can it detect motion? Can it determine shapes?" This set of questions would put the blind-man-explaining-vision in the proverbial blind-men-with-elephant situation (each feels a different body part and each has a different idea of what an elephant is). The blind man might conclude that vision was not one thing but was made up of separate detecting systems that handled various detection tasks. This sort of piecemeal understanding would bring him closer to the underlying mechanism of visual perception in the brain than was Marr with his sighted person's intuition of a unitary process. Although the blind man would have no appreciation of the personal experience (the qualia) of seeing, he would have some notion of how the brain actually performs the task. Lacking access to the myriad unconscious processes of our brain, we are nonetheless blessed with the illusion of introspection. Sight makes us blind to vision.

Modern, sophisticated engineering takes place in big laboratories. Discovery proceeds in reasoned steps according to quarterly plans. The old engineering was, to paraphrase Edison, all inspiration and perspiration. This is the engineering of tinkerers and hackers, who take discarded bits and pieces of machinery and cobble them together so as to make them do things they were not originally meant to do. When it doesn't work, the tinkerer bends and hammers and makes it work. The workman's ideal is always to have the right tool for the job. The tinkerer is cursed with the wrong tools, the wrong materials, the wrong job.

This discrepancy between available material and the exigencies of the task is also the plight of the evolutionary process. For example, a gill is not well suited to life on land. Since water must be moved across the gill in order to replenish oxygen, the gill needs to be exposed externally. However, gills must also be kept wet at all times. Keeping wet becomes a problem when we move the gill onto dry land. Although the modern lung looks great, early versions would have been crude hacks that managed to barely satisfy these contradictory needs: invaginating the gill to prevent drying, while exposing it enough to prevent asphyxiation.

2.7 The neural code

The brain denies the philosopher's, the mathematical modeler's, and the guy-on-the-street's desire for clarity and simplicity. Since there is no single overarching task for the brain to do, different facets of brain function must be studied separately. This does not necessarily mean that there are no unifying principles. There may well be basic neural codes that are used throughout the animal kingdom. However, the discovery of neural codes will no more free us from the need for further research into different brain areas than the discovery of the genetic code revealed all the functions of all enzymes and structural proteins.

The analogy between the search for the genetic code and the search for a neural code has been highlighted by Francis Crick, discoverer of the former and pursuer of the latter. To doubters, he points out that the quest for a simple genetic code seemed quixotic to anyone who considered that the complexity of the natural design encompasses enzymes and organs, growth and development. Of course, the discovery of a simple genetic code did not in any way provide an understanding of all the things that are coded for. It did, however, provide a powerful new tool for exploring these things. Similarly, the discovery of neural code or codes will not tell us how any part of the brain works, but will enable us to start to understand what we see when we amplify electrical signals from different parts of the brain.

Several neural signals are well established. However, some of these signals probably carry no information at all, while other signals carry information that is not used by the brain or body. For example, the electroencephalogram (EEG) is a very well studied signal that is emitted by the brain. There is information in the EEG that permits an outside observer to determine whether a brain is awake or asleep or even, after some signal processing, whether the brain is hearing clicks or seeing a flashing checkerboard. These field signals are generally an epiphenomenon, a side effect that has no functional relevance. These signals are not used within the brain under normal circumstances, and are too weak to be used for telepathy, no matter how close you put the two heads together. There are some cases where the field is used or misused. Some neurons in the goldfish communicate internally with such field effects. Field effects are used to communicate between individuals in the weakly-electric fish (the strongly electric fish use their fields to stun prey). Field effects are also responsible for pathological signaling in cases of epilepsy and multiple sclerosis. However, in general the EEG can be considered an information-carrying neural signal that is not used internally as a neural code.

Various signals are used directly by the brain and therefore can be considered to be codes. For example, the rate of spiking of neurons carries information that determines how powerfully a muscle will contract. This is a code that has been cracked: the nerve tells the muscle "squeeze ... squeeze harder." It appears likely that similar rate coding is also used in the central nervous system. Rate coding has also been suggested to be the primary code in parts of sensory cortex. Neurons in visual cortex spike fastest when presented with oriented bars of a certain configuration, and auditory cortex neurons will spike faster in response to particular sound frequencies.

There are an enormous number of electrical and chemical signals that influence neuron firing. Many of these can be considered to have a coding function as well. Most neurons use chemical synapses to communicate. The presence of neurotransmitter is a coding signal at these synapses. Synapses are typically viewed as passive information conduits connecting complicated information-processing neurons. An alternative view is that a synaptic complex may itself be a sophisticated information processor. Neurotransmitter concentration may vary and be a relevant signal in some cases. Within the postsynaptic cell, ions and molecules function as second and third messengers in cascades of chemical reactions. These chemical reactions can be very rapid. It may be that sequences of chemical reactions are as important as electrical activity for neural information processing.

2.8 The goals and methods of computational neuroscience

Conferences in computational neuroscience often feature energetic debates about what constitutes the correct approach to the problem of understanding brain function. Generally, it's biologists against computationalists, bottom-uppers versus top-downers. To caricature, the most rabid biologists believe that a model that deviates from the details of known physiology is inadmissibly inaccurate. Meanwhile, the computer scientists, physicists, and mathematicians feel that models that fail to simplify aggressively do not allow any useful generalizations to be made. The view presented here is one of compromise. Both perspectives are in part correct. Leaving out biological details will lead to models that can no longer make the connections with physiological experiment. However, failure to simplify at all can produce models that may not generalize at all. For example, it is possible to model a specific experiment with such fidelity to detail that one just has another copy of the experiment. Such a model would not be able to generalize so as to explain other related experiments in the same brain area. Also duplicating the system will not by itself give you any insight into how the system works.

In addition to the inherent intellectual tension between dry computers and wet biology, there are also historical tensions between traditional applied mathematics and the newer computational approaches. Traditionally, applied mathematics and theoretical physics were done with paper and pencil. The resulting formulations embedded complex physical phenomena in simple attractive formulae that could be disseminated by T-shirt and coffee cup. The Maxwell equations and $E = mc^2$ are examples that have been translated into both of these media. Although these equations are mysterious to most people, their elegance and aesthetic appeal is evident. They look like a key to the mysteries of the universe. Computer modeling, on the other hand, has little of the elegance and none of the generality of the traditional great equations. Although it is possible that neuroscience may someday yield clear-cut defining equations of this sort, it seems to me more likely that it will not. Just as with wet biology experiments, the results of computer simulations are rarely definitive and perhaps never canonical in the way of the great physics equations.

Computer modeling or simulation can be considered to be experimental mathematics. Simulations are themselves so complex that they must be studied by using virtual experiments to try to understand them. The simulated complex system, like the original, shows emergent behaviors whose origins and implications are not immediately obvious and must be explored experimentally. Traditional mathematics provides clean translations of reality. Simulation provides an alternative reality with advantages of manipulability and accessibility.

Simulation is used to assess large sets of complex mathematical formulae that cannot be solved by traditional analytic (paper-and-pencil) means. Since the single simulation never unequivocally represents the biology, it is often necessary to cross-check results among several simulations that represent the same system with different levels of detail or scale or simply with different choices for undefined parameters.

On the bright side, simulation also produces a variety of very nice benefits. Simply transforming a notion about how something works into an explicit computer model requires a complete accounting for all system parameters. Compiling this list often reveals basic, critical aspects of the system that are not known. Sometimes this is simply because no one ever bothered to look. Additionally, running computer simulations permits one to test specific questions about causality that can only be guessed at in paper-and-pencil modeling. Finally, working with computer simulations provides a way of getting a very intimate view of a complex system. The next time you take a commercial airliner flight, consider that this may be your pilot's first flight in this aircraft type, since many airlines now do all step-up training on a simulator. Just as flight simulators provide an intuitive feel for flight, neural simulators can provide intuition and understanding of the dynamics of neural systems. If I swim with the neurons long enough, maybe I'll learn to think like a neuron.

2.9 Summary and thoughts

I have presented this brief history of brain thoughts partly to present my own view and place it in perspective. The view in this book is particularly contrasted with Marr's ideal separation of ends from means. The present evolutionary view of implementation entangled with task is comparable to the mainstream programming practices of Marr's era. In Chap. 5, I present basic computer science through exploration of computer practices from that era, when hacking was necessary to perform complex computations despite hardware limitations. These practices have been lost with the growing power of computers and increasing sophistication of programming tools.

This book focuses on the interface between task and machine, where tricks and shortcuts, hacks in software parlance, are used to optimize a function on a particular architecture. The assumption is that the brain uses a thousand tiny hacks, each cleverly evolved to do some little task very well.

3
Basic Neuroscience

3.1 Why learn this?

I come from a biology background. For this reason, I am sometimes surprised that there are people in the field who do not care where the thalamus is. On the other hand, those who come from a physics background would be shocked to hear that I don't use Hamiltonians.

So there is a debate. If you want to understand the brain, do you need to know what the thalamus is, or the amygdala, or even the cortex? The most theoretical theorists would say no. Their belief is that fundamental theories will emerge independent of the details of evolutionary history that gave us a cortex and gave birds a wulst (a cortex-like structure). They will note that understanding the emergence of parabolic trajectories from Newton's second law is only obscured by close study of the principles of cannonry, with its attention to the details of combustion and barrel hardening. If the basic principles of neural computation are already known, as is assumed by some theorists, then close attention to brain details may not be necessary. If this is the case, then our current understanding of brain function is an adequate framework to build upon for future research. My guess is that the current intellectual framework (the paradigm) is basically sound. However, I wager that it will be so heavily built upon that it will be barely recognizable 100 years from now. Even if we don't suffer a paradigm shift, we may still get a major paradigm face-lift.

The viewpoint of this book is that new concepts of computational neuroscience will largely emerge from study of the brain. If readers of this book intend to study the brain directly or to study the research of those who study it, they must know standard terminology, coordinate systems, and concepts. This chapter provides an initial introduction to these. It is hoped this will be sufficient so that the interested reader can segue painlessly to neuroscience textbooks.

3.2 Microscopic view of the nervous system

Living tissue is made up of cells. A cell has a fatty membrane and is filled with liquid and proteins known as cytoplasm as well as smaller functional parts called organelles. In the bodies of humans and other multicellular organisms, these cells are typically specialized and organized. Livers are made of liver cells and brains are made of brain cells. There are two major types of brain cells: neurons and glia. Neurons are usually identified by their ability to produce action potentials, but not all neurons can produce action potentials. It is believed that neurons are the principal elements involved in information processing in the brain.

In the classical model, a neuron has dendrites, a cell body, and an axon. According to Cajal's century-old "neuron doctrine," information comes into the dendrites of the neuron. Signals then travel to the cell body, which in turn activates the axon, which can send a signal out of the neuron. Note that this is a classical view that may be true for many but certainly not all neurons.

Fig. 3.1 shows a typical pyramidal cell of the mammalian cortex. These cells are considered principal cells because of their large size and long projections to other areas. The cell body or soma (small oval), which would be the bulk of the cell in other cell types, is dwarfed by the dendrites, which extend out in all directions. In this cell type, there are lots of small dendrites and then one major long dendrite with only a few branches. This major, apical dendrite can be a millimeter in length. The axon, which is thinner and much longer than the dendrites, is shown here coming out from the lower right. It would continue off the page. The axon is typically much longer than the dendrites — several of them go from your lower back to your big toe, a distance of about a meter. Axons also branch to connect with multiple targets. An axon branch from another cell is shown at the upper left forming a synapse on the cell (rectangle). The synapse is shown enlarged below the label, illustrating the presynaptic bouton at the end of the axon and a spine on the dendrite. The terminology of presynaptic and postsynaptic defines the direction of signal flow. Transmitter is released presynaptically, floats across the synaptic cleft, and activates receptors postsynaptically. The two neurons are not directly connected but communicate via this cleft.

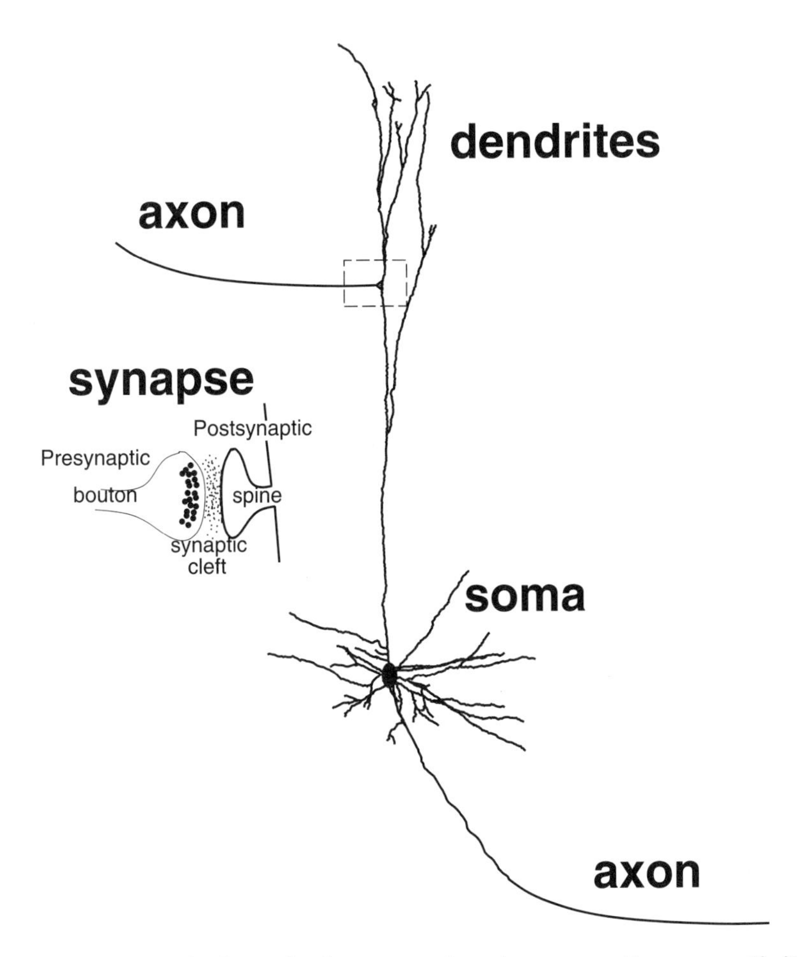

Fig. 3.1: A classical cortical neuron showing synaptic connectivity. The neuron is explored in more detail in Chap. 11.

The information in the neuron is in the form of electrical potentials (voltages) across the membrane. Information is conveyed via the synapse through arrival of neurotransmitter on receptors. This triggers postsynaptic potentials (PSPs). These can be excitatory postsynaptic potentials (EPSPs) or inhibitory postsynaptic potentials (IPSPs). These are graded potentials, meaning that they vary in size. Fig. 3.2 shows a tracing of potentials in a model neuron. The trace starts at a resting membrane potential (RMP) of about −70 mV. An IPSP pushes the membrane potential down to more negative values and away from its firing potential. A pair of EPSPs of increasing size move the potential in a positive direction, toward firing threshold. Finally a larger EPSP reaches the firing threshold and the cell fires a spike (action potential, AP). This spike is all-or-none. It is of stereo-

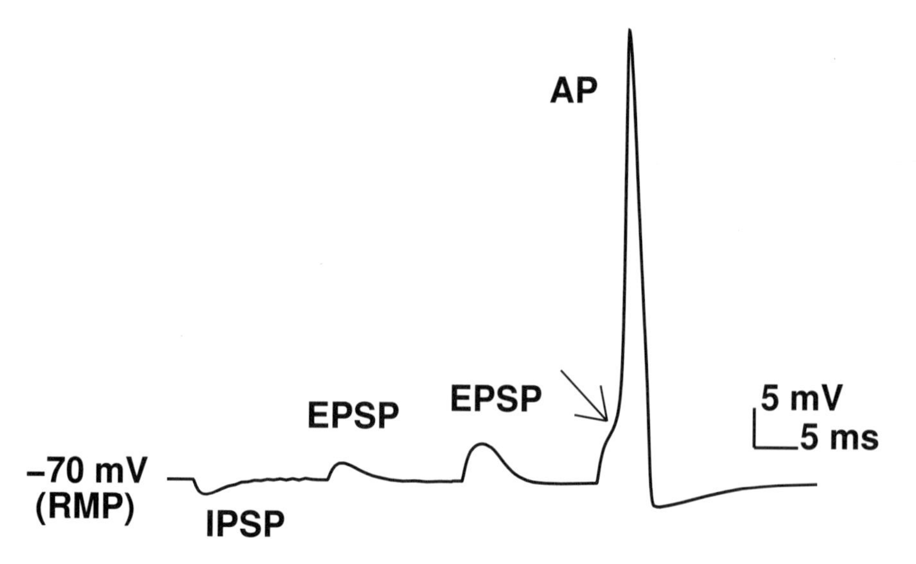

Fig. 3.2: Postsynaptic and action potentials. Arrow shows where 3rd EPSP triggers the AP.

typed shape and amplitude and is not graded. The action potential is the signal that can be sent down the axon to create a PSP in another neuron.

Most synapses are chemical: neurotransmitters are released from one neuron and picked up by receptors on another neuron, generating an electrical signal in the latter neuron. The classical synapse is axodendritic; it connects an axon to a dendrite. There also exist dendrodendritic synapses, which connect dendrites to dendrites. In addition to chemical synapses, there are electrical synapses with specialized channels (gap junctions) that allow current to flow directly from one neuron to another. Complementing these synaptic mechanisms are various nonsynaptic interneuronal signals. These signals includes volume transmission, where transmitters are broadcast through extracellular space to a lot of neurons. Glia, as well as neurons, may be involved in volume transmission. There is also ephaptic transmission, where an electric field is generated by one cell and influences another. As our understanding of the nervous system continues to grow, more nonsynaptic transmission will need to be included in models.

Glia are presumed to be supporting cells but they may also play a role in information processing. Long regarded as passive onlookers, it is now appreciated that glia also have electrical potentials and chemical receptors. (Actually, pretty much all cell types, including liver cells and kidney cells, have electrical potentials and chemical receptors, so this doesn't necessarily mean much.) Glia selectively take up and release neurotransmitters and extracellular ions. In doing this they cannot help but influence neuronal activity. It is not clear whether this influence could play a part in information processing. Astrocytes are the major glial type that take up and release chemicals. There are also other types of glia. Oligodendroglia are notable since they provide the myelin sheaths that insulate axons and speed up action potential transmission. Microglia protect the brain against bacteria and other invaders.

3.3 Macroscopic view of the nervous system

The nervous system can be structurally subdivided in several ways. There is the peripheral nervous system, which reaches out to hands and feet, and the central nervous system (CNS), which lies within the bony protection of spine and skull. The CNS is divided up into forebrain, brainstem, and spinal cord. Within the CNS one can distinguish gray matter, made up of primarily of cell bodies and dendrites, and white matter, made up of axons. The white matter is white due to the fatty white myelin. Gray matter at the surface of the brain is cortex. Gray matter masses deep in the brain are called nuclei.

In addition to dividing the brain by appearance and location, we can also divide it into different areas that have shared or common connectivity and presumably shared function (e.g., thalamus, basal ganglia). We can also discuss different directions in which the brain can be sliced.

Slicing the brain

To describe a structure, it is helpful to have a coordinate system. Unfortunately, the brain, and the body in general, has two major coordinate systems that are often used together in describing a single slice. The first anatomical coordinate system is based on the external world, using the common terms *left, right, lateral, medial, anterior*, and *posterior.* "Up" and "down" are the other cardinal directions in this system, but they are not used. The other coordinate system is based on the body axis and uses the terms *rostral, caudal, ventral, dorsal, lateral,* and *medial.* In the context of the nervous system, we use the neuraxis. The neuraxis is the axis of the CNS as it curves around from tail (caudal) to head (rostral). The front part of this curve is the ventral (belly) part and the back part is the dorsal (back) part. Because of the curvature, ventral and anterior are synonyms at spinal cord levels. However, ventral is downward at the brain. Thus, the two coordinate systems have different correspondences at different locations.

Left and right would seem to be easy. However, in the context of neuroscience, left and right mean anatomical left and anatomical right. These coordinates always refer to the subject's viewpoint and not to the viewer's viewpoint. It is assumed that the subject is facing the viewer so that the

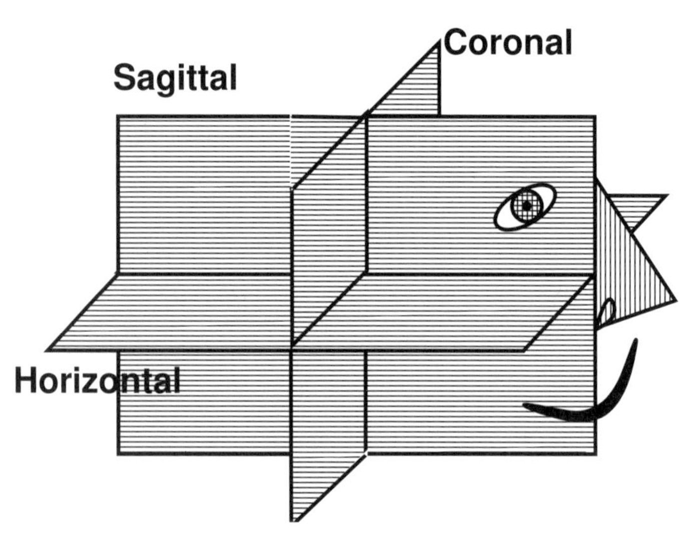

Fig. 3.3: The three planes of body sectioning.

right side is on the viewers left. When asked to look for something on the left side of a sliced brain (or a CAT scan or MRI), look to your right.

There is a standard anatomical position that differs depending on the animal. It's generally a standing position. In the case of our favorite animal, the human, it is standing up, face forward, with hands hanging at the side, palms forward. From there, we can conceptually (or actually) slice the body in three orthogonal planes (Fig. 3.3): horizontally (a plane horizontal to the ground), sagittally (a vertical plane passing from belly to back), and coronally (a vertical plane running from ear to ear).

Each plane has its own pertinent coordinates from the three-dimensional body structure. I'll refer to each of these primarily in the context of the brain though they also apply to the spinal cord, which has different coordinate axes due to CNS curvature. In the horizontal plane, directions are anterior vs. posterior (toward or away from the nose), left vs. right, and medial vs. lateral (toward the middle or toward the edge). For the coronal plane, left vs. right and medial vs. lateral also pertain. The other direction would be up and down, but the neuraxis coordinate system is usually used. For the brain coronal, up is dorsal and down is ventral. In the sagittal plane, we have anterior to posterior and dorsal to ventral. Sagittal planes can be mid-sagittal or parasagittal (off to the side). Coronal planes can be more anterior or more posterior. Horizontal planes are more dorsal or more ventral in the brain but are more caudal or more rostral in the spinal cord.

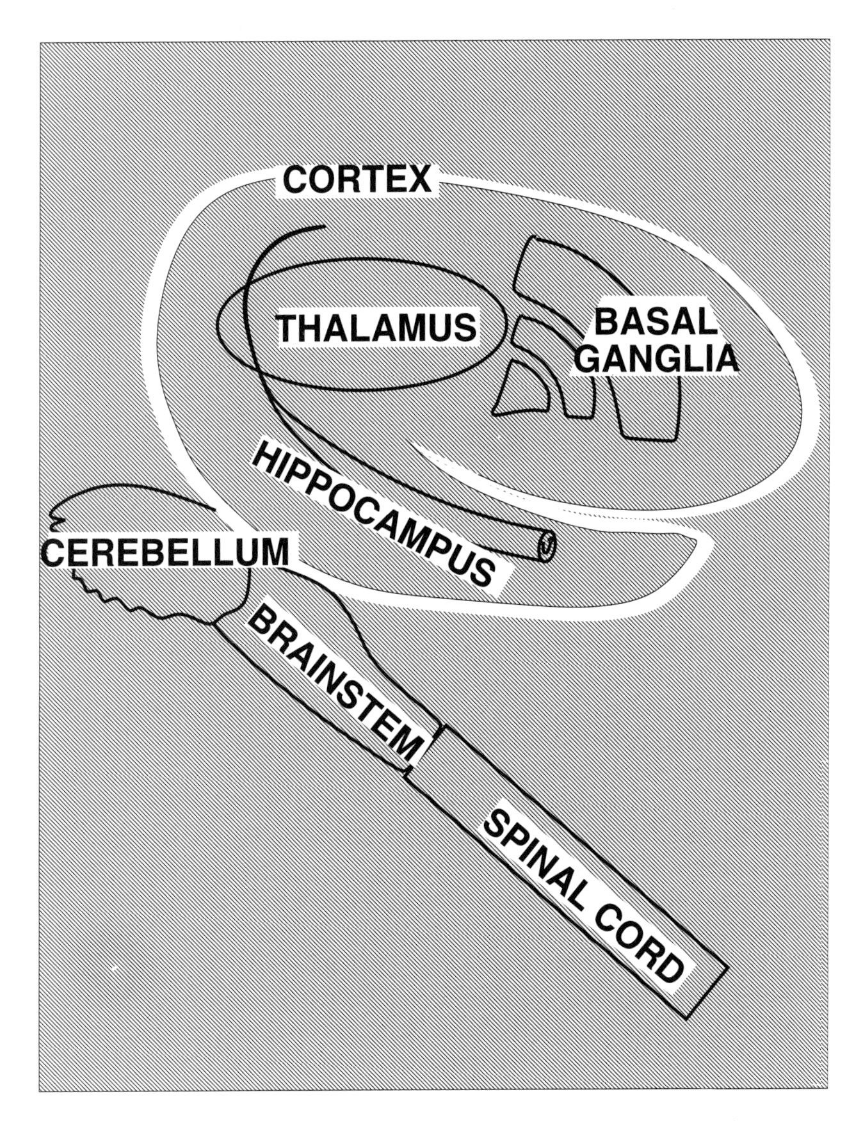

Fig. 3.4: Labeled schematic of the brain in sagittal section.

3.4 Parts of the brain

The brain is itself subdivided into an outer rind called the cortex and multiple internal nuclei (Fig. 3.4). The cerebellum is an additional "mini-brain" stuck on behind; it has its own nuclei and cortex. Major deep nuclear complexes are the thalamus and the basal ganglia. Along with the wiring to and from the cortex, these fill up most of the brain. The nuclei and cortex are gray matter, the subcortical connections are white matter.

Computational neuroscience seeks to answer organizational and functional questions about different areas of the brain. This goal is often frustrated by the fact that, for many sections of the brain, basic functions are still a mystery. As an analogy, imagine that, in the far future, technical know-how has been lost after one of the usual apocalypse-movie scenarios. Someone finds a radio, an artifact of ancient civilization. The effort to understand what it does will be much frustrated by the fact that when turned on it just hisses, having no radio transmissions to pick up. Similarly, many brain areas are in the midst of a great system of other brain areas of equally unknown function. We take one of these areas out to study, shock one end, and look at the mysterious hissing coming out the other end.

For this reason, many of the best-studied parts of the brain are those that interact directly with the environment. The retina is such a part. We know that it developed to perceive light, and we can control the amount of light that hits it. By contrast, although we know that the hippocampus has something to do with memory, we don't really know what it does. It is connected to structures on either side whose function is even more mysterious. Therefore, as we study the basic anatomy and physiology of the hippocampus, we need to hypothesize global as well as local functions.

The cortex is one of the most popular brain bits for humans since it is a major part that makes us different from other creatures. The major parts of the cortex, called neocortex (new cortex), form the outer layer of the brain. The cortex is heavily folded, giving it greater surface area. Human cortices are considerably more folded than those of most monkeys. This is important; babies who are born with smooth, unfolded brains (lissencephaly) suffer serious retardation.

Moving across the cortex, there is evidence of functional organization into columns laid out in a two-dimensional array. Moving down into neocortex, six layers of interconnected axons, dendrites, and cell bodies can be identified. Cortex can also be subdivided into different areas that are responsible for different aspects of sensory processing, motor control, and cognition. Occipital cortex, in the back, does vision. Frontal cortex, in the front, does socializing and judgment. Left temporal cortex (lateral and ventral) does speech recognition. Right parietal cortex (more dorsal) does spatial orientation. Grossly, the left brain is literate and numerate and the right brain is good at pictures. This might well remind you of phrenology, the old science of feeling the skull to figure out personality. In fact, areas of the brain may increase in size depending on their usage so that highly musical people will have a larger auditory area. Cortical areas are generally represented as forming a hierarchy from the simple perceptions of primary sensory areas, through perceptual integration, up to the multimodality representations of association areas. However, cortical areas typically feature strong recurrent loops between connected areas, and there is much evidence to suggest that even lower areas participate in high-level perception.

In addition to neocortex, there are other cortical areas that are regarded as more primitive because they are found in lesser creatures and because they appear to be more simply organized. Two of these areas are favorites for modeling: the piriform cortex and the hippocampus. Piriform or olfactory cortex is responsible for smell. Olfactory cortex is the only sensory cortex that receives its input directly from the end organ rather than receiving sensory information via the thalamus (described below). In addition to its simplicity, the closeness of olfaction to the periphery makes piriform cortex particularly popular for studying structure–function relations.

The hippocampus is one of the most heavily studied and modeled areas of brain. It is the area responsible for episodic memory. Episodic memory means memory for specific incidents and can be contrasted with procedural memory (memory of how to do things like ride bikes and play simple games) and semantic memory (how many states in the U.S.). The hippocampus is characterized by largely feedforward excitatory connections and the presence of long-term potentiation, a form of synaptic weight change suggestive of the types of changes used in artificial neural networks. Additionally, the hippocampus is involved in certain types of epilepsy, making it a favorite modeling locus for clinically oriented studies as well.

Another favorite area for modeling is the thalamus. The thalamus is thought of as the gateway to the cortex. All sensory information except smell comes into the brain through individual nuclei of the thalamus, which are called relay nuclei. One of the best studied of the relay nuclei is the one for vision, the lateral geniculate nucleus (LGN). In addition to sensory nuclei, there are other divisions of the thalamus that are involved in memory and with movement. In general, the thalamus and neocortex are heavily interconnected with excitatory feedback loops. There are also a number of inhibitory feedback loops within the thalamus. Overall, the thalamus's relay function remains obscure. One hypothesis is that the thalamus may be involved in directing attention to a particular stimulus. In addition to its role in sensation, the thalamus is also heavily involved in sleep. Again, its role here remains obscure, partially because the role of sleep remains obscure.

The basal ganglia is a central motor region in the brain. It appears to be involved in initiation and planning of movement. Diseases of the basal ganglia include Parkinson disease, which causes people to move too little, and Huntington and Tourette diseases, which cause people to move too much. Unusual features of basal ganglia organization include the facts that it is largely feedforward, that projections are largely inhibitory, and that there is massive convergence from cortex. The basal ganglia is part of a big loop involving motor cortex → basal ganglia → thalamus → cortex.

The cerebellum is also generally regarded as a motor area, though it may better be thought of as being a site of sensorimotor organization. It gets a lot of limb and body position information and helps coordinate movement and posture. The cerebellum, like the basal ganglia, is notable for the dominant role of inhibition. Purkinje cells are massive inhibitory cells in the cerebellum. They are arrayed in long rows and look like power stanchions with 10,000 wires running by them making synapses. This very regular organization has made the area a favorite for modeling as well.

Other, more remote provinces of the nervous system can also be modeled. The autonomic nervous system, with both central and peripheral branches, is largely responsible for nonvoluntary activities such as heart function, blood flow regulation, and pupillary response to light. These systems generally maintain the level of some physiological variable such as blood pressure or light hitting the retina, just as a thermostat maintains temperature in a house. Such negative feedback systems are well explored in engineering, and these models have been applied to the autonomic nervous system.

The enteric nervous system runs the gut. Although the enteric nervous system has connections that allow it to communicate with the brain, it can function independently. An isolated gastrointestinal tube, disconnected from the body, will still eat and defecate quite happily. Fancifully, I think of the enteric nervous systems as a separate creature within (i.e., an alien), with the approximate sentient capabilities of a large worm. The vague abdominal notions of fear, anger, or satiety are perhaps the closest we can come to understanding what a far simpler organism feels and knows about his surroundings: the worldview of a worm.

The brain is a complicated place. Here are a few of the many other brain areas that I haven't mentioned: cingulate cortex, claustrum, amygdala, Forel's field H3, red nucleus, substantia nigra (black stuff), locus ceruleus (blue place), nucleus solitarius, nucleus ambiguus. Some of these have an approximately known function, or at least known affiliations with areas of approximately known function. Others remain enmeshed in utter puzzlement and enigmatic mystery.

3.5 How do we learn about the brain?

In the above section, I emphasized how little we know about the brain. In this section, I want to talk about how much we do know and how we know it. As discussed in the first chapter, the levels to be investigated range from studying the flow of ions through membranes to studying cognition. In computational neuroscience, major areas of interest are information, representation, and calculation. It's likely that many of the levels of investigation will turn out to be relevant to understanding these areas. Other levels may not be relevant to these functional issues but will still be of interest as we try to understand neural dynamics. For example, our understanding of epilepsy has benefited greatly from computer models of brain activity.

Although any and all levels are of potential interest, a few levels get most of the attention. These are generally the lower levels, from the gene to the single cell. Like the parable of the drunk looking for his keys (under the light rather than where he dropped them), researchers look where they can see. Researchers are even more strongly motivated than drunks: they have to get grants.

The good news is that our light on the brain expands as new techniques are developed. The development of physiological probes such as positron emission tomography (PET) and functional magnetic resonance imaging (fMRI) have permitted the viewing of activity in the living, thinking human brain. However, there still remains a large gap in the scale at which activity is accessible. It is possible to record from a single cell in an animal brain but not usually in a human brain. It is generally not possible to record from a neuronal ensemble (hundreds to hundreds of thousands of neurons).

An important distinction to be aware of in discussing research methods is the distinction between anatomy and physiology. Anatomical methods provide static measures of structure: classical slice-and-dice man-in-the-pan methods. Anatomy does not show activity. Physiological methods do show activity. In general, when we are interested in information transfer, we are more interested in physiological measures. However, anatomical measures can help us with the wiring and basic layout of the brain.

Anatomical methods

Major anatomical methods involve using some sort of radiation, mostly visible light, infrared, radio, x-ray, and electrons. Anatomy is the science of figuring out how to look and then figuring out what has been seen. The major anatomical tool is the microscope. Another major type of tool is the imaging device. This includes computer tomography (CT or CAT scan), which uses x-rays, and magnetic resonance imaging (MRI), which uses the nuclear magnetic resonance (NMR) that results when radio waves are applied to atoms in a strong magnetic field.

Light microscopy gives a nice view — one can zoom in from the naked eye to about 1000-fold magnification. Light microscopy is very convenient since we have an organ that can directly detect and image radiation in this range. The limitation to light microscopy is due to the wavelength of light, about half a micron. It is impossible to clearly resolve anything much smaller than a micron. Cell somas are about 5 to 30 microns across and can be easily seen under a microscope. Synapses, however, are tiny structures of about 1 micron and cannot be seen in any detail with light.

Another problem with most types of microscopy is that the raw brain doesn't provide much contrast — it's mostly white on white with some gray on gray. To get contrasting images, various stains are used. The Golgi stain is historically important because it was used by Cajal in his turn-of-the-20th-century studies that identified the neuron and its role. Ironically, this stain permitted Cajal to see the separation between neurons and correctly counter Golgi's contention that neurons formed a continuous syncytium of tissue. The Golgi stain is usually picked up by only a few cells in a slice of tissue and turns these cells black. Some modern techniques involve direct injection of dyes though electrodes placed in specific cells. Since these electrodes can also be used to measure electrical signals, this offers a nice opportunity to correlate electrical activity with cell shape.

Electron microscopy (EM) works at a shorter wavelength and permits photography of objects down to about 10 nanometers. A disadvantage is the inability to zoom in and out so as to identify the connectivity of the structure seen. This means that it is not always obvious whether one is looking at a piece of a neuron or a piece of some other cell. For example, it took researchers a long time to pick synapses out of the many little circles and lines seen on the black and white images provided by EM. Additionally, EM requires difficult staining, embedding, and cutting procedures.

Imaging methods, CT and MRI, are used to look at macroscopic structures. Presently these methods cannot be brought down to microscopic levels except in small pieces of tissue. These imaging views can be extremely clear, sometimes better than looking at sliced brain after death. Of course, they have the further advantage of not requiring either death or slicing. These are *in vivo* techniques, used in the living organism, to be contrasted with *in vitro* techniques, used in tissue that has been removed and placed in a dish. These techniques are particularly useful clinically since pathology such as brain tumors and stroke can be seen. However, since current imaging technology does not permit visualization down to microscopic levels, there remains a large gap between what can be seen in life and what can be seen after death

In addition to the neuroanatomical imaging techniques, there are also physiological techniques that give some measure of activity in the brain rather than just recording structure. The most prominent of these techniques are PET and fMRI. These can be used to measure ions or metabolites in a living animal or person.

3.6 Neurophysiology

Neuroanatomy is the study of form. Neurophysiology would like to be the study of function. Unfortunately, it only occasionally rises to this ideal. We are often so clueless as to what's going on that neurophysiology becomes more a description of the dynamics than an explanation of its meaning or intent. Physiology measures things that change. In neurophysiology, a major thing that changes is electrical potential; hence, much of neurophysiology is electrophysiology. Other things that are measured include the flux of specific ions such as calcium, the uptake of nutrients such as glucose, and the binding of neurotransmitters such as glutamate.

The techniques of electrophysiology generally involve the use of microelectrodes that measure either voltage or current (see Chap. 16, Section 16.6 for description of electrical concepts, and Chap. 11 for details of electrophysiology). The electricity measured in the nervous system is generated by the equivalent of a battery that produces voltage across the cell membrane. This battery keeps the inside of the membrane at a negative potential (potential and voltage are synonyms), called the resting membrane potential (RMP). Deviations from the RMP are electrical signals in the neuron (Fig. 3.2). Signals that make the membrane more negative inside are called hyperpolarizing. Hyperpolarization means further polarizing the membrane beyond its already negative polarization. These signals are considered inhibitory, although they do not always reduce cell firing (see Fig. 12.9). Signals that make the membrane more positive inside are called depolarizing (they relieve some of the negative polarization) and are usually excitatory. The action potential or spike is a brief (1 millisecond) duration depolarizing signal that briefly reverses membrane polarity, making the membrane positive inside. The action potential can travel along the axon and produce release of chemical (neurotransmitter) at a synapse.

The most direct way to measure potential is to insert a glass microelectrode through the membrane and compare the potential inside to the potential of a wire placed outside. This is intracellular recording. Electrodes used for intracellular recording are tiny hollow straws of glass called micropipettes. The glass itself doesn't conduct electricity but the hollow tube is filled with a salt solution that does. Alternatively, metal electrodes are placed near the cell for extracellular recording. Although these electrodes do not directly measure a neuron's intrinsic voltage, they do detect electrical fields that closely reflect this voltage. If an extracellular electrode is close enough to a cell, it can record activity from that cell alone. This is called single-unit recording. If you pull the electrode further away, it will pick up signals from multiple cells — multiunit recording. The signals from the different cells will have slightly different shapes that make them distinguishable from one another. If you pull the electrode still further away, all of the signals blur into an average field potential. Field potentials can be strong enough to be detected from outside the head. This is the potential measured by electroencephalography (EEG), which detects very small potentials by using electrodes glued to the head.

3.7 Molecular biology and neuropharmacology

Molecular biology has in recent years been the most rapidly advancing field of biomedicine. The capability exists not only to knock out specific genes but in some cases to knock out specific genes in specific brain areas at specific times in development. The well-known "central dogma" of molecular biology is that DNA makes RNA makes protein. By creating a "knockout," a particular gene is inactivated, permitting us to evaluate changes in an animal's behavior that reflects the function of that gene. The problem with this approach from the modeling perspective is that these behavior changes are unlikely to be a direct and obvious consequence of function. By way of a hoary example, if you stick a screwdriver into your radio and it starts making a high-pitched whine, this does not necessarily imply that you have cleverly ablated the "whine-suppressor" for this radio.

Neuropharmacology is the study of the effects of drugs on the nervous system. Neuropharmacology is mostly the study of receptors — figuring out where in the brain different compounds bind and what their binding activates. In general there are many different subtypes of receptor for each endogenous ligand and there are usually many different non-endogenous compounds that also bind to specific receptor subtypes. Some of these non-endogenous compounds are synthetic, having been developed by drug companies to try to treat disease. Many more, however, are natural products that one organism makes in order to kill or disable another organism. General examples of this are apple seeds, which contain cyanide, and fungi that make penicillin to kill bacteria. Neuroactive compounds are quite common poisons since the nervous system is a particularly vulnerable spot in animals. The puffer fish makes tetrodotoxin, a sodium ion channel blocker. Bees make apamin, a potassium channel blocker. Scorpions produce charybdotoxin, another potassium channel blocker. The deadly nightshade plant makes belladonna, deadly at high dose but formerly used at low doses to make a woman beautiful by increasing the size of her pupils.

3.8 Psychophysics

Psychophysics uses precisely defined stimuli and then asks how they are perceived by a person, often the investigator himself. Early triumphs in this area involved the use of specific colors to probe the way that the eye perceives light. Even before detailed molecular and cellular studies could be done, this led to the understanding of the three different types of color receptors in the retina, each tuned to a different frequency. The relation between these frequencies explains the effects of color combinations: why blue and yellow appear green. More recent psychophysical research has used computer-generated images to probe higher levels of visual function. For example, it has been shown that rapid projection of images can lead to breakdown in binding of stimulus attributes such that the color and shape of an object are not correctly correlated. For example, rapid presentation of a red square and a blue circle can lead to perceptual error such that a blue square and a red circle are seen (illusory conjunction).

3.9 Clinical neurology and neuropsychology

Medical study of people with brain disorders has played a central role in our understanding of normal brain function. Damage to the brain produces in the victim a very personal experience of the dependence of mind on brain. Brain damage can produce peculiar states of depersonalization and derealization. Patients with obvious brain pathology are treated by neurology, while those with grossly normal brains are classified as having mental disorders and treated by psychiatry. A disease like stroke is easily seen to be a brain disease; if you take the brain out, you find a hole in it. In a disease like schizophrenia, abnormalities in brain structure are much more subtle and are only now being discovered. Schizophrenia was traditionally viewed as a mind disease, due to having a bad mother rather than a bad brain. In addition to the increasing awareness of brain abnormalities in psychiatric disease, there is also increasing appreciation of the changes in personality and thinking that occur in association with brain diseases such as multiple sclerosis and Parkinson disease.

The object of clinical research is to alleviate illness rather than to understand human function. Nonetheless, studies of patients have contributed substantially to our understanding of the brain. Disease can be regarded as an experiment of nature that may reveal important insights about neural organization. From this perspective, ablative and intrinsic brain diseases represent different kinds of experiments. Just as the design of an experiment determines what can be and what is discovered, the kinds of brain diseases that occur have had a large influence on our view of the brain.

Ablative diseases

An ablation knocks out a piece of brain. Stroke, tumor, and trauma are major ablative disorders. These disorders come from outside of the functioning brain itself. They are imposed on the brain. War injuries were the first brain diseases studied in the early days of neurology, during the Civil War. More recently, brain dysfunction after stroke has been heavily studied. A stroke occurs when a blood vessel is blocked off, starving a part of the brain and killing it. A stroke of the left middle cerebral artery typically causes an aphasia, a disorder of language. Such aphasias have been widely studied to learn how the brain processes language. However, the specific patterns of aphasia may have as much to do with the particular patterns of middle cerebral artery organization as with the brain's organization for language production.

Since ablative diseases are imposed on the brain from outside, these insults can produce idiosyncratic effects that do not reflect attributes of brain organization. In general, the modular nature of ablative disease has suggested a modular view of brain function. After a stroke, an area of brain is lost and the patient has a particular dysfunction. It is then natural to conclude that the piece of brain that was lost was the "center" for the function that was lost. From there, one builds up a view of the brain as a series of these centers that pass information to one another. Before the development of brain imaging, such modular brain theories were very useful for the localization of strokes. Modular theories continue to be highly influential in brain science as well as in neurology. Certainly, the brain is not equipotential: some brain areas are dedicated to vision, while others are used for hearing. However, while there is merit to modular theories, they tend to be oversimplified and can tell us only a limited amount about how the brain functions.

A remarkable set of stroke-related mental disorders are those seen with damage to the right hemisphere. Neglect is a neurological syndrome that occurs frequently, albeit often transiently, following large strokes in that area. The neglect is of the left side of the world and the left side of the patient's body. Patients ignore people and things situated to their left. Most remarkably, they may be unable to identify their own left arm. If pressed, they confabulate absurdly, concocting ever more elaborate stories to explain why a strange arm would be found in such close contact with their body. One patient, for example, was asked so frequently about his arm that he developed the peculiar habit of throwing one cigarette across his body whenever he smoked. When asked, he explained that his brother was sharing the bed with him and he wanted to share the cigarette with him.

The bright side of ablative diseases is that they are not progressive. They damage the brain and stop. The brain can then start to recover. It is remarkable how well the brain can recover. A few months after a stroke, a person may be left with no noticeable deficit, despite having a hole in his brain that can be seen on MRI. The process that leads to recovery of function is similar to the process of functional organization during development and to the normal process of cortical reorganization that occurs with learning. In all of these cases, modeling has suggested that an initial phase of altered dynamics is followed by synaptic plasticity. The brain is a dynamic structure, with functional circuits that are repeatedly organized and reorganized.

Intrinsic diseases

Diseases caused by alterations in cellular organization, metabolism, neurotransmitters, or electrical conduction are intrinsic diseases. Correlation of mental dysfunction in these diseases with underlying cellular or chemical abnormalities will likely provide important insights into how the brain works. For example, schizophrenia produces peculiar patterns of thought that are also seen between seizures in some forms of epilepsy. These syndromes are likely to involve widespread abnormalities in intercerebral communication. Intrinsic diseases are typically progressive, making them degenerative diseases

Intrinsic disease are generally harder to study than ablative disease. As experiments of nature, they correspond to more recent microscopic experiments that classify various cellular and molecular components of the nervous system. Study of these diseases tends to suggest a global interconnected view of brain function. Intrinsic diseases of the brain are caused by disorders of the elements of brain organization: synapses, ion channels, or particular neuronal types. The prototypical intrinsic diseases are those involving intoxication with neurotransmitter agonists or antagonists. Disorders of mentation associated with various drugs are examples of this.

Parkinson disease is another example of an intrinsic disease. Loss of a particular type of cell results in a decline in release of dopamine, a neurotransmitter. This has profound effects on movement and thinking, causing a marked slowing as well as other problems. Giving a dopamine precursor as a drug can restore these functions. This finding suggests a diffuse network of interacting units somehow glued together by use of a common neurotransmitter.

3.10 Summary and thoughts

Even without considering computational neuroscience, one finds many different ways of thinking about the brain. These come from the different types of experimental techniques available, the different kinds of training of the people doing the experiments, and even from the varying types of diseases that people get.

Much of the research that has been done has gone into elucidating the chemical and electrical properties of individual neurons. Far less has been done at higher levels. In particular, the technology does not exist to determine detailed neuronal connectivity or to assess the activity in large neuronal ensembles in behaving animals or people. Historically, notions about the highest levels of perception and behavior have mostly been deduced from strokes and personal intuition. The recent development of functional imaging should give more solid information about brain activity during mentation.

Now that techniques for studying most of the levels of organization are becoming available, it is tempting to look forward to the day when we can just look at all levels of neural activity and see how the brain works. With the experimental means to identify everything everywhere all the time, we could build the perfect brain model. It would do everything that the brain does in just the way the brain does it. Even with this, understanding would still elude us. We still wouldn't understand any more about *how* the brain does what it does. Complex systems like the brain show emergent properties, properties that cannot be explained by just knowing the properties of the constituent parts. The classic example is the emergence of the laws of thermodynamics from the bulk behavior of very many gas particles, each of which follows Newtonian rules. Newton's laws (or quantum mechanical laws) do not directly suggest the thermodynamic laws.

Similarly, a small ganglion in the lobster runs the lobster stomach with only about 30 neurons. These neurons have all been studied thoroughly; their responses mapped. Given the simplicity of the network, it would seem that knowledge of the elements would yield understanding of the network dynamics. This has not been the case. Computer modeling is being done to understand the emergent properties of this simple network. There are several Web sites devoted to this thing — look for stomatogastric ganglion.

Neuroscience subfields like neurophysiology, neuroanatomy, and neurochemistry each cover a particular area of neuroscience understanding. Explicit computer models build bridges across the gaps between these fields. Every researcher is a modeler. Without a model, whether held in the head or written out in words, pictures, or equations, data do not confer understanding. The brain is a place of myriad wonders and many different models will be needed to explain it.

Part II

Computers

4
Computer Representations

4.1 Why learn this?

In the 1950s, biologists were homing in on the chemical source of heredity. Skeptics pointed out the enormous variety of things that had to be coded for: enzymes and structural proteins, the shapes of faces and bodies, the number and the form of limbs, the structure of the brain, and the structure of the kidney. It seemed inconceivable that all these representations could be reduced to any single process or principle, much less to a single code or a small set of symbolic chemicals. The skeptics were wrong. Turns out that there are four chemicals that are arranged in triplets to code for 20 amino acids plus stop and start codes. Admittedly, this hasn't solved the problem of understanding where an arm comes from. However, it's remarkable that at some level everything can be expressed with an alphabet of four letters, word lengths fixed at three, and some awfully long sentences, all of which can now be read.

Many people are skeptical that any simple code will be extracted from the brain that will allow us to congeal our understanding of vision, audition, volition, and respiration, to name just a few. Actually, I'm skeptical. However, some of the pioneers who unraveled the genetic code subsequently switched over to the search for the neural code. Perhaps their experience in genetics taught them something of the possibilities in impossible problems.

The classical behavioralists held that, as far as the brain was concerned, "There is no there there": [1] everything could be explained by simple loops connecting a stimulus with a response. If something happened out in the world, the brain would respond appropriately; there was no need to postulate complex internal representations or internal processing. I haven't read much behavioralist literature, but even behavioralists must have admitted that people and animals can remember things. Anyway, nowadays it's generally accepted that there are representations in the brain. At some level, the brain has to represent or model the world out there. A lot of research in computational neuroscience, cognitive neuroscience, and neurology has to do with hypothesizing what these representations might be and then coming up with tests to see if these representations can be validated.

In this chapter, I discuss basic computer representations. Representations in the brain most likely differ considerably from computer representations. Nonetheless, learning about computer representations is valuable for a couple of reasons. First, these representations are very influential in neural network research. For example, most neural network models assume that a neuron can be approximated by a unit with a scalar state (a scalar, a single number, is contrasted with a vector or array of numbers). Some go further and suggest that this scalar state will only take on two values (i.e., a binary device — like a transistor).

Another reason to study computer representations is that these representations are the best-quantifiable examples of complex information storage strategies that we have access to. Note the qualifier "quantifiable" here. One would expect language, poetry, music, and painting – complex products of the free-range brain — to be closer to brain representations. Linguists argue that the structure of language is a good indication of the structure of the brain and that commonalities among different languages reflect this structure. At some level this is likely to be so. However, language is a rare, unusual brain capacity seen only in one species. The human brain has probably developed far in a unique direction to perform this clever trick. I suspect that language will turn out to involve a peculiar mapping, remote from any essences of neural representation.

Representations in a computer are called data structures. They are designed to be efficient for some purpose. Such design always involves tradeoffs. In this chapter, we look at typical computer representations for the alphabet, for numbers, and for pictures. In the next chapter, I discuss computer design and the design of programs, demonstrating how a program is just another type of representation. The design of appropriate data structures is a key aspect of the art of computer programming. Similarly, competing needs and pressures have formed representations in the brain as well.

---

[1] Gertrude Stein's comment about Oakland, her hometown.

4.2 Calculator or typewriter

Many people think of computers as overgrown calculators. However, the major thing most people do with a computer is word processing, which is not calculation at all. The computer, a mysterious object for many, thereby seems all the more mysterious: a glorified calculator or a glorified typewriter? This has also been a problem for computer experts. A lot of debate about computer architecture had to do with dual design goals: data processing for business markets, and number crunching for scientific and engineering markets. When the founder of International Business Machines (IBM) made the famous prediction of a world market for only five or six computers (1958), it reflected the difficulty of seeing any advantage to using an expensive, and at that time unreliable, machine as a substitute for file cabinets and punch cards.

The central concept that reconciles the computer's calculating and word processing functions is the notion of representation of information. Ideas about representation come as much or more from business as from mathematics. Computers emerged as tools for business and government under the auspices of IBM. Historically, the early "information revolutions" came out of business. (And perhaps prehistorically, counting and writing probably developed for trade.) In the early Renaissance, double-entry bookkeeping developed as a data management technique to organize debits and credits in an accessible data structure. Other early revolutions included both well-known high technologies such as printing, and more obscure low technologies such as pigeonholes and alphabetical filing.

The epitome of turn-of-the-20th-century information processing technology is the often caricatured large room filled with clerks, each of whom has an in-box and an out-box. This technology was widely used for "statistics," a field that originally meant keeping track of information about the state or nation. At the time, this labor-intensive technology was proposed as a way to do weather forecasting, using the same algorithms as those now used by supercomputers for this purpose. The idea was that each clerk would be provided with information about air temperature, pressure, and winds at a particular part of the globe for which he was responsible. By comparing his information with that of neighboring clerks responsible for neighboring areas of the globe, he could determine whether air would be flowing into or out of his zone, and thereby determine weather. This was an early discovery of the finite element algorithm that is used today. The algorithm was never implemented using office workers.

4.3 Punch cards and Boolean algebra

At the end of the 19th century, standardized data forms and punch cards came into use, introduced in order to tabulate counts for the United States census. Early data processing machines were developed to count holes in punch cards. These were simple mechanical computers but differed from modern electronic computers in that they were not user programmable. Their programming was built in, and was generally fairly straightforward — e.g., count the number of times that a hole in row 15, column 5 is present along with a hole in row 23, column 73.

A simple version of this basic computing device can be built with index cards. Prepare each card by punching 10 holes uniformly along the top. Binary information can then be entered by cutting out the top of a hole to indicate "true" and leaving it closed over to indicate "false." For example, let's say you have a file of recipe cards. Some recipes require sugar, some require flour, and some require broccoli. You assign each of the punched holes to a particular ingredient. If that ingredient is present in the recipe, you cut open the hole to the edge of the card. If the ingredient is not used in that dish, you leave the hole alone. The intact hole (the default) then means false (represented as F or 0 in Boolean algebra); the open hole represents true (represented as T or 1).

If a dish requires sugar and flour, but no broccoli, the sugar and flour holes should be opened up to indicate that it is true that the recipe requires sugar and it is true that the recipe requires flour. Now comes the clever bit. You have a stack of recipe cards and you have discovered that you have broccoli on hand, but no sugar or flour. You take a knitting needle and pass it through the "sugar" hole and the "flour hole." You pick up the stack by the two needles and shake slightly. Those cards that require sugar or flour will fall out. You put aside the cards that fell out. You now do the same thing with the broccoli hole, but in this case you keep the cards that fall out. This is the set of recipes that require broccoli but do not require sugar or flour. In Boolean notation, you have performed the logical operation "(not (sugar or flour)) and broccoli." which can also be written "(not sugar and not flour) and broccoli." In Boolean notation, using the first letter of each word as the symbol, this is $(\sim (S \vee F)) \wedge B$ or, equivalently, $(\sim S \wedge \sim F) \wedge B$. The needle always pulls out cards that don't use that ingredient (False) and leave behind cards that do use it (True). Complex Boolean operations can be performed using the proper combination of needles.

This recipe card scheme was marketed as a practical database management technique in the 1960s. Although it has been made obsolete with the increased availability of databases on electronic computers, the algorithm remains a viable one for some database implementations. Multiple true/false fields can be stored in a single word with each true/false choice taking up only a single bit. These can then be combined using Boolean operators such as "and," "or," and "not."

Any information storage strategy has advantages and limitations. Different techniques will differ in the type of information most readily stored, the storage capacity, the ease of entering new information, the speed of accessing old information, and other factors. Trade-offs depend on what aspect of data manipulation needs to be optimized. For example, a database that has to have frequent changes will be optimized far differently than one that gets set up once and then accessed many times. One well-studied trade-off is that between space and time: big memory-consuming data structures typically offer faster access to information. Although the specific data storage issues are likely to be far different in the brain, the general trade-offs may well be similar. In particular, we might expect that brain areas that require rapid access to information might be bulkier than those where speed is not important. This might be a contrast between the memory functions of cerebellum and hippocampus.

A common data format that illustrates the time/space trade-off is the scantron test form that's used for the Scholastic Aptitude Test (SAT) and other standardized tests. These forms have been optimized for ease of input by the untrained user. Although it always feels like they take a lot of time to fill out, they are actually rather quick when one considers the training that would be required for use of a more complex, more compact data structure. At the beginning of these tests, you fill in your name and student ID with the famous number 2 pencil. Each circle on the scantron form represents a single bit. It can either be filled in or not. If it is filled in, it represents a 1 and can be considered a *set bit.* If it is empty, it represents a 0 and can be considered a *clear bit.*

The bits on the scantron form represent not only binary numbers but also represent your name, social security number, and answers to the test questions. This is where the choice of representation comes in. One needs to represent the many digits and many possible values of each name field using bits. Let's take the binary representation of a last name as an example. Each letter of the name is represented by a set of 26 bits (in a column), out of which one bit is set and the rest are clear. The name is represented across a fixed number of columns, usually 20. This is a representation of your name in binary. It uses $20 \cdot 26 = 520$ bits of storage.

The scantron representation of your name is inefficient but is an easy one for untrained people to understand and quickly master. This representation is therefore a good one for the person–machine interface. Since we know that the first letter of the name is a capital and the rest are small letters, we only need the 26 letters and can simply translate the first letter into its capital form and the others in the small form. Including all 52 large and small letters would be a redundant representation for this purpose. This is an example of how the information content of a message can't be viewed in isolation but is also dependent on the knowledge base of the transmitter and receiver of that message.

A more efficient name representation would be a direct bit code for each letter. Given that there are only 26 letters, only five bits would be required to store a letter in binary since $2^5 = 32$. As we see below, the standard computer encoding for a letter actually uses seven to eight bits. Given that your last name is probably less than 20 characters in length, bits are being wasted there as well. If we instead wanted to use a five bit code for each letter of the alphabet, the user would have to look up each code in a table. This is poor design for a human interface; there would likely be many errors. An additional inefficiency comes from having a fixed-length representation of 20 characters for the last name. This inefficiency represents a limitation of the medium: paper can't grow and shrink as needed. Although hard disks and memory also don't physically grow and shrink, many data structures do so. If you want a data structure to be arbitrarily sized, you typically use a marker to indicate where the field ends or a variable counter to give the field size for each instance. A compact code for my six-letter last name might use 36 bits: six bits to give the number of letters (this allows counts up to 63) and $5 \cdot 6 = 30$ bits for the name itself. This is much smaller than the 520 bits of the scantron form.

4.4 Analog vs. digital representations

The word *digital* comes from the word *digit*, meaning finger. Though this would suggest that it refers to the base 10 number system, the word is used to refer to any equipment that uses discrete rather than continuous states. Thus, the modern computer is a digital computer that uses binary representations. This is likely to be a major difference from brain function. Although Von Neumann and other early computationalists took the all-or-none nature of the axonal spike as evidence that the brain was primarily a binary digital device, this viewpoint is not currently popular.

Binary means 2-valued: each element (a binary digit or bit) takes on either a value of 0 or a value of 1 (see Chap. 16, Section 16.3). One can produce a discrete system using any number of symbols. For example, a computer could use decimal instead of binary by dividing up the 5 volts into 10 ranges instead of 2, each of which would then represent a digit from 0 to 9. Discrete representations can be readily expressed in symbols. Each symbol of a set of symbols corresponds to a particular value. For example, the two physical values of the binary digital representation used in a computer are nominally 5 volts and 0 volts. These values are symbolized as HIGH vs. LOW, True vs. False (using positive true), T vs. F, or 1 vs. 0. Using binary symbols is a good idea for discussing computer systems since the underlying signal, voltage, is actually an analog measure. Because of this, a computer has to accept a range of values for each of the symbolic values: typically anywhere from 3.5 to 5 volts for 1 and anywhere from 0 to 1.5 volts for 0.

A digital representation uses discrete values. An analog representation uses continuous values. A property of a continuous value is that there is always another value between any two values that you pick. For example, there are temperatures between 53° and 54°. In fact, there are an infinite number of temperatures between these values. Like temperature, other measurable macroscopic values in the world are analog values, although quantum values are, by definition, discrete. Voltage is the main quantity used to encode information in the computer. Digital information is extracted from this analog signal by thresholding. Discrete representations have the advantage of being precise. The field of information theory is based on the ability to "count" the amount of information by translating it into binary representations.

Analog representations are more complicated, since they can theoretically contain infinite amounts of information, limited by the ability to read out the information, determined by measurement resolution and noise. This infinite precision doesn't imply good accuracy, however. Early analog computers were used to model the trajectories of artillery shells. Since the equations that described the behavior of the electrical components in the computer are identical to those that describe the behavior of cannon shells, these machines could theoretically produce perfect simulations. However, the electronic components were temperamental. Their parameters would change with change in temperature and this would change the predictions. The behavior of the artillery shells would also change slightly with temperature change but this change was not be as great and perhaps not even in the same direction.

4.5 Types of computer representations

I now describe some typical representations used in the computer. The point is to illustrate that the nature of the computer's design dictates the forms of its data structures. The most substantial single requirement is that everything must be reduced to binary. In the computer, binary provides a single common representation from which all other representations are built. Because of this, all representations are interconvertible. A picture on the computer screen can be mistakenly printed out as gibberish text. It can also be executed as a computer program, which will most likely crash the machine. This interconvertibility of representations is a significant difference between computer and brain. In the nervous system, information is stored in a variety of chemicals and electrical forms. It is unclear how these different representation systems can be related or if they may in part operate largely independently. The complexities of neurons presented later in the book make computer data structures appear simple and elegant.

As we saw in the case of the scantron, computer representations may be designed to allow the computer to interact with people and the outside world or to communicate efficiently with each other, with other data structures, and with programs. In addition to the environmental interactions required by input and output to people, computers may also have to directly sense or manipulate the physical world. Development of such programs, the realm of robotics, is of great interest for our understanding of brain function. Getting a machine to do something practical in the outside world often reveals the limitations of representations that otherwise seemed adequate. For example, a digitized photograph seems a pretty good representation for planning navigation around a room. However, extracting the needed information about the three-dimensional world from this two-dimensional representation is a difficult task that has perplexed researchers for years. One trick that brains appear to use is to actively seek out sensory information on an as-needed basis rather than depending on the passive absorption of information that happens to hit receptors. In the context of seeing, this is called "active vision." Your eyes are always moving around, building and rebuilding your image of the world rather than just waiting for something interesting to drift into view.

Representations are also specialized depending on extraneous factors having to do with communications media or conditions. For example, long-distance communication is expensive and makes a virtue of brevity in the representation. This has led to development of data-compression algorithms. Under circumstances where security is a concern, such as communicating credit card numbers over the Internet, brevity is be sacrificed in order to assure obscurity and prevent eavesdropping. This is the realm of cryptography. Under conditions where transmission lines may be unreliable, redundancy is used to ensure accuracy of transmission using error-correcting codes and check sums.

4.6 Representation of numbers

It may seem funny that one has to worry about the representation of numbers on a computer, a machine that was designed to handle numbers. Even here, however, compromises must be made to balance the demands of input/output (here communication with the human user) and internal functionality (arithmetic operations). Arithmetic and logic operations are performed in the computer with strings of zeros and ones. Modern computers are designed to do this as efficiently as possible. To achieve this goal various number representations have been developed to cleanly handle negative numbers, fractions, decimals, complex numbers, logarithms, and exponents. Operations range from addition of integers to transcendental functions on complex numbers.

Early on, some computers were designed to work directly in base 10. This meant easy communication with users but inefficient representations with many wasted bits. It was also difficult to program these machines to do arithmetic. Since the underlying hardware uses binary values (0 volts, 5 volts) on the transistors, it became clear that everyone would be much happier if all internal operations were done in binary. All modern digital electronic equipment uses binary. When you use a calculator to multiply two numbers, the calculator translates the base 10 numbers that you enter into bits.

Addition and subtraction are the most basic arithmetic operation. Subtraction is, of course, just the addition of a negative number. The development of various representations of negative numbers in order to permit efficient subtraction illustrates the use of a machine-dependent hack to optimize performance. In particular, the now-standard representation, two's complement, uses to advantage an architectural quirk that could otherwise be regarded as a limitation on the computer's ability to do arithmetic. As far as I know, every computer nowadays uses two's complement.

Two's complement uses the fact that binary numbers are stored in physical memory locations of finite length. Computer memory is divided up into words and bytes. Depending on the specific computer architecture, words can be of different sizes. However, a byte is now pretty much standardized at 8 bits in length. A word will typically be 4 or 8 bytes, hence 32 or 64 bits. Using straight binary storage an 8-bit byte can store up to 11111111_2 (FF_{16}) which is $100000000_2 - 1$ or $1 \cdot 2^9 - 1 = 255$. If we needed to store bigger numbers, we could just take bigger and bigger chunks of memory. Of course, we would then need some kind of additional tag to indicate how many bytes were being used to store the number. Perhaps the number of bytes used could be stored as the first byte of the data structure. This data structure would be fairly efficient in terms of memory use but would make it difficult to do additions and subtractions since the numbers would have to be aligned correctly first. This representation also could not readily handle negative numbers. Since this sort of elastic representation is not used, numbers must fit within the limited space allotted. This means that numbers that are too large cannot be represented. In particular, if you add two big numbers, you will get a number that won't fit. This process will lead to an overflow condition, and bits will be lost. As we will see, this is the machine limitation that is used in the two's complement representation.

First let us look at a couple of alternative negative number representations that have been used in computers. One way to represent negative numbers is to take one bit away from the numerical representation and make it a sign bit. This is called the sign magnitude representation. With this representation, a byte only has 7 bits to represent the number, up to 127_{10}. If the first bit of the byte is 0, that would indicate a positive number and if 1, a negative number. This method has the advantage of relatively easy translation for entry or print-out since the presence or absence of a negative sign translates directly into one bit. Unfortunately, addition and subtraction are not particularly easy using sign magnitude.

Another representation is called one's complement. This representation takes advantage of the fact that it is easy to design hardware that will invert all the bits of a number, changing all the 1s to 0s and all the 0s to 1s. This inversion of all bits is called the bit-wise complement of a number. The bit-wise complement can be used as the negative of a number. Since the bit-wise complement creates a 1 wherever there was a 0 and vice versa, any number plus its complement will yield a byte with all 1s: 11111111. Since a number plus its negative is zero, this means that 11111111 will be a representation for zero in one's complement. Since the complement of 11111111 is 00000000, this is another zero. The former is considered -0 and the latter $+0$. Having two zeros is a bit of an inconvenience for input and output. An additional complexity arises in the need to add back the overflow when doing subtraction with one's complement. As an example, let's subtract $7_{10} - 1_{10}$ using 4-bit one's complement. The bit-wise complement of 0001 (1_{10}) is 1110. 7_{10} is 0111. We add $7_{10} + -1_{10}$ in one's complement by doing $0111 + 1110 = 10101$. We are using a 4-bit representation and the extra bit on the left represents an overflow bit. Using one's complement, the overflow is added back to the 4-bit result to get the final answer: $0101 + 1 = 0110$ or 6_{10}.

An improvement on the one's complement scheme is two's complement. Under two's complement, there is only one zero value. There is also no need to add back overflow bits. Instead, overflow can be discarded. The two's complement is formed by adding one to the one's complement. Remember that the sum of a number and its one's complement is 11111111. Since the two's complement is the one more that the one's complement, the sum of a number and its two's complement is $11111111 + 1$. Since this produces an overflow condition, $11111111 + 1$ gives 00000000, which is the unique zero. This overflow situation is familiar from experience with car odometers. These can only represent a certain number of digits and will eventually overflow to go back to all zeros. On an odometer, we can consider the two's complement of any positive number to be an equal negative excursion on the odometer (i.e., driving backward). If we start at that negative point and then drive forward the same distance, the odometer will register zero again.

Let's look at the previous problem, $7_{10} - 1_{10}$, using two's complement. As before, the bit-wise complement of 0001 is 1110, so the two's complement is 1111. Then $7_{10} + -1_{10}$ is $0111 + 1111 = 10110$. We throw away the left-most bit and get the desired answer: $0110 = 6_{10}$.

To give another example, the calculations for $6 - 5$ compared:

	One's complement	Two's complement
5	0101	0101
negation	complement	complement + 1
−5	1010	1011
6	0110	0110
sum: low bits	0000	0001
sum: overflow	1	1
overflow handling	add back	discard
final result	0001	0001

Representation of letters and words

Many of us have had our primary exposure to computers through the use of a word processor. In this case, the primary representations used are those for letters and symbols. Ancillary codes are used for things like formatting and choice of font. The standard code for letters and symbols is called the Ascii code. This assigns a unique 7-bit number for each letter, digit or symbol (e.g., , . ; + =, etc.) as well as codes for non-printing symbols like "newline" (the end of the line where you start up again at the next line).

Computer representations are generally designed for efficiency but the standard for efficiency may differ widely depending on the purpose of the representation. The computer I am writing on has a 64-bit architecture. Since there are 8 bits to the byte, my computer has 8 bytes to a word. The first sentence of this paragraph (italics) happens to be 160 characters long, including the period. It will therefore fit into 20 words of memory. We can therefore lay the sentence out as it would appear in the memory of my computer (Fig. 4.1, top). This is a bitmap: each bit in a region of memory is shown with a black square for a 1 and a white square for a 0. Each of these squares is considered a pixel or picture element. As in the paintings of the pointillists, if you make the pixels small enough they blend together and produce a picture. In this figure, each word of memory is read from left to right and consecutive words are read from bottom to top. The sentence begins at a word boundary.

With a little effort, it is possible to read the text off of the bitmap. At the lower left corner 01000011 is the capital C in the word computer — the string of four 0s gives the horizontal black bar four squares long. It is easy to see the layout of letters because all of the symbols of printable Ascii begin with a 0 — Ascii is really a 7-bit code that is commonly stored in the 8 bits of a standard byte. The vertical black bars in the bitmap, representing 0s that line up in every word, show where characters begin. Spaces are easily found because they begin with 00, as does punctuation, for example the period in the upper right hand corner.

Since printable Ascii is basically a 7-bit code, we could store it in less space. At the bottom of Fig. 4.1, I stored the same sentence 7 bits at a time. This saves about two and a half words. There is 0-padding at the top right

Fig. 4.1: Two computer memory representations of the italicized sentence in the text.

and lower left. The C starts at the first white pixel from the left on the lowest row. The final period appears in mid-row at the top. Most printable characters have a 1 in the first bit. Nonprinting Ascii representations are used for monitor control, clearing or flashing the screen, or resetting the cursor, and are available from the keyboard. For example, the lowest Ascii values are used for control characters (entered from the keyboard by holding down the control key and then pressing one of the alphabetic keys): CTL-A is 1 (0000001) and CTL-C is 3 (0000011). The 7-bit codes for the letters A to Z and a to z all begin with 1. Just using the lower 6 bits is an adequate code for letters, numbers, and some punctuation.

If you want to start a new religious cult, take some text from a major religious work and turn it into a bitmap that happens to look like something, preferably a face. You can use Ascii or any other arbitrary encoding. I tried doing this with the Book of Genesis (the one from the Bible) but wasn't able to come up with anything remarkable.

4.7 Representation of pictures

Above, we have used a bitmap as a way of displaying the contents of memory. More commonly, bitmaps are used to paint pictures on the screen. For example, the pictures one sees on the World Wide Web are color bitmaps,

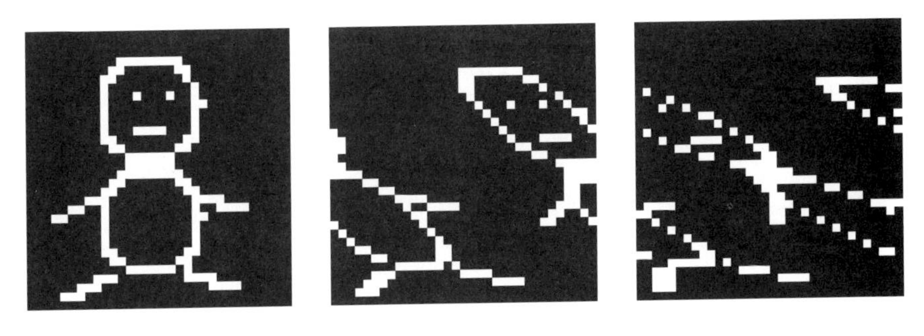

@@@@@@N@@@@@x@Cp@C‘G‘@@C}@@@C@p@@A‘F@@@P@‘@
@D@H@@A@B@@XP@p@Ad@Ix@G@Cp@@XA‘@@C@p@@@_x@@@
C|@@@@ @@@@pL@@@XA‘@@D^H@@A@B@@@P@‘@@D@L@@AD
R@@@P@‘@@F@P@@@pL@@@G~@@@@@@@@@@@@@@@@@@

Fig. 4.2: A bitmap picture of a baby represented in a 32×32 bitmap. Two incorrect bitmap decodings are shown: a 33×33 bitmap and a 34×34 bitmap. Bottom: the same image decoded as Ascii.

with a value for each pixel, which gives its color. This is more easily appreciated in a black and white picture, the way pictures are reproduced in newspapers. To turn a real-world scene into a pixilated picture, one divides the picture into a grid and then thresholds each grid location for luminosity. For black and white, a simple bit code suffices, e.g., 0 for white and 1 for black. Copying the values of each pixel row by row turns a picture into a bitmap or bitstring.

The integrity of a bitmap depends on knowing its proper dimensions. For example, the bitmap picture of a baby (Fig. 4.2) is 32 bits by 32 bits for a total of 1024 bits. Displayed incorrectly as a 33 by 33 bitmap, the picture is obscured (middle bitmap) and it is unrecognizable in a 34 by 34 format (right bitmap). If read out as Ascii (below in 6-bit groups left-padded with 01) instead of as a bitmap, all identity is of course lost.

A vector plot is an alternative to bitmaps for picture representation. In a vector representation, the picture is laid out using a set of commands that describe where lines are to be drawn. Strictly speaking, a vector plot draws everything as stick figures made up of line segments. The individual commands tell where each line segment is to be drawn. Most vector languages include more complex commands in addition to line segments. For example, the vector language used to produce Fig. 4.2 uses circle, rectangle, and filled-rectangle commands in addition to a line command. The following commands were used to produce Fig. 4.2.

```
//      32  32  -1
circ    15  23  6
circ    15  10  6
frect   12  15  17  17
line    4   1   11  4
line    19  4   25  2
setbit  13  24
setbit  17  24
line    3   10  9   13
line    20  13  27  11
line    13  20  16  20
```

The first line gives the size of the bitmap to draw: 32×32. Subsequent lines tell what to draw using a coordinate system starting at pixel $x = 0, y = 0$ in the lower left-hand corner and going up to 31,31 at the upper right-hand corner. For example "circ 15 10 6" says to draw a circle centered at $15, 10$, about halfway across and one-third of the way up in the bitmap. The circle has radius 6. This circle is the lower circle in Fig. 4.2. Note that the outline of the circle isn't very circular. At this resolution, the bitmap is too coarse to draw a very smooth curve. The other commands are also fairly obvious: "line" draws a line between two points, "frect" fills in a rectangle between two points. "Setbit" just changes a single bit from 0 to 1, black to white in the picture.

The above vector description is a simple graphical program that serves as an alternative to the bitmap representation of the picture. Bitmaps and vector maps are complementary and are both used extensively in computer graphics. Vector maps have the advantage of usually providing a much smaller representation. A bitmap has to represent all of the 0s as well as the 1s. Line drawings like this one are usually sparse. Vector maps give data to represent only what is to be drawn, not what is in the background. A vector map scales well and is well suited to animation. Unlike the bitmap representation of Fig. 4.2, the commands will still function correctly if the size of the bitmap is misconstrued. Also the whole picture can be expanded by simply multiplying the values given as arguments to the commands by a number. Vector languages are commonly used for printer languages such as Postscript. It's critical to use a vector language when working with a vector device such as a plotter (a device that moves a pen across the paper to draw graphics). Although it is easy to generate a bitmap from a vector language, it is very difficult to go the other way and figure out what vector commands were used to make a particular bitmap.

Although bitmaps are generally bigger than vector maps, they are easier to print out to screen or paper. To present a bitmap on the screen, the bits can be drawn out in the sequential order visited by the scanning electron beam on a television or cathode-ray computer monitor. A vector map, however, requires interpretation and does not provide all of the bits in order. For example, the command "circ 15 23 6" requires that the program calculate radial locations using the equation for a circle ($r^2 = x^2 + y^2$) and then round off where each of these positions falls on a square grid.

Description length is one concept that is used to consider the relative advantages of various representations. Description length takes into account not only the size of the data structure for the information but also the size both in space and in time for the algorithm needed to decode and recode the information. The complexity of this algorithm will depend not only on the complexity of the representation, but also on the design of the input and output organs and on the design of the computer itself. As we've noted above, bitmaps draw directly onto a screen, while vector maps translate directly into plotter commands. For the vector language given above, the program for interpreting the language requires about 100 lines of programming code. By contrast, a simple program to interpret the corresponding bitmap takes only four lines of code. The vector language interpreter requires both greater space for storage of the program and greater time for its execution.

As you can see, the choice of bitmap or vector plot as a data structure depends on the requirements of the program that will use the data structure. In modern computers, memory is cheap. Graphics software commonly maintains both representations so that they are immediately available for interactive use. For example, an animation program will use vector representations for its primary store to make it easy to move objects around independently. However, it will need to calculate a bitmap for each image in order to present it on the screen.

4.8 Neurospeculation

Just as complex computer representations are all built on top of the low-level binary of the transistor, so brain codes are compounded on top of low-level codes suited to the chemicals and voltages available. Then there are mid-level codes such as Ascii and two's complement. In the nervous system, these codes involve spikes and synaptic function: rate coding, burst coding, or synaptic co-transmitter codes. Vector-plot graphic representations are examples of high-level data structures in the computer. In the brain, this would involve circuitry-level codes such as cell assemblies. All codes will be adapted for ease of manipulation, for rapid translation from stimulus to response, or for compact storage. Pressures of multiple use may also dictate hybrid or redundant data structures.

It is interesting to speculate how biological data structures might be fitted to the needs of the organism. In industrial robots, the machine sees with a scanner or CCD (charge-couple device) camera. In either case, the natural representation is a bitmap. However, to move an arm toward a visually identified target, the bitmap has to be translated into a vector representation that maps direction in space. Then this vector has to be translated to a different coordinate system that provides angles for the joints of the arm. There may also be a sensory feedback system that reports back whether the arm actually arrives in position. In the biological context, this is an issue of sensorimotor integration.

The concept of description length may provide insight into possible strategies for brain storage. If there is a lot of information to be stored, then it makes sense to develop compact data structures even at the expense of utilizing a lot of brain to construct and deconstruct it. However, in some cases the need for survival will require that speed take precedence over size in the space–time trade-off. This means choosing representations that are close to those of primary receptors or primary effectors so as to translate in or out rapidly.

Language is an example where massive storage requirements compete with the demand for speed. Neuropsychological evidence suggests that language is stored in a variety of ways for different modalities of input and output. There are phonemic representations for rapid auditory comprehension. There are facial motor representations for manipulation of tongue, lips, and palate. There are visual representations for reading and additional motor representations for writing. None of these representations even touches on real language use: semantics, grammar, meaning. With regard to word meaning, there is evidence for different semantic storage schemes depending on the type of word being stored. Words for objects that are primarily defined by their appearance (e.g., animals) are stored separately from words that are primarily defined by their use (e.g., tools).

Given the profusion of representations, it is necessary to have a lot of translation programs. There are two approaches to producing translators. One is to develop a central master data structure. The format of this data structure doesn't have to be useful for anything in particular, but should contain information or pointers that make it easy to translate into all of the other data structures. The alternative approach is to provide many little translators to go between pairs of representations. Neuropsychological evidence suggests that the brain takes the latter route for language. There are a variety of acquired dyslexias (reading disorders) that are seen after stroke. These provide evidence for separate grapheme (written letters and words) to phoneme (pronunciation) and grapheme to meaning pathways. On the other hand, it is probably the case that direct translation is not necessary for some representations, allowing translation to take place via a shared central format. For example, it is fairly easy to read aloud. This suggests that grapheme to phoneme translation is immediately available online. It is hard, and rarely necessary, to go straight from reading to writing. Therefore, grapheme to orthography might go via a central phonetic representation.

A similar case, in terms of the requirement for massive online data storage with rapid access, is the motor system. Movements must be produced fast and unerringly: riding a bike, throwing rocks at tigers, etc. Also, once learned, they are not forgotten. Motor programs are often used inside of other motor programs, a situation that in computer science would dictate the use of subroutines and pointers. It is not clear whether or how the brain could make use of such strategems.

Episodic memory, memory of things that happened to you over the years, also has massive storage requirements, more and more massive as the years go by. However, this database does not have to be online — you do not have to access a full description of last year's July 4th picnic in one or two hundred milliseconds. This would suggest the use of compression formats, which would save space but take longer to store and access (space–time trade-off). There is evidence to suggest that the hippocampus is involved in preparing these representations and perhaps coordinating their transfer to other storage sites in neocortex.

By contrast, some sensory systems may have relatively little need for storage, but need to optimize rapid processing strategems. Most of us, as nonmusicians, probably have little raw auditory signal storage requirement or storage ability. However, we must still process auditory signals quickly. Music comes in at high rates. Primary auditory decodings must be passed quickly to language centers.

An enormous amount of brain is devoted to vision. It is known that the brain breaks up vision into many different processing pathways and different representations used for different purposes. One major pathway seems to be responsible for detecting and mapping motion in the visual field. It would seem natural to use some sort of vector representation for this. Other systems process shape, color, and other attributes of the visual scene.

In both vision and audition, the design of internal structures is likely adapted to the needs of input circuitry. However, there is enormous variation in the ability of individuals to store and reproduce visual or auditory data. Perhaps certain people are gifted with internal data structures that are well suited to store particular sensory system patterns. This would presumably reflect some mixture of nature and nurture. Given the enormous plasticity of brain, it seems plausible that individuals develop their data structures in slightly different ways during development. Visual cortex is built somewhat differently from auditory cortex. It is not optimized to process sound information. Yet functional imaging shows that congenitally blind people stuff extra auditory processing into their unused visual cortex. Somehow, their brains have built auditory data structures and auditory data processing structures in this alien cortical area.

4.9 Summary and thoughts

In this chapter I have presented common computer representations: binary number encodings, Ascii text encoding, bitmap, and vector-plot picture encodings. Each is a tool suited to the needs of use. However, all of these representations are at heart determined by the structure of the computer — we ultimately get down to bits. All computers now use binary number encodings for numerical calculations, allowing them to run directly on the machine. At the other end, the highest-level encoding that we looked at was the vector-plot graphical language. Like other computer programs, this is a compound encoding built upon the basic underlying binary coding schemes. The vector language used in this chapter is stored as Ascii, but just decoding Ascii would not tell you anything about the picture being represented.

This layering of representations provides an indication as to why it may be difficult to figure out the representations in the brain. We can probably assume that they have been pretty well optimized, but we can't be sure exactly what they've been optimized to do. A single common code in the brain would have the advantage of permitting easy communication between different brain areas. On the other hand, it seems more likely that different brain areas that must perform very different tasks have each developed their own complex codes suitable for these tasks. The many different circuit designs used in the brain (thalamus, basal ganglia, cortex) also suggest that different representations will be used as information is passed around from area to area. Different circuits will also have different roles with regard to speed of processing. Circuits such as thalamic nuclei get only a single brief shot at the data as they relay masses of information on to other areas. Other circuits can take their time, chewing over information in protracted loops of recurrent activity. Perhaps the phenomenon of coming up with new ideas during sleep is due to such slow brain rumination.

5
The Soul of an Old Machine

5.1 Why learn this?

Computer science pioneers had an exciting idea: human intelligence was a result of sophisticated information processing and could be emulated by any machine that could handle enough information speedily enough. Artificial intelligence was the effort to produce software that would validate this hypothesis. Forty years later, they're still trying. I don't believe they'll ever find this particular grail, because I believe that there is something special about brain design that allows it to display intelligence. Theologians are typically dualists, believing that this special thing is immaterial. I think that it's a set of clever design features.

If brain function, intelligence, memory, thought, etc. are the results of design features, then we wish to understand both how the brain is designed and how design constraints are likely to influence function. To put this in a computer science context, we want to understand the hardware and then see how hardware constrains software. This approach differs substantially from Marr's top-down approach, which started with function and worked down to implementation. However, it will not be possible to simply start with implementation and infer function. Only by probing from the top and the bottom simultaneously are we likely to catch something interesting between our pincers.

In this chapter, I want to make two points: 1) hardware molds and constrains software, and 2) a hacker can exploit the idiosyncrasies of a machine in order to push the machine beyond its design limitations. Extending on the example of two's complement in the previous chapter, I want to show how hardware has molded software in the modern digital computer. Programmers exploit computer design to obtain surprising results in programs. My contention is that nature has similarly made the most of design constraints, utilizing tissue in new ways to obtain novel results.

5.2 The art of the hack

In computer lingo, a hack is a trick. A trick performed with mirrors takes advantages of optical anomalies to produce unexpected images. A hack uses computer design anomalies to make programs behave in unusual ways. Nowadays, computer hackers are generally bad guys. The primary reason for using hacks is subterfuge, to break into or simply break a computer system by making it do something other than what the designers intended. In the early days of computers, hacking was a standard programming skill, as the limitations of the computer hardware made it impossible to solve many problems in a straightforward way. The primary limitation was that of space; programs had to be short. Old machines were also limited due to their small instruction sets. These computers could not directly perform basic tasks like multiplication. A lot of programming had to be done in order to perform simple but necessary tasks.

With limited commands and limited space for both data and programs, programmers looked for shortcuts and tricks, many of which took advantage of quirks of a particular machine. This made for very obscure, hard-to-read programs that could not be readily moved from one machine to another. It also made for programs that were not very robust. For example, the Y2K problem arose as the year 2000 approached because many old programs had only alloted two bytes to store the year, assuming that the two digits would always be preceded by 19. Although this saved bytes, it led to confusion and the need for massive reprogramming as 2000 neared. Now, programmers have switched them all over to four digits. There will be even more hassles when the year 10000 rolls around.

Below, I introduce a simple machine that is modeled on the historic PDP-8. I explain how it works. Then I show how hacking was used to extend the machine. The first example is the two's complement hack given in the previous chapter. Then I show how a program can alter itself while it is running. Finally I give a simple example of a computer virus. Such minimalist early examples of virtual reproduction inspired the study of artificial life.

5.3 Software and hardware

Before the invention of programming languages such as FORTRAN and BASIC, programs were written in machine language or assembler language. Machine language is the pattern of bits that the machine works with. It reflects the way that transistors are put together to perform specific simple binary operations such as addition. Machine language sits in memory. It is loaded into an instruction register and from there to the central processing unit (CPU), where the individual bits are interpreted. For the purpose of this book, I generally represent machine language in octal, base 8, rather than binary simply for ease of representation (see Chap. 16, Section 16.3). I also use assembler language. Assembler language is the same as machine language but uses symbolic names as stand-ins for commands. For example, in the language we will explore, the name "ADD" is used in assembler as an obvious label that translates directly to the number 0_8 ($= 0_{10}$, the little "8" tells that the number is given in base 8). This is 000_2 in machine language. The assembler language for a particular computer has exactly the same commands as the machine language for that computer. Different computing machines will have different machine languages, so that something written in machine language on one will not run on the other.

By contrast with the one-to-one translation of assembler to machine language, single lines in a higher-level language, such as FORTRAN, C, or Pascal, are translated, or compiled, into many steps of machine language. As we will show, the statement $x = x + 5$ (augmenting the variable x by 5) may take many machine language steps, and therefore many machine cycles, to execute. An advantage of a high-level language is portability. FORTRAN compilers have been written for almost all computers and a FORTRAN program can therefore be easily ported (moved) from one computer to another.

Translation of a high-level language into machine language is called compilation. Another way that a higher-level language can run on a computer is through interpretation. With compilation, the higher language is translated once and machine code is produced. This machine code can then be run again and again without making reference to the original program. If the original program is changed, then it has to be compiled again. With interpretation, the program does not get translated into machine code. The program itself does not run directly on the machine. Instead an interpreter, which has been either written in or compiled into machine code, runs on the machine. The interpreter reads commands from the high-level language one by one and simply does what each command tells it to do. Interpreters do not translate programs but only interpret what the program says. A previously compiled program no longer requires the compiler in order to run, but an interpreter must always be present for an interpreted program to run. Common interpreted languages include BASIC, JAVA, and Matlab.

5.4 Basic computer design

Representations in the computer all boil down to groupings of 1s and 0s in patterns that the CPU can make some sense of. Whether the 0s and 1s are a compiled program or a document written on a word processor, the computer just sees bits laid down in consecutive main memory. When a program runs, memory is further manipulated as a result of the machine instructions. If a word processor is used to edit a manuscript, binary Ascii code is replaced and rearranged. On the machine there is no intrinsic distinction between the binary of the machine language program and the binary of the paper's Ascii. If pointers get mixed up, as sometimes happens, the machine will execute the Ascii as if it were a program or will display the machine language of the program on the monitor as if it were Ascii. Either way, the results will be highly unsatisfactory.

At the lowest level, computers are organizations of switches. Once these were vacuum tubes; now they are transistors. These switches maintain an electrical lead at either a high or a low voltage. The low voltage is interpreted as a 0, the high as a 1. Binary representation is thus only slightly abstracted from the reality of high and low voltages. Having boiled down higher-level computer representations to groupings of 0s and 1s, we can then probe more deeply into computer design and ask about the machine language, the way that specific patterns of 0s and 1s will be interpreted by the computer.

To explore this low-level "mind–body" interface of a computer, we will look at a simple computer loosely modeled on the PDP-8, a computer manufactured by Digital Equipment Corporation (DEC) starting in the mid-1960s. The PDP-8 is often used as an example machine in computer science because of its simplicity. I have simplified it still further by paring away some design details to make a few core points. I refer to this imaginary machine as the PDP8– to distinguish it from the real PDP-8. A language emulator for the PDP8– is available on the Web site. There is a lot of information on the real PDP-8 available on the Web, including full instruction sets and copies of old programming manuals from DEC.

A new breed of small machine, the PDP-8 was dwarfed at the time by the giant calculating behemoths sold by IBM. Calling the machine a parallel data processor (PDP) instead of a computer was a marketing notion. Computers at that time were immense machines that required a large staff to run and maintain. DEC wanted to distinguish the PDP as a machine that didn't require dedicated rooms or dedicated teams of technicians. Its price made it affordable to a small company or university. PDPs were the first step towards personal computing, and the PDP-8 was the smallest, simplest machine in the PDP series. By modern standards, it was not very powerful, with a complexity comparable to that of modern programmable calculator. Nonetheless, some of these machines remained in use for decades. I spotted one in 1992 in the betting parlor of a jai alai fronton in Tijuana.

Pointers come from computer memory design

Although the CPU is the brain of the computer, the design of computer memory was the big breakthrough that determined the course of modern computing. John von Neumann, the pioneering Hungarian mathematician, formalized the design of the stored program computer by noting that internal memory could be used to indistinguishably store both commands and data to be accessed by the CPU. This basic design feature, present in almost all modern computers, is now known as the "von Neumann architecture," although his collaborators deserve equal credit. This design is also called the "stored-program concept."

The von Neumann architecture determines much of computer programming strategy. Computer memory is referred to as main memory, core, core memory, or, as here, random-access memory (RAM). RAM is where programs and data reside. Random access means that any location in memory can be retrieved at any time. It is contrasted with the sequential access of some storage media, such as magnetic tape. Although RAM is laid out in sequential words, any of these words can be addressed using a pointer. Each word has a fixed number of bits. In the case of the PDP-8, each word has 12 bits. Words are numbered sequentially from 0 up to the total size of memory. A modern computer will have millions of words of memory (several megabytes where a word will be 2 to 4 bytes). A typical PDP-8 had about 4000 words of main memory. A word of RAM is accessed by its address, which is simply its number in the sequence from 0 to the number of words of memory. In the PDP8–, I have only allowed 24 (30_8) words of memory, numbered from 00_8 to 27_8 (remember how to count in octal: 25,26,27,30,31, ...).

Pointers are a prominent feature, explicitly or implicitly, in all programming languages because of the need to address locations in RAM. For example, $x = 5$, an assignment, creates a pointer named x to a location in memory that stores the number 5. Pointers are fundamental to the way programs are written because they reflect the way computers are built. At the machine level, a pointer is the address of a location in memory. Of course, most memory aids that we use in everyday life employ some sort of pointer to access information. For example, the alphabetical ordering of files in a file cabinet is a pointer system.

Sequential algorithms come from computer control flow

The PDP-8 has only a few major components (Table 5.1): a CPU doing the computing, RAM storing the program, a bus connecting the two, and a set of specialized registers used by the CPU as local memory. The CPU does the actual calculating; it is the computer within the computer. Any programs and all data for programs are stored in RAM. Registers are separate words of memory that are not random access, but are instead hard-wired to be

Abbrev	Name	Description
CPU	Central processing unit	The computer in the computer — performs the command passed from the IR.
RAM	Random access memory	Also called "core memory" or "main memory." Contains program and data accessed by the CPU.
BUS	Bus	The backbone with the wires that connect CPU to RAM and external computer components.
Registers		
ACC	Accumulator	Used to process (accumulate) numbers.
PC	Program counter	Points to the next instruction to be executed.
IR	Instruction register	Holds word from RAM indicated by PC and passes it to CPU for execution.

Table 5.1. Computer parts

used for particular purposes. In the PDP-8, three registers are important: the program counter (PC), the accumulator (ACC), and the instruction register (IR).

The program counter is a memory pointer that indicates the location in memory that will be loaded into the instruction register. When the program counter points to a word of random-access memory, it will be fetched and used as a command. However, when this same word of RAM is pointed to by *another word* in RAM, it will typically be considered to be data. There is no intrinsic difference between data and command in the structure of the computer. This, as we will see, is an essential feature utilized by hackers. The accumulator is a register that is used as a temporary storage point for doing arithmetic. The instruction register is a register used as a temporary storage point for commands before they are accessed by the CPU.

In the normal cycle of operation, the PC gives an address whose contents are moved from RAM into the IR. In the PDP8–, we start running most programs at memory location 0. This is similar to the way the bootstrap sequence works on most computers: a memory block from location 0 of the disk is loaded into location 0 in RAM and execution is then started from there. After the word is fetched into the IR, a command component is split off from an address component. The first 3 bits, a number from 0 to 7, is the command. The other 9 bits are generally an address that the command will act on. The CPU will then interpret and execute the command. The execution of the command will change the state of the machine, usually by changing a word in RAM or changing a register. After this, the PC is automatically incremented by 1 and the cycle repeats.

Computer programs can be easily laid out in algorithms — step-by-step instructions similar to the recipe in a cookbook. Algorithms are a major focus in computer science. Algorithms may or may not be fundamental to the brain, but it is apparent why they are fundamental to the von Neumann computer architecture: the machine is designed to proceed in sequential steps. Thus the machine cycle not only determines control flow in the computer but also defines the way that programs are written and the way that computation is regarded.

CPU: machine commands

As a 12-bit word is moved out of the IR into the CPU, it is broken into two parts. The first 3 bits are the command. In the real PDP-8, some commands used the lower 9 bits of memory to microprogram other commands, but here I only discuss the PDP8–, in which the lower 9 bits are either an address or else are irrelevant and are ignored. Since 3 bits are used to define the command, there can be only eight commands: 000, 001, 010, 011, 100, 101, 110, 111 (0–7 octal). The PDP8– instruction set that corresponds to these commands will be given the following mnemonic names: 0 — ADD, 1 — DEC, 2 — INC, 3 — SKP, 4 — JMP, 5 — CLA, 6 — LDA, 7 — HLT. These names are meant to be easy to remember: add, decrement, increment, skip, jump, clear accumulator, load accumulator, halt. When using these names instead of the numbers, one is using assembler language. Thus the only difference between assembler language and machine language is that the former is human-readable and the latter machine-readable. Otherwise, they are the same language. We are now about three steps removed from the machine representation: the machine uses voltages, which we symbolize with binary, and then translate into octal and now further into assembler. The language of the PDP8– is summarized in Table 5.2.

5.5 Programs and hacks

Of the eight commands, three commands alter RAM (LDA, DEC, INC), two alter the ACC (ADD, CLA), and two alter the PC (SKP, JMP). The re-

Inst.	Code	Description	Usage
ADD	0	Add address contents to ACC	ADD [address]
DEC	1	Decrement address contents by 1	DEC [address]
INC	2	Increment address contents by 1	INC [address]
SKP	3	Skip next instruction if address contains 0	SKP [address]
JMP	4	Jump to specified address	JMP [address]
CLA	5	Clear the accumulator	CLA – – –
LDA	6	Load accumulator to address	LDA [address]
HLT	7	Halt the program	HLT – – –

Table 5.2. Assembler and machine code definition. '– – –' means that the bits are ignored

maining command, HLT, stops processing but does not change any memory. Program 1 is an addition program using this instruction set.

ADDR	ASSM	OCTAL	BINARY	COMMENT
00	CLA 00	5000	101000000000	clear ACC
01	[ADD 05]	0005	000000000101	add contents of 05 (13) to ACC
02	ADD 06	0006	000000000110	add contents of 06 (5) to ACC
03	LDA 05	6005	110000000101	put ACC contents in 05
04	HLT 00	7000	111000000000	stop
05	ADD 13	0013	000000001011	data: x
06	ADD 05	0005	[000000000101]	data: 5
07	ADD 00	0000	000000000000	not used

Program 1: Addition

The command at each address (ADDR) is shown in assembler (ASSM) as well as in machine language, showing the number in both octal (e.g., 5000) and binary (e.g., 101000000000). In the machine's main memory, the transistors will hold high or low voltages according to the values shown for each bit in this word.

The accumulator is the location where arithmetic is done. Before doing any arithmetic operation, it is good to clear the accumulator in case a previous program left a number there. This is done with the CLA command. The address argument 00 is in this case meaningless. The next two steps are the heart of the program. ADD 5 superficially looks as if it would add the number 5 directly to the accumulator. This is not correct since all of the commands use pointers to other locations in memory. Therefore, the ADD

5 command instructs the CPU to access memory location 5. The contents of address 05 is ADD 11, which appears to be another command rather than a number. This apparent paradox is resolved when we remember that data and commands are both represented by binary numbers in memory — there is no fundamental difference between a datum and a command. The content of memory location 05 is $13_8 = 11_{10}$. Notice that the contents of locations 01 and 06 are identical. In location 01, I have bracketed the assembler command name to indicate that this is being used as a command. In location 06, I have bracketed the binary to indicate that this is being used as data. In step 02, ADD 6 adds the contents of address 6 to the accumulator, yielding $13 + 5 = 20_8$. Step 03 copies the accumulator back into location 05, overwriting the contents and replacing it with the number 20_8. Finally, in step 04, HLT 00 halts the processing. As with the CLA command, the 00 argument is unused.

If the halt command weren't there, the data in locations 05 to 07 would be taken as commands, adding the pointed-to contents to the accumulator in each case. Notice that location 07 is empty; it is not being used as either data or program. Nonetheless, it can be run and will be handled as a command that will add the contents of location 00 to the accumulator. In the absence of a HLT statement, the machine will continue executing until it encounters an illegal command (e.g., a command that points outside of memory, a so-called segmentation error), or runs out of memory. If there is a jump command that loops control back, the machine will enter an infinite loop, repeatedly executing these inconsequential commands.

This program, without the halt, would be typical of compiler output for a higher-level language expression such as $x = x + 6$ (or $x = x + y$ where y was previously set to 6). The compiler would perform a first pass where it would find a word of memory to assign for x and another word where the 6 was stored. It would create a symbol table to keep track of where these variables and fixed numbers were located. The compiler would then be able to translate the algebraic expression $x + 6$ into steps 0 through 2 in Program 1 and translate the assignment $x =$ into step 3. The name of the first compiled language FORTRAN, means FORmula TRANslator because it was able to translate algebraic formulae into machine code in this fashion.

Conditionals

Memory addressing dictates the use of pointers in computer languages. The sequentially incrementing program counter dictates serial processing. Another key feature of computer languages emerges from the SKP command, or, more fundamentally, from the association between binary representation and Boolean algebra. Boolean algebra is the arithmetic of truth and falsehood, represented in the computer as 1 and 0, respectively. In the context of a computer language, truth and falsehood dictate the fundamental usage of branching of control flow in computer programs. The true path will be the road taken and the false path will be the road not taken.

The high-level computer language manifestation of this aspect of machine representation is branching at a conditional statement. In high-level programming languages, the conditional is usually written as an if-then statement. For example, one frequently sets the value of one variable based on the value of another variable: if x equals 5, then set y to 3 else set it to 1. In the C computer language, this could be written `if (x==5) y=3; else y=1;` or `y=(x==5?3:1);`. Note the use of a double equal sign. This represent the normal meaning of equal, querying the truth or falsehood of an equality: 5==5 is true. The single equal sign in C represents the assignment of a number to a variable: $x = 5$ makes x a pointer to 5 in memory. Every computer language uses different symbols for these two very different operations. Let's compile these conditional assignments into our machine language (Program 2). We'll say that x is located at address 10 and y is located at address 11. We'll also need to provide storage space for the constant numbers 5, 3, and 1: let's store these in locations 12 to 14.

The assignments of y to 3 or 1 are easy. Each of these just involves copying memory from one location to the other. In each case, this requires three steps at the machine level: 1) clear the accumulator, 2) copy from memory to the accumulator, and 3) copy from the accumulator to memory. The conditional itself is more complicated. The only conditional command we have in this machine language is the SKP command (the actual PDP-8 had a few more). It checks whether a given memory address contains a 0. There are two ways that we can turn a 5 into a 0 with our instruction set: either add −5 or decrement by 1, five times. We'll look at both techniques.

Program 2 shows a conditional program using decrementing.

ADDR	ASSM	OCTAL	COMMENT
		(x == 5)? section	
00	CLA 00	5000	clear to start
01	DEC 10	1010	decrement unknown value
02	DEC 12	1012	decrement the counter
03	SKP 12	3012	done when counter is 0
04	JMP 01	4001	otherwise do again
05	JMP 20	4020	finished: skip over data section
		Data section	
10	ADD 05	0005	x — the "unknown"
11	ADD 06	0006	y to be set
12	ADD 05	0005	5 for decrementing
13	ADD 03	0003	number 3 (for $y = 3$)
14	ADD 01	0001	number 1 (for $y = 1$)

⋮

y assignment section

20	SKP 10	3010	x==0? (i.e., did x start out as 5)
21	JMP 24	4024	if $x \neq 0$ (if x not equal to 0)
22	ADD 14	0014	if x==0 ACC = 1
23	JMP 25	4025	set y and finish
24	ADD 13	0013	if $x \neq 0$ ACC = 3
25	LDA 11	6011	y = ACC (set y)
26	HLT 00	7000	stop

Program 2: Conditional program: if (x==5) y=3; else y=1;

The main point of this example is to show how the simplest line of a program becomes horribly complicated and convoluted when it actually gets performed by the computer. Similarly, we may be disappointed as we look for the nervous system to do things in simple, straightforward ways.

In Program 2, we set up x in location 10, y in 11, and the test number 5 in 12. Program 2 does not show binary but instead includes comments in pseudo-code (similar to but not identical to the C programming language). I have also not listed all of memory here, only those words that are used. Steps 1 and 2 decrement the unknown x and the known 5 in tandem. When the decrementing reaches 0 in location 12 (the known value), control flow jumps over the data section to location 20, having set location 10 (the unknown value x) to $x - 5$. At address 20, location 10 is evaluated to see if it contains 0 (which would mean that x was 5 originally). If it contains 0, control jumps to address 22, and the contents of address 14 are put in the accumulator. Otherwise, the contents of address 13 are put in the accumulator. The accumulator is then copied to address 11.

This is an inefficient way to write the program. The loop to do a comparison by decrementing could take a while if the number was very large instead of being 5. However, since the machine does not have a subtraction command, decrementing would be the obvious way to do subtraction. As discussed in Chap. 4, an efficient, direct way to do subtraction is to use two's complement as the representation of a negative number, using machine architecture to do something that cannot be easily done in a straightforward way.

In Program 3, we use the two's complement overflow hack to represent -5 as 7773_8. This allows us to write our conditional program in a slightly more compact fashion that would run much faster if a large number was given. Effectively, instead of performing decrements in steps 1 and 2, we use an appropriately pre-decremented value in address 12 to complement the unknown x in address 10. Then we can add the numbers and proceed with the conditional as before.

ADDR	ASSM	OCTAL	COMMENT
00	CLA 00	5000	clear to start
01	ADD 10	0010	"unknown" x value
02	ADD 12	0012	add in the −5 from address 12
03	LDA 10	6010	$x = x - 5$ store result in x location
04	CLA 00	5000	clear accumulator again
05	JMP 20	4020	skip over data section
		Data section	
10	ADD 05	0005	x — the "unknown"
11	ADD 06	0006	y to be set
12	HLT 773	7773	−5 in two's complement
13	ADD 03	0003	number 3 (for $y = 3$)
14	ADD 01	0001	number 1 (for $y = 1$)
		Test and set — same as Program 2	
20	SKP 10	3010	$x == 0$? (i.e., did x start out as 5)
21	JMP 24	4024	if $x \neq 0$ (if x not equal to 0)
22	ADD 14	0014	if $x == 0$ ACC = 1
23	JMP 25	4025	set y and finish
24	ADD 13	0013	if $x \neq 0$ ACC = 3
25	LDA 11	6011	$y =$ ACC (set y)
26	HLT 00	7000	stop

Program 3: Conditional program with two's complement

5.6 Pointer manipulation

In modern computer programming technique, the distinction between program and data should always be maintained. A program should not operate on itself as if it were data. Modern virus scanning programs look for exactly this kind of hack. Early programmers did not have room in memory to maintain this clean separation. It was common for a program to alter its own body. This is shown in Program 4, which adds a list of numbers.

The numbers to be added are in memory locations 10 to 12. The counter for the number of words to be added is in location 07. The program alters its own text by using the code at location 03 (INC 01) to increment the address of the ADD command at location 01. This moves the pointer for this ADD command so that it points to the next location in memory. This pointer will be incremented as long as the counter in 07 is being decremented. Each of the sequential locations will be added into the accumulator. The incrementing of a pointer is a standard technique that is still used today. The archaic aspect of this program is that the pointer and the command are all part of the same word so that the program itself has to be modified in order to access the sequential data.

ADDR	ASSM	OCTAL	COMMENT
00	CLA 00	5000	clear
01	ADD 10	0010	this command (pointer) will be altered
02	DEC 07	1007	decrement the counter
03	INC 01	2001	increment address in 01 (the hack)
04	SKP 07	3007	if counter==0 finished
05	JMP 01	4001	else go back to 01 for next number
06	HLT 00	7000	stop
		Data section	
07	ADD 03	0003	the counter (three numbers)
10	ADD 33	0033	value #1
11	ADD 27	0027	value #2
12	ADD 42	0042	value #3

Program 4: Add list of numbers in locations 10 to 12

A kludge

A really egregious hack is called a kludge (with the long vowel of "clue"). An example of this is the notorious and archaic programming technique whereby a program modifies itself while running and then loops back and runs the modified code. When another programmer tries to change such a program later, the altered software would behave in bizarre ways since the modifications would also alter this unseen program that is created on the fly. This type of hack is often highly machine-dependent since it takes advantage of the specific numbers used for the instruction set. An example of this that we will now show is to use overflow or underflow on an instruction to create a different instruction.

ADDR	ASSM	OCTAL	COMMENT
00	ADD 13	0013	value #1
01	ADD 22	0022	value #2
02	ADD 17	0017	value #3
03	CLA 00	5000	start here
04	ADD 02	0002	add/halt statement
05	DEC 04	1004	manipulate the program
06	JMP 04	4004	loop back

Program 5: Kludge to add list of numbers

In Program 5, I managed to condense a program to add a list of numbers into a three-step loop, as compared to the five-step loop of Program 4. In the two's complement example we saw how numeric overflow could be used to manipulate numbers. It is also possible to use numeric overflow or underflow to manipulate commands. In this example, the hack uses underflow.

Decrementing 0000 gives an underflow condition that changes the word to 7777. If this word is being used as a command, the underflow changes the command from an ADD to a HLT statement. The decrementing of the address to change the pointer would be considered reasonable programming practice in the PDP-8 era; continuing the decrement to an underflow would not.

The functioning of Program 5 depends on the data being stored at the beginning of memory. Although the program code is relocatable, the data section is not. After the accumulator is cleared (step 03), the program adds in the contents of location 02 (step 04). This is the pointer, which is then decremented in step 05. The jump takes us to the next iteration at 04, which adds in the contents of location 01. On the next iteration, the contents of 00 are added in. The next decrement step changes the ADD 00 (0000) command to the HLT 777 command (7777). The lower, address, bits in this case have no meaning. The jump to location 04 halts the program, leaving the summed contents of locations 00 to 02 in the accumulator.

This program is meant to be started at 03. This is a small adding routine that would be used by other code. Numbers to be summed are placed in the data section, followed by a JMP to 03. If someone mistakenly JMPed to 00 and began execution in the data section, the program would still work fine for small positive numbers. The commands at 00 to 03 just point to numbers elsewhere in memory that get added to the accumulator; the CLA at 04 clears the accumulator anyway. On the other hand, if large or negative numbers were entered to be summed, one of the new numbers could be a less innocuous command that would crash the computer or intermittently produce weird results.

This hack uses a machine-dependent trick: a halt and an add command happen to lie only one arithmetic bit apart. Similar hacks can be designed to use different commands since adding a multiple of 1000_8 will switch from one command to another.

A computer virus

Program 6 is a simple but defective prototype of a common malicious hack — a virus. A virus is a piece of code that copies itself into a different location in memory. On the Internet, such viruses not only copy themselves on one machine but then propagate themselves to other machines as well. After clearing the accumulator, the next two steps (ADD and LDA) shift consecutive memory from RAM to ACC back to RAM at a higher address. Each of these two commands is decremented to access the entire program from high to low addresses. The program will stop after an underflow at address 01 after six iterations that have copied the program from addresses 00 – 05 to addresses 12 – 17.

This particular virus is defective. For the virus to run properly in its new location, it would have to be updated so that the ADD, DEC, and

JMP commands all correctly address the new location of the code. Also, a proper virus would transfer control to its new location instead of halting. Trying to hack programs that will do this sort of thing is an entertaining sort of puzzle, which is one of the reasons that this activity is so popular among misguided youths.

ADDR	ASSM	OCTAL	COMMENT
00	CLA 00	5000	clear
01	ADD 05	0005	move code into ACC — becomes HLT
02	LDA 17	6017	move code from ACC to higher memory
03	DEC 01	1001	alter the address to copy from
04	DEC 02	1002	alter the address to copy to
05	JMP 00	4000	loop back to get another address

After running:

ADDR	ASSM	OCTAL	COMMENT
00	CLA 00	5000	unchanged
01	HLT 777	7777	ADD turned into a HLT
02	LDA 06	6006	now points to below relocated code
03	DEC 01	1001	unchanged
04	DEC 02	1002	unchanged
05	JMP 00	4000	unchanged
06	ADD 00	0000	
07	CLA 00	5000	**program relocated commands here**
10	ADD 01	0001	
11	LDA 11	6011	
12	DEC 01	1001	
13	DEC 02	1002	
14	JMP 00	4000	

Program 6: Relocating code — before ... and after.

5.7 Neurospeculation

The brain is sometimes referred to as a "black box." A black box, in system engineering jargon, is a component that is bought from an outside vendor and just plugged in. The system engineer doesn't have to know how it works. He is content to know that it takes available inputs and provides needed outputs. For example, if you're building a Boeing 747, you might pick up your inertial guidance system from Lockheed. They send you a box. It might be black. You go to the cockpit and plug it into the navigation system. Then off you go.

The brain has long been mysterious in this way. It was purchased from an outside vendor. It fits neatly into a slot at the top of the case. It picks up signals from the rest of the system and provides output signals that are used to navigate and maneuver the entire device. In a classic "Star Trek" episode, a nasty alien stole Spock's brain and plugged it into a central control station so that it would run the air conditioning or something. Complex plot twists led to the immortal line: "Brain, brain ... what is brain?" This question resonates at two levels. We generally know neither what the circuits are doing nor how they are doing it. There is no instruction manual. A good black-box instruction manual tells not only what the whole box is for, but also what kind of signals go into each input plug and what kind of signals come out of each output plug.

In electrical engineering jargon, the next step is to "open the black box." As we poke around inside the brain, we have to start with the assumption that systems are performing calculations similar to the ones we are familiar with. This is not always a good assumption. A brain system is likely to be designed very differently from comparable artificial technologies. The visual system does not work like a camera or video camera, or like an electric eye. Despite this concern, below I'll look at some brain analogies for various computer concepts and components: the software/hardware dichotomy, the distinction between CPU and memory, and the use of a central bus.

The notion of a hardware/software distinction is often used by linguists and cognitive psychologists who do not want to open up the box. This is a version of the Marr approach that I have been complaining about. My contention is that the hardware develops so that it readily performs certain algorithms in particular ways. However, it seems possible that we could consider natural language to be the brain's software.

Above, we discussed the distinction between compiled and interpreted software. A compiler translates programs into machine language so that they can run on the machine, while an interpreter leaves the program as it is and performs the steps of the program one at a time. Code compiled onto a PDP-8 would look like the machine code programs presented in this chapter. Compilation is done when you need speed but don't care about being able to change the program. Once it is compiled, you can even throw away the original higher-level code, making it nearly impossible to alter the program in the future.

In the brain, motor learning would seem to be a kind of compilation. Motor tasks are compiled so that they can run with optimal speed, in real time. Once the task has been compiled down into the brain, it is extremely difficult to alter. This is most clearly seen in the case of speech. After learning, as a child, the pronunciation patterns of one's own language, it is very hard to learn new speech patterns in order to correctly pronounce a foreign language learned as an adult. Of course, brain compilation goes beyond computer compilation. In computer compilation, a program compiles down to the machine level but cannot alter the machine itself. In brain development and learning, the machine is changed, and underlying circuitry is reorganized. The brain program is compiled into machine circuits rather than into a predetermined machine language.

The brain is also an interpreter. Natural language (e.g., English) is the programming language that the interpreter processes. Thought is the process of interpretation. In computer science, the interpreter is itself a compiled program that runs in machine language. (It is possible to write an interpreter inside of another interpreter. With time to waste, I did this for an old Hewlett-Packard computer and then waited 47 seconds to add 5 to 3.) Because the interpreter itself runs in machine language, I would contend that it is still necessary to understand the brain to truly understand natural language.

There are some languages, like LISP, that are both interpreted and compiled. In LISP, when you write a subroutine, you initially use the code in its interpreted form while debugging it. Then, once you've got it right, the subroutine is compiled. This way, you don't have to do repeated compilations while making many changes to the program. Once it's finished, you can put it in a compiled form and it runs faster. This process is analogous to the shift from interpreted to compiled that is seen when learning a motor task. Take the example of learning a tennis serve. You start out with a natural language description, an algorithm, for stepping through the process. As you say the algorithm to yourself and attempt to execute it, you are interpreting the code. You are also very clumsy. Over time, you eventually compile the code directly onto the machine. After this is done, you can throw away the old program. You no longer have to think about doing the task, it is fast, automatic, and hard to change. The discrete steps that were used to learn it are forgotten. Subsequently, you may have to reconstruct the algorithm by monitoring yourself while you serve in order to teach it to someone else.

In years back, CPUs were designed to either be RISC (reduced instruction set chip) or CISC (complex instruction set chip). In a RISC architecture, the CPU can perform only simple instructions but performs them fast. On a RISC, basic routines have to be written in software. Although the PDP-8 far predates the RISC/CISC distinction (it has a comically meager instruction set by modern standards), the RISC concept is illustrated by the need to write a multiplication instruction rather than have one built into the chip. By contrast, a CISC performs a lot of complicated processing on-chip. Using this analogy, I would suggest that brain CPUs are likely to be CISCs. They develop massive built-in instruction sets. In the motor system particularly, learning builds long hardware routines that can be triggered and then allowed to play out without conscious intervention.

In a computer, the bus is a set of shared wires that carries information between different components: RAM, CPU, disks, etc. This is an efficient design since the same wires can be used for different signals. It is not necessary to have dedicated wires that just run from RAM to CPU and from

RAM to hard disk. The downside of this design is that signals on the bus must include markers or labels indicating their destination address. In general, everything that is connected to the bus must look at every signal, reading the address to see if that signal is theirs. This extra reading represents a time inefficiency. The need for labels on the signals expands the data structure and represents a coding inefficiency. These minor inefficiencies are worth it since the savings on physical wire is enormous.

The brain appears to use primarily labeled lines rather than labeled signals. Labeled lines are wires that are dedicated to carrying only certain information between specified sites. Brain wires, axons, run primarily one way (although they can be back-fired), and are not easily shared (although there are gap junctions between axons). Pyramidal cells are the large cortical cells whose projection axons connect remote parts of the central nervous system. If the brain has a bus-equivalent, these wires are it. It is possible that the brain could use both labeled lines and labeled signals. In that case, a projection axon (or set of projection axons) would be conceived as connecting cortex A to cortex B rather than simply connecting cell a to cell b. This would require that the projecting pyramidal cell show resonance at the cellular level. The receiving region could show resonance at either the circuit or cell level. In this scenario, a particular pattern of activity in A would activate the projecting pyramidal cell in a distinct manner, causing it to fire a distinct spike sequence. This spike sequence would be the labeled message. At the other end the spike sequence would be read out through its tendency to activate a particular neuron or particular subcircuit. It is even possible that these labeled messages could be multiplexed, allowing the triggering of more than one target circuit. Although this model is highly speculative (i.e., there is not an ounce of data to support it), the pattern of broad convergence and divergence in cortex would make such a model possible.

Modern computers define a hierarchy of memory storage based on accessibility and speed. At the very top are the CPU registers, such as the program counter and accumulator. A modern CPU will have many more of these registers, allowing it to temporarily store and repeatedly run brief machine-level subroutines on-chip. At the next level of accessibility will be a cache. This is particularly useful for graphics chips that need to maintain screen information in order to update the screen quickly. The third level of memory is RAM. This is considered main memory. As in our PDP-8 model, this is where programs and data reside before use. At the fourth level is the disk. This also offers random-access but is much slower than RAM; data retrieval depends on the physical movement of an arm that samples the magnetically charged surface of the spinning disk. At the fifth level is some kind of slow offline media used for backups — paper punch tape in the case of older PDP-8s.

So what are the layers of memory management for the brain? In the previous chapter I alluded to the widely shared concept of the hippocampus as an encoder of episodic memory for subsequent storage in other cortical areas. However, for most of the brain we encounter a problem that we've been skirting all along — the brain does not generally distinguish between data representation and data processing. The two are ubiquitously intertwined. This has analogues in various types of software data structures that use programs or mix executable code and data. The vector plot data structure presented in the last chapter was an interpreted program. As another example, an industrial robot might utilize a data structure that lists common target locations in both a vector format and a camera-centered bit format along with the actual code that is needed to move the arm to that location. Minimal description length describes the trade-off between processing description length and data description length. In general, the existence of some kind of processing technique is implicit for any data structure.

So rather than have levels of memory, the brain can be viewed as having levels of memory/processing. The brain typically divides a sensory signal shortly after its arrival in the central nervous system (CNS). One track leads to nearby cells optimized for speed. These cells mediate reflexes that immediately alter behavior to improve survival. Slower tracks take the input to more advanced processing centers, arrayed both in parallel and in series. When you stub your toe, lower centers register the mishap and convey a sense of unease to conscious processing before higher centers identify the misery location. Higher processing pathways are presumably also divided up according to speed needs. For example the "magno" pathway that mediates visual motion perception is faster than the "parvo" pathway that mediates object recognition. This may reflect a need to know "where?" before you know "what?"

An additional problem faced by the brain is that it receives much too much data, most of which is redundant and unneeded. So in addition to dividing the input into streams according to different processing needs, the CNS will also want to split off input that can simply be discarded.

Above, I've mostly thought about the situation on the sensory side. Comparably, the motor side must often initiate movement before it's entirely worked out the details of how the movement will end. In a reach-to-grasp task you start the arm movement before the fingers are configured for the grab.

There are several other aspects of computer design that can be considered in the context of the brain. These include the existence of a central clock, use of pointers and other addressing schemes, and the use of stacks, queues, heaps, and other standard data structures. One could go on and on, whether fruitfully or fruitlessly.

5.8 Summary and thoughts

In this chapter I've presented the procedures and process of hacking in the context of its original value before its modern-day descent into evil. I may have indulged myself a bit or two; the programs were fun to write. I encourage the reader to get hold of my software and try it out.

Although different computer programming languages have different features, the fundamental design constraints of the underlying hardware appear again and again. The design of the data processing machine determines how we use and how we can use that machine. These computer design features include 1) pointers, 2) conditionals, 3) control flow, and 4) equivalence of program and data.

A message of this chapter is that hardware molds software. The other message is that cleverness can overcome apparent hardware limitations. Development of various types of channels, neurons, and patterns of connectivity has served to impose a higher algorithmic order on the basic underlying architecture. Consider a jellyfish or a sponge. These animals have simple nervous systems that don't do a heck of a lot more than the nervous system of the venus flytrap, a plant. Imagine that you are trapped on a desert island, surrounded by these creatures. Now, starting with jellyfish and sponges, using a penknife and a magnifying glass, build a robot that can see, hear, walk, talk, and think.

Part III

Cybernetics

Cybernetics is an old, perhaps archaic term for developing machines that can behave like nervous systems. It dates from a time when hardware, rather than software, was preeminent. The focus of cybernetics was on system analysis and system control, closer to the current study of robotics than to neural networks. Nowadays, the term *cybernetics* is perhaps more common in science fiction than in science: the borg of "Star Trek" or the cyborg of *The Terminator*.

Cybernetics is a relatively broad term that denotes studies of the man–machine or animal–machine interface. I use this term in preference to *neural networks* because it is broader, and because it emphasizes the contrast between the machine and system perspective, and the biological perspective. Cybernetics is a top-down approach, in that one looks for ways to design machines to do a particular function. The word *cybernetics*, like the word *artificial* in *artificial neural network*, emphasizes the aggressive rationality of the approach, as distinct from the detailed, deductive thinking of biological research. The fictional construct of the nasty unfeeling borg or cyborg embodies this contrast between cybernetics and biology.

Using the term *cybernetics* also brings into focus one of its central tenets. This is cybernetic's goal of developing a general, domain-independent theory that can be used to understand any large system, from robots to spacecraft to ecosystems to brains to social networks. This perspective, or academic field, is sometimes called general systems theory. It is very big-picture and quite seductive, since it would be nice to be able to explain pretty much everything with a few well-chosen concepts. My feeling, though, is that this universalist impulse to dissolve academic boundaries trivializes every field that it touches. Abstractions, theories, and concepts have their own aesthetic, which can lead to theory dominating reality. For example, back-propagation, discussed in Chap. 9, was a major advance in neural networks. However, it was taken too seriously by many of its early proponents. They suggested that it was such a good algorithm that it had to be used by the brain and insisted that neurobiological research should focus on demonstrating its existence.

In Part IV, I switch to a primary biological focus and approach again this contrast between the mindset of cybernetics and biology, mindsets that must be reconciled in individual minds as the field moves forward.

6
Concept Neurons

6.1 Why learn this?

We now embark on the study of one of the two major strands of computational neuroscience: artificial neural networks (ANNs). This is the more popular strand, since it touches more closely on the big questions of learning and memory. The ANN movement is heir to cybernetics and to AI (artificial intelligence), the attempt to use standard programming techniques to develop machines that could compete with human intellectual activities. In some areas, like chess, this effort paid off. In 1997, an IBM computer defeated Garry Kasparov, the reigning chess champion. In other areas, such as doctoring, lawyering, and translation of natural language, AI programs have had only limited success. Still less successful have been efforts to replicate the more elementary aspects of human, and animal intelligence, such as understanding visual scenes (i.e., seeing). Indeed, although humans are justifiably proud of higher cortical functions such as language and thought, visual areas make up about two-thirds of our cortex.

We begin our study of both artificial neural networks and realistic neural networks at the level of the neuron and its behavior. In the 1950s, two neuroscientists, Warren McCulloch and Walter Pitts, proposed a basic neuron model. Although McCulloch and Pitts were not the first to consider neurons as calculation elements, they were pioneers in their attempt to formally define neurons as computational elements and to explore the consequences of neural properties. Neurobiology has advanced enormously since the 1950s. The original McCulloch-Pitts neuron was a binary element.

The modern artificial neural network unit is no longer binary but is otherwise little changed. The artificial neuron remains an insubstantial shadow of a real neuron. These simplified neural models can be considered concept neurons, explicitly designed based on a concept of how neural processing takes place. The words *unit* and *neuron* are often used interchangeably in discussing neural networks, although some biologists object that these units are so abstract that it's unfair to call them neurons.

In this chapter, we emphasize the differences of implementation and interpretation between various ways of modeling neurons. We show how concept neurons differ from real neurons. Although a recitation of these differences makes it look like these are lousy models, they are not. The concept neurons are attempts to get to the essence of neural processing by ignoring irrelevant detail and focusing only on what is needed to do a computational task. The complexities of the neuron must be aggressively pared in order to cut through the biological subtleties and really understand what is going on. As more information comes to light, some of these models will turn out to be right, in that they capture some critical aspect of neural information processing, while others will turn out to be wrong, meaning that they are so far off base that they obscure more than they illuminate. Even the wrong models can be valuable, however. They illustrate different approaches to simplifying the neuron and can be used as meta-models, models of how to build, and how not to build, models.

6.2 History and description of McCulloch-Pitts neurons

In their 1943 paper, McCulloch and Pitts considered the computational power of simple binary units. Their paper is more mathematical than it is biological and was therefore more influential in computer engineering than in neuroscience. John von Neumann, one of the great mathematicians of the 20th century, became intrigued by the brain and by McCulloch and Pitts's view of it. This neural model had a substantial influence on the thoughts and studies that led to modern digital computer design.

McCulloch and Pitts knew that neurons had spikes (action potentials) that involved sudden, transient shifts of membrane voltages from negative to positive. These spikes somehow carry information through the brain. McCulloch and Pitts assumed a simple coding scheme for this information carrying: each spike would represent a binary 1 (or Boolean true), and the lack of a spike would represent a binary 0 (or Boolean false). They showed how these spikes could be combined to do logical and arithmetical operations. From the perspective of modern computer design, there is one glaring problem with the design of these circuit elements — the brevity of the spike. A duration of 1 millisecond (ms), compared with inter-spike

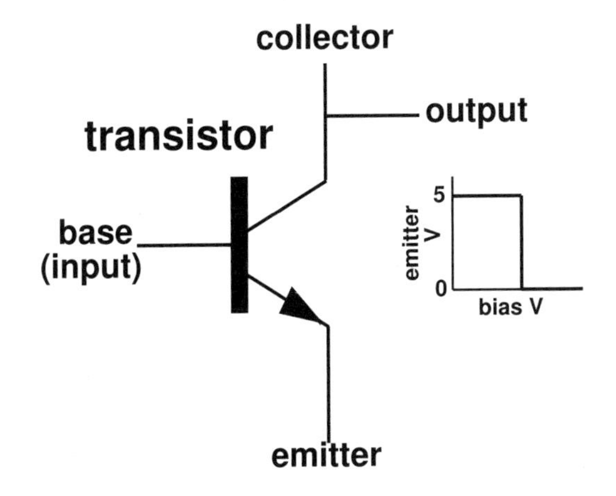

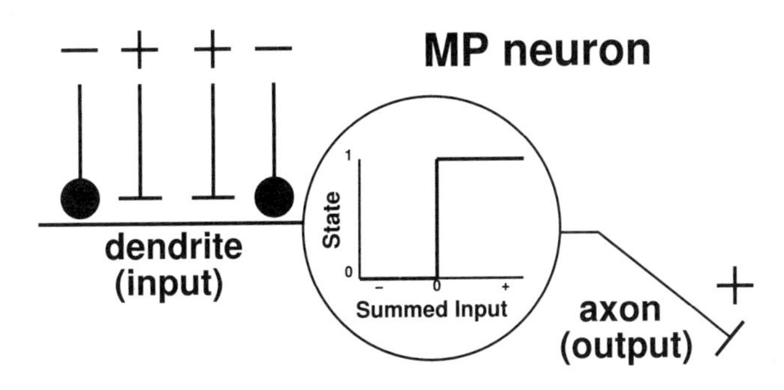

Fig. 6.1: Comparison of transistor and McCulloch-Pitts neuron. Standard symbols are used: a perpendicular line segment is excitatory (+) and a filled circle is inhibitory (−).

intervals of 50 ms or longer, means a duty cycle (signal duration as percent of the period) of only 2% (1 ms/50 ms). This means that it is easy to detect the signal for 0, which sticks around for 49 ms, but hard to pick up the signal for 1, which is there for only this small percent of the time. There is a high risk of missing or failing to process such a short signal.

A McCulloch-Pitts neuron functions much like a transistor (Fig. 6.1). At the time of McCulloch and Pitts, computers used vacuum tubes or electromechanical switches. I will briefly discuss the transistor, the modern analogue of the vacuum tube. A transistor has a "base" lead that controls the flow of current between the other two leads. Generally, a high voltage is considered a logical one or true (*positive true* in computer engineering) and a low voltage a logical zero or false. If the base is activated so as to permit current flow between emitter and collector, the voltage on the emitter will go to low and the transmitter output will be logical false (0). If the base does not permit this flow, the voltage on the emitter will remain at high. The base can thereby set or clear (turn on or off) this single bit device, acting as a switch with which to change the output from 1 to 0. A certain voltage, called the bias, is required to trigger the base. The bias is a threshold that changes the output state of the transistor. Since the transistor is controlled through voltage applied to the base, voltages can be routed from the collector of one transistor to the base of another transistor in order to build up complicated logical or arithmetical operations based on a domino effect: highs and lows from one transistor produce highs and lows in the following transistors. In the example shown, the transistor performs a logical NOT operation: input 1 produces output 0; input 0 produces output 1.

Like the transistor, the original McCulloch-Pitts neuron had a threshold that needed to be reached in order to activate the unit. Like the transistor, the output was completely stereotyped, being a binary 1 if threshold was reached and remaining at a resting value of binary 0 if threshold was not reached. In an artificial neural network, these inputs come from the outputs of other McCulloch-Pitts neurons, just as the inputs to a transistor come from other transistors.

Transistors are typically organized as sets of switches. The output from one transistor will turn a follower transistor on or off. The McCulloch-Pitts neuron, like the transistor, handles incoming inhibitory input as a switch: any inhibitory input is sufficient to shut down the McCulloch-Pitts neuron. This is called veto inhibition, a highly nonlinear process. By contrast, excitatory inputs to a McCulloch-Pitts neuron just add up linearly.

6.3 Describing networks by weights and states

In neural networks, a critical but sometimes confusing distinction is that between weights and states. A *weight* is the strength of the connection between two neurons. Biologically, this can have several interpretations, but is thought to most closely correspond to the size of a synaptic conductance change. We use the biological term *synapse* to refer to the location of a connection between units. The term *presynaptic* refers to the unit projecting into that connection. *Postsynaptic* refers to the unit receiving that connection. Since most biological synapses are unidirectional, it makes sense to speak of pre- and postsynaptic units.

A *state* is the degree of activation of a single neuron. The state is an artificial neural network abstraction. Biologically, a neuron cannot be adequately described by a single scalar state. However, to make comparisons between real neurons and artificial neural network units, it will be useful to figure out how best to reduce the complexities of real neuron activity to

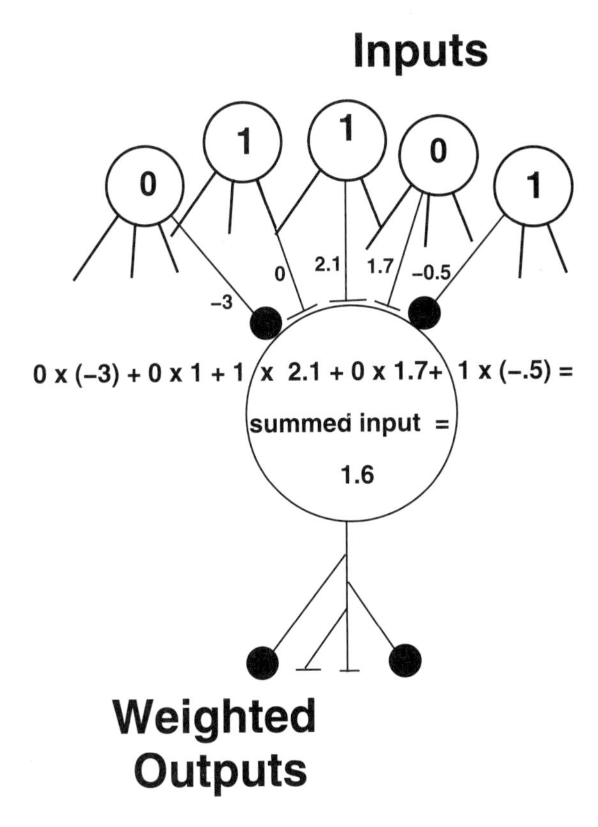

Fig. 6.2: Calculation of input into a single unit. Line segments represent connections between presynaptic units (small open circles) and this unit (large open circle). Filled circles = inhibitory connections (negative weights); perpendicular lines = excitatory connections (positive weights). Weight values are analog and will be determined by a learning algorithm.

a single value. As we see in Chap. 11, average membrane potential or spike rate would be two typical choices for a scalar state in a real neuron.

In an artificial neural network, the state of a neuron is just a number that serves as the output of that neuron. An *update rule* determines how the input to a unit is translated into the state of that unit. Generally the update rule will requires two steps. First, determine the unit's *total-summed-input* based on presynaptic states. The total-summed-input to a unit is the summation of products of presynaptic states with corresponding presynaptic weights. Second, determine the state based on the total-summed-input. The first step of the update rule, the summation of products, is fairly standard across a wide variety of network models. The second step, determination of states, differs between models.

In Fig. 6.2, we depict binary units that take on values of either 0 or 1. The numbers in the circles at the top are state values of units that are presynaptic to the unit represented by the large circles in the center. Each of these state values serves as an output from that unit. Each output is then multiplied by the corresponding weight value, which is represented in the figure by a number to the side of the line connecting them. Negative numbers represent negative weights that are inhibitory — activity in a presynaptic neuron operating through a negative weight will tend to reduce activity in the postsynaptic neuron. The weight values are analog. Those shown were chosen arbitrarily. In an artificial neural network, weights will initially be randomized and will then be adjusted by a learning algorithm. The main unit in Fig. 6.2 projects to other units via the weighted outputs shown at the bottom of the figure. Specific weight values are not given in the figure, but the symbols indicate that both negative and positive projections are coming from the same unit. This is another nonbiological feature; real neurons are strictly excitatory or strictly inhibitory (Dale's principle).

Looking at the leftmost input, the weight is irrelevant, since the presynaptic cell is inactive (i.e., its output is zero). Conversely, the cell second from left is active but it also has no effect on the neuron at the center since the weight connecting to this cell is zero. The only cells that do have an effect in this case are the center cell and the rightmost cell. In this example, activity in an active cell is represented by a state value of one, so that the weights are all multiplied by either 1 or 0. The calculation is shown at the center of the figure with the multiplications by zero explicitly shown.

Calculating total-summed-input by dot product

Arithmetically, each input coming into a unit is a multiplication of state and weight. A shortcut for representing this state times weight multiplication comes from the dot product of linear algebra (see Chap. 16, Section 16.4). Linear algebra uses scalars, vectors, and matrices (matrices is the plural of matrix). A scalar is just a single number. A vector is a one-dimensional array of numbers. A matrix is a two-dimensional array of numbers that is described by the number (M) of rows and number (N) of columns.

An arrow over an italic letter (e.g., $\vec{x}$) is used to indicate that the variable being represented is a vector, rather than a scalar. In the standard vector, numbers are arranged from top to bottom in a column. This is called a column vector. In the other vector orientation, numbers are listed from left to right. This is called a row vector. (A vector can also be represented by a magnitude and angle without worrying about whether it's a column or a row, but I won't use that representation.) The *transpose* is the operation that switches a column vector to a row vector or *vice versa.* Transpose is symbolized by a superscript T. Therefore, $\vec{x}^T$ is a row vector. In a spreadsheet, a single row can be thought of as a row vector, a single column as a column vector, and the entire spreadsheet as a matrix.

The dot product (also called inner product) is the scalar result of multiplying a row vector times a column vector. This scalar result is created by taking the sum of the pairwise products of the elements of the two vectors. The dot product can be represented by a dot ($\cdot$) or by specifying the orientation of the vectors by using the transpose symbol and leaving out the dot. For example, for $\vec{x}^T = \begin{pmatrix} x_1 & x_2 \end{pmatrix}$ and $\vec{y} = \begin{pmatrix} y_1 \\ y_2 \end{pmatrix}$, $\vec{x} \cdot \vec{y} = \vec{x}^T \vec{y} = x_1 \cdot y_1 + x_2 \cdot y_2$. Note that the subscripted variables, x_1 and x_2, are scalars that are arrayed to make the vector $\vec{x}$. Using numbers, with $x_1 = 2$, $x_2 = 5$, $y_1 = 3$ and $y_2 = 7$, then $\vec{x} = \begin{pmatrix} 2 & 5 \end{pmatrix}$, $\vec{y} = \begin{pmatrix} 3 \\ 7 \end{pmatrix}$, and $\vec{x} \cdot \vec{y} = 2 \cdot 3 + 5 \cdot 7 = 6 + 35 = 41$, a scalar.

In our application, the two vectors will be the weight vector and the state vector. We create the weight vector by simply producing an ordered row of numbers that are the weights coming into the neuron in question. Similarly we list all of the corresponding states of presynaptic units in a column to make the state vector. It is critical that the ordering be the same so that the proper weight lines up with the corresponding state. The dot product is defined as the sum of the pairwise products of the elements of the two vectors. Therefore, the dot product of the state vector and the weight vector is the total-summed-input to the unit.

For example, in Fig. 6.2, the weight vector $\vec{w}^T$ is $\begin{pmatrix} -3 & 0 & 2.1 & 1.7 & -0.5 \end{pmatrix}$ and the state vector $\vec{s}$ is $\begin{pmatrix} 0 & 1 & 1 & 0 & 1 \end{pmatrix}^T$. Note that I've written out both of the vectors in row form to save space. In the case of the state vector, the transpose symbol is used next to the row of numbers to indicate that this is actually a column vector. Now we can do the dot product $\vec{w} \cdot \vec{s}$

$$\begin{pmatrix} -3 & 0 & 2.1 & 1.7 & -0.5 \end{pmatrix} \begin{pmatrix} 0 \\ 1 \\ 1 \\ 0 \\ 1 \end{pmatrix} = 1.6$$

The dot product is commutative: $\vec{w} \cdot \vec{s} = \vec{s} \cdot \vec{w}$ or $\vec{w}^T \vec{s} = \vec{s}^T \vec{w}$. However, multiplying a column vector times a row vector is NOT the same as multiplying a row vector times a column vector. Therefore, to reduce confusion, we will do the multiplications in one way: weight times state ($\vec{w} \cdot \vec{s} =$ *total-summed-input*). This may seem counterintuitive since when diagrammed, state comes before weight. The reason for this order will become clear when we introduce the weight matrix.

Calculating state

The first step of the update rule was the calculation of total-summed-input. Depending on how many inputs there are and how large the weight numbers, total-summed-input could be a very large number. The state of a single unit is restricted to lie within a certain range. The *activation function* is used to convert the total-summed-input to unit state. The activation function is sometimes called a *squashing function* because it takes a large range of values and squashes them down to fit within the required state range.

Different artificial neural network models use different squashing functions to produce either an analog or a binary state. An analog state can be any value between two endpoints. The range of a particular squashing function will determine a range of possible state values. A binary state is restricted to being one of only two values. The extrema of the squashing function will be the two possible state values. For example, allowed states for an analog unit might be real numbers between 0 and 1, while a comparable binary unit would take on one of two states: inactive (0) or active (1). It is also common to used ranges of -1 to 1 or binary states or -1 and 1. Generally, the squashing function will be monotonically increasing (i.e., only go up). In Chap. 9, I use a squashing function that provides analog states between -2 and 1. The state of a unit serves as the output from that unit. This output gets multiplied by weights to provide inputs to follower neurons.

To produce a binary unit, the squashing function uses a sharp threshold that determines whether an input value is translated to a state value of 0 or 1 (Fig. 6.3). The discontinuous function shown here can be readily defined on the computer: `if (total_summed_input <0) then state=0 else state=1`. This particular function, named after Oliver Heaviside, is called the *Heaviside function* or the step function. Given an input of 1.6, the unit state will be 1. Alternatively, to produce an analog state, a continuous function is used, as shown to the right of Fig. 6.3. This *sigmoid* curve produces a continuous value between 0 and 1: for the total-summed-input of 1.6, the unit state will be 0.7.

Now we can summarize the update rule for a single unit: $s_i = \sigma\left(\vec{w} \cdot \vec{s}\right)$, where s_i is the state of unit i and $\sigma(x)$ is the squashing function applied to a scalar x. This is the standard generalization of the cyborg neuron. It differs from the McCulloch-Pitts neuron since inhibition is added in rather than having veto power. It is a generalization since the squashing function can be chosen to yield either binary or continuous values. As a shorthand, I call this basic unit the sum-and-squash unit.

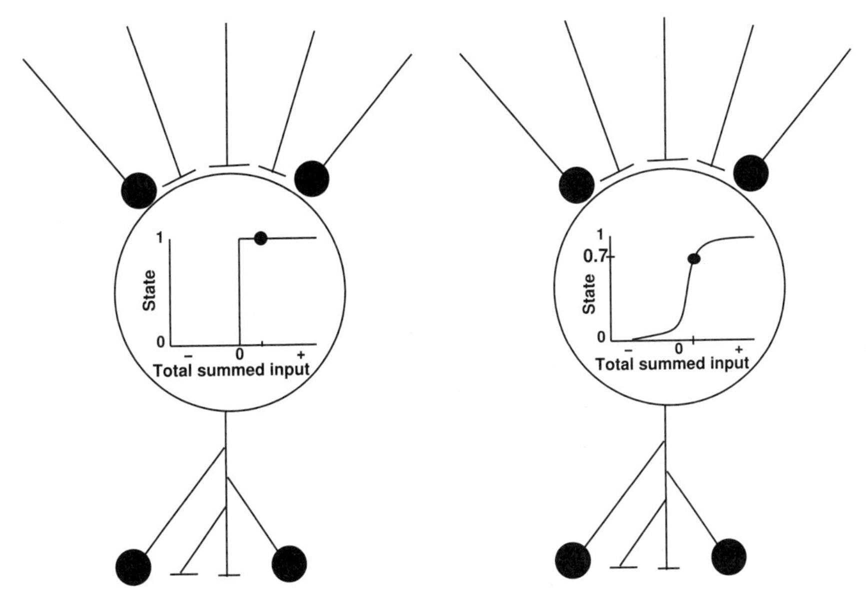

Fig. 6.3: Calculation of state of a single unit. The input calculation is the same as in Fig. 6.2. The state calculation differs depending on the squashing function chosen. A sharp threshold gives a binary state of 0 for negative inputs and 1 for positive inputs. In this case the state is 1. A sigmoid function gives an analog state between 0 and 1. In this case the state is 0.7.

6.4 From single unit to network of units

In the previous section we considered a single unit and its inputs. For one unit, introducing the dot product to do such a simple calculation may seem unnecessary. However, when we put together a lot of units and start connecting them up, the notions of linear algebra become valuable. This is particularly true when it comes time to enter these neural networks into a computer, which can readily store a spreadsheet of numbers but not a diagram of circles and line segments.

This simple update rule for the single unit ($s_i = \sigma(\vec{w} \cdot \vec{s})$) allows us to generate an update rule for the entire network. Instead of using the weight vector $\vec{w}$, we use a weight matrix W, which represents all of the weights in the network. The update rule for the state vector is $\vec{s}^{\,t+1} = \sigma(W \cdot \vec{s}^{\,t})$. Superscripted t and $t+1$ indicate passage of time by one time step. Applying a function (σ) to a vector is simple: the rule is to simply apply the function to each element of the vector. For example, if we have a function $f(x) = 5 \cdot x + 2$ and a vector $\vec{a} = \begin{pmatrix}3 & 5 & 7 & 1\end{pmatrix}^T$, then $f(\vec{a}) = \begin{pmatrix}17 & 27 & 37 & 7\end{pmatrix}^T$. We simply multiply each element by 5 and add 2.

Each element of the W matrix is the value of a weight between two units. $W \cdot \vec{s}$ (multiplication of the weight matrix times the state vector) gives a vector of inputs into each element. This works because the weight matrix is a pile of weight vectors, and matrix multiplication produces a pile of dot products:

$$W = \begin{pmatrix} \vec{w}_1^T \\ \vec{w}_2^T \\ \vec{w}_3^T \\ \vec{w}_4^T \\ \vdots \end{pmatrix} \quad \text{hence} \Rightarrow \quad W \cdot \vec{s} = \begin{pmatrix} \vec{w}_1 \cdot \vec{s} \\ \vec{w}_2 \cdot \vec{s} \\ \vec{w}_3 \cdot \vec{s} \\ \vec{w}_4 \cdot \vec{s} \\ \vdots \end{pmatrix}$$

If there are nine units, the weight matrix will have nine rows and there will be nine dot products (between each row and the state vector) giving a vector of length 9. Each element of this vector is the summed input into a unit. This vector is then squashed to get the state of each unit after one update step.

To see an example, let's look at a circle-and-stick drawing of a simple network (Fig. 6.4). This is a fairly small network by artificial neural network standards and is very sparse, meaning that of the possible $9^2 = 81$ connections shown in the table below Fig. 6.4, only a few are present. While it is easy enough to examine the circle-and-stick diagram in this small network, the scribble can quickly become overwhelming, making a table a more easily read representation.

With the table (and matrix) laid out in this way, each row represents convergence onto a particular unit. For example, unit c (third row), has 3 non-empty entries, indicating a convergence of 3 inputs onto unit c, which receives projections from presynaptic units a, b, and f. Similarly, each column represents divergence from a single unit. For example, unit f (column 6), has a divergence of 3 with projections to postsynaptic units c, e, and i. I have placed black squares along the main diagonal. These are the locations of self-connects. Units are sometimes connected to themselves in a neural network, as they may be in biological networks. In this case there are no self-connects so the diagonal is all zeros. Given the simplicity of Fig. 6.4, all of the connections can be readily confirmed by looking at the circle-and-stick diagram.

We are using the formula $\sigma(W \cdot \vec{s})$ as our update rule. $\vec{s}$ will be a column vector. Each row of W will represent the ordered sequence of weights converging onto the unit with that row number. Equivalently, each column of W will represent the ordered sequence of weights diverging from the unit with that column number. Any row, column number pair i, j will give the strength of the connection from unit j to unit i. Therefore, the connectivity table above can be directly copied into a weight matrix W (also called the connectivity matrix). Each number is located in the same place in W as it

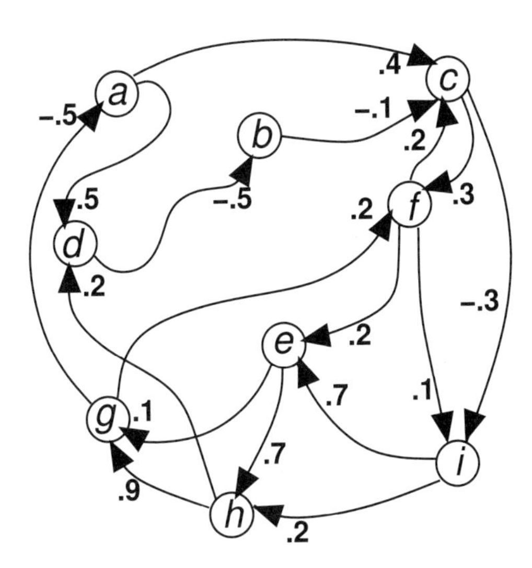

Fig. 6.4: A nine-unit sparse recurrent network. Each neuron is labeled to make it easy to keep track of their identities. I have used arrows for connections instead of circles and lines to indicate that the sign of these weights is changeable with network learning. This diagram can be translated into the table below:

FROM ⇒ / TO ⇓	a	b	c	d	e	f	g	h	i
a	■						-.5		
b		■		-.5					
c	.4	-.1	■			.2			
d	.5			■				.2	
e					■	.2			.7
f			.3			■	.2		
g					.1		■	.9	
h					.7			■	.2
i			-.3			.1			■

This table can then be copied directly into a weight matrix W:

$$\begin{pmatrix}
\mathbf{0} & 0 & 0 & 0 & 0 & 0 & -.5 & 0 & 0 \\
0 & \mathbf{0} & 0 & -.5 & 0 & 0 & 0 & 0 & 0 \\
.4 & -.1 & \mathbf{0} & 0 & 0 & .2 & 0 & 0 & 0 \\
.5 & 0 & 0 & \mathbf{0} & 0 & 0 & 0 & .2 & 0 \\
0 & 0 & 0 & 0 & \mathbf{0} & .2 & 0 & 0 & .7 \\
0 & 0 & .3 & 0 & 0 & \mathbf{0} & .2 & 0 & 0 \\
0 & 0 & 0 & 0 & .1 & 0 & \mathbf{0} & .9 & 0 \\
0 & 0 & 0 & 0 & .7 & 0 & 0 & \mathbf{0} & .2 \\
0 & 0 & -.3 & 0 & 0 & .1 & 0 & 0 & \mathbf{0}
\end{pmatrix}$$

is in the table. Zeros are used in the matrix where there is no connection between corresponding units.

Matrix times vector multiplication is done by taking the dot products of each row (here the convergent weights onto a unit) times the column vector (the convergent states of presynaptic units). (This is the right-hand side matrix multiplication rule — the vector is on the right. There is also the left-hand side vector times matrix multiplication using a row vector on the left and resulting in a row vector. I will only use right-hand side multiplications in this book.) Because each row is simply the list of projecting weights, the dot product gives the input for each unit as illustrated in Fig. 6.2 and the corresponding dot-product equation. By doing each of these dot products in turn, we generate an input vector where each element corresponds to the input for one unit. If we then take this input vector and squash it, we have the state for each unit after one time step.

To simulate the network, we have to set not only the weights but also a starting point for the network state. This set of values is referred to as initial conditions. Let's start with unit c active (set to 1) and all other units inactive. We use sharp thresholding at 0: any input greater than 0 becomes a state of 1, and any input less than or equal to 0 becomes a state of 0. The first multiplication step is:

$$\begin{pmatrix} 0 & 0 & 0 & 0 & 0 & 0 & -.5 & 0 & 0 \\ 0 & 0 & 0 & -.5 & 0 & 0 & 0 & 0 & 0 \\ .4 & -.1 & 0 & 0 & 0 & .2 & 0 & 0 & 0 \\ .5 & 0 & 0 & 0 & 0 & 0 & 0 & .2 & 0 \\ 0 & 0 & 0 & 0 & 0 & .2 & 0 & 0 & .7 \\ 0 & 0 & .3 & 0 & 0 & 0 & .2 & 0 & 0 \\ 0 & 0 & 0 & 0 & .1 & 0 & 0 & .9 & 0 \\ 0 & 0 & 0 & 0 & .7 & 0 & 0 & 0 & .2 \\ 0 & 0 & -.3 & 0 & 0 & .1 & 0 & 0 & 0 \end{pmatrix} \begin{pmatrix} 0 \\ 0 \\ 1 \\ 0 \\ 0 \\ 0 \\ 0 \\ 0 \\ 0 \end{pmatrix} = \begin{pmatrix} 0 \\ 0 \\ 0 \\ 0 \\ 0 \\ 0.3 \\ 0 \\ 0 \\ -0.3 \end{pmatrix}$$

This multiplication is very simple since a single active unit will simply project forward to its postsynaptic cell. Algebraically, this is equivalent to saying that the lone 1 in the state vector will pick out the divergence values for that unit. Indeed, it's easy to compare and see that the resultant vector (the right-hand side of the equation) is identical to the 3rd column of the matrix. This result serves as the input to each of the units. Repeated application of the update rule (matrix multiplication to produce inputs and thresholding to produce states) produces a sequence of vectors:

$$\begin{pmatrix}0\\0\\1\\0\\0\\0\\0\\0\\0\end{pmatrix}_{\vec{s}^{\,0}} \overset{W\cdot\vec{s}}{\rightarrow} \begin{pmatrix}0\\0\\0\\0\\0\\0.3\\0\\0\\-0.3\end{pmatrix}_{\vec{i}^{\,1}} \overset{\sigma(\vec{i})}{\Rightarrow} \begin{pmatrix}0\\0\\0\\0\\0\\1\\0\\0\\0\end{pmatrix}_{\vec{s}^{\,1}} \overset{W\cdot\vec{s}}{\rightarrow} \begin{pmatrix}0\\0\\0.2\\0\\0.2\\0\\0\\0\\0.1\end{pmatrix}_{\vec{i}^{\,2}} \overset{\sigma(\vec{i})}{\Rightarrow} \begin{pmatrix}0\\0\\1\\0\\1\\0\\0\\0\\1\end{pmatrix}_{\vec{s}^{\,2}} \overset{W\cdot\vec{s}}{\rightarrow}$$

$$\begin{pmatrix}0\\0\\0\\0\\0.7\\0.3\\0.1\\0.9\\-0.3\end{pmatrix}_{\vec{i}^{\,3}} \overset{\sigma(\vec{i})}{\Rightarrow} \begin{pmatrix}0\\0\\0\\0\\1\\1\\1\\1\\0\end{pmatrix}_{\vec{s}^{\,3}} \overset{W\cdot\vec{s}}{\rightarrow} \begin{pmatrix}-0.5\\0\\0.2\\0.2\\0.2\\0.2\\1\\0.7\\0.1\end{pmatrix}_{\vec{i}^{\,4}} \overset{\sigma(\vec{i})}{\Rightarrow} \begin{pmatrix}0\\0\\1\\1\\1\\1\\1\\1\\1\end{pmatrix}_{\vec{s}^{\,4}} \overset{W\cdot\vec{s}}{\rightarrow}$$

$$\begin{pmatrix}-0.5\\-0.5\\0.2\\0.2\\0.9\\0.5\\1\\0.9\\-0.2\end{pmatrix}_{\vec{i}^{\,5}} \overset{\sigma(\vec{i})}{\Rightarrow} \begin{pmatrix}0\\0\\1\\1\\1\\1\\1\\1\\0\end{pmatrix}_{\vec{s}^{\,5}} \overset{W\cdot\vec{s}}{\rightarrow} \begin{pmatrix}-0.5\\-0.5\\0.2\\0.2\\0.2\\0.5\\1\\0.7\\-0.2\end{pmatrix}_{\vec{i}^{\,6}} \overset{\sigma(\vec{i})}{\Rightarrow} \begin{pmatrix}0\\0\\1\\1\\1\\1\\1\\1\\0\end{pmatrix}_{\vec{s}^{\,6}}$$

Since the thresholding is at 0, all of the inputs that are less than or equal to 0 become 0 and any that are greater than 0 become 1. In this sequence, states are subscripted as $\vec{s}$ and total-summed-input vectors as $\vec{i}$. The result of matrix multiplication $W \cdot \vec{s}$ is shown as $\vec{s} \rightarrow \vec{i}$, while the result of the squashing (thresholding) function $\sigma(\vec{i})$ is shown as $\vec{i} \Rightarrow \vec{s}$. The time step (iteration number) is shown as a superscript.

The same sequence of states are shown in the circle-and-stick diagrams in Fig. 6.5. Units c and f are mutually excitatory, so activity starting in c spreads to f after the first time step and then spreads back to c after the second time step. In the absence of other inputs, the c–f excitatory loop allows activity to bounce back and forth between these two units. This is seen from time 0 to 3 in this example. At time 2, activity also spreads down from f to e and i. However, c inhibits i so the activity of c at time 2

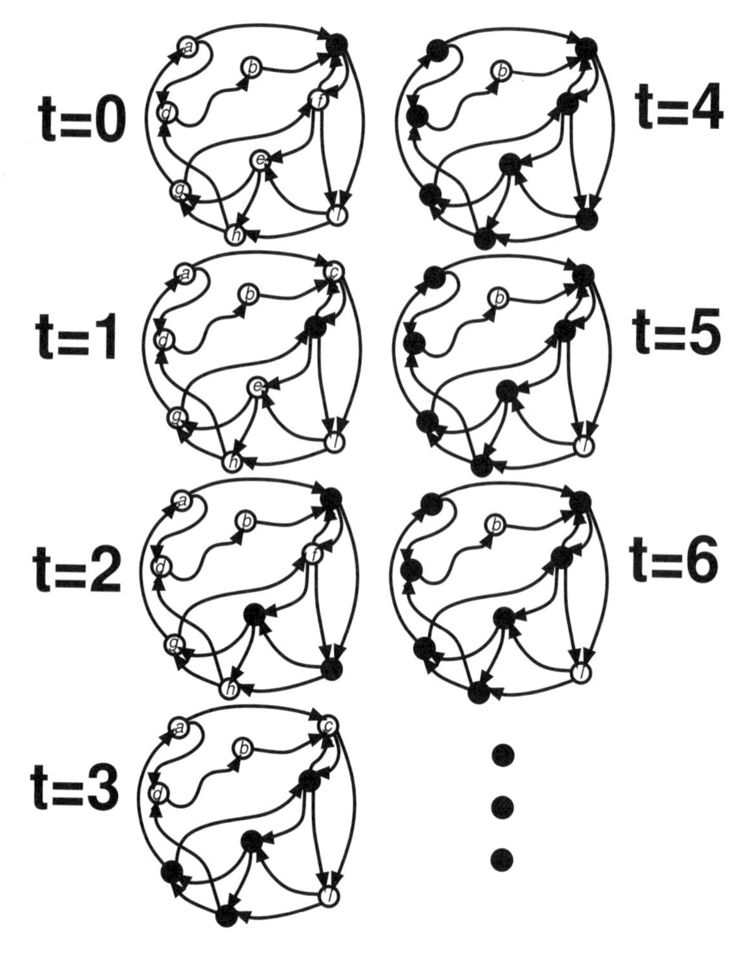

Fig. 6.5: Sequence of network states from time $t = 0$ to $t = 6$. 0 is represented by a white unit and 1 by a black unit.

leads to i turning off again at time 3. It turns on again at time 4, due to the activity in f at time 3. By time 4, activity has percolated throughout the network, excluding only b, which receives only inhibitory inputs. In the final state both c and f are active. Since $c \rightarrow i$ inhibition is greater than $f \rightarrow i$ excitation, i is turned off at time 5. It's pretty hard to keep track of activity as it roams around in this tiny network. Imagine trying to follow activity patterns in a network of 1000 or 10,000 units. To do this, it would be helpful to discover rules that describe common propagation patterns. A general taxonomy of network state evolution has not yet been developed.

Some networks will change indefinitely over time. In Fig. 6.5, the network stops changing. A network that reaches a final constant state in this manner is said to arrive at a steady state. The state of the network, as represented

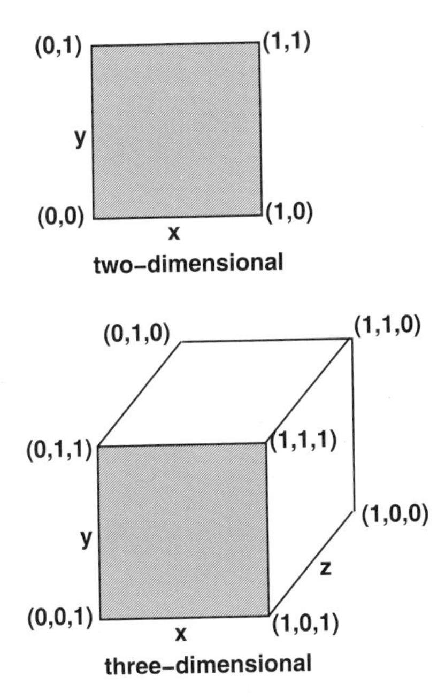

Fig. 6.6: A two-dimensional binary vector can be mapped as a point at one of the vertices of a square. If a 2-D vector took on analog values between 0 and 1, it could be mapped as a point inside the square. Similarly, a 3-D binary vector is mapped on a vertex of a cube; here (0,0,0) is on the hidden vertex. A 3-D analog vector is a point inside the cube. The concept generalizes to hypercubes in higher dimensions.

by the state vector, will not change further with time. In the case of the network of Figs. 6.4 and 6.5, the final state shown at $t = 6$ in Fig. 6.5 is not only the steady state for this particular initial condition, but also the final state for any initial condition. This makes it the sole *attractor* for this network. Since this attractor is a steady state, it is a *point attractor*. For most initial conditions, the network will end up at this attractor. The other attractor in this network is $\vec{s} = (0 \ldots 0)^T$. Wherever the network starts, it will end up in one of these two attractor states.

The concept of an attractor comes from physics, where it refers to the behavior of an object in an energy field, such as gravity. An object in a gravitational field will tend to roll downhill until it reaches a state of lowest potential energy. Then it will stay there. The fact that it goes toward a goal makes that goal an attractor. If the dynamics made it run away from a point, that point would be a repellor.

State space is the multidimensional space in which you can map the entire state vector (Fig. 6.6). You can map a two-dimensional vector on a square, a three-dimensional vector on a cube, and a nine-dimensional vector on a nine-dimensional hypercube (not shown). An attractor that is a single spot in state space is a *point attractor.* We will revisit point attractors when we discuss memory (Chap. 10). If the system ends up switching state repeatedly among two or more state vectors, the attractor is a limit cycle.

6.5 Network architecture

The term *architecture* refers to the connectivity of a network. We can describe the architecture of Figs. 6.4 and 6.5 as being recurrent and sparse. Recurrent means that there are loops in the connectivity, for example, the loop between $c \leftrightarrow f$ and longer loops like the circumferential $c \rightarrow i \rightarrow h \rightarrow g \rightarrow a \rightarrow c$. Note that a network would be considered recurrent even if there were only inhibitory loops — the strength or sign of the connection is not important. As for sparse, we noted above that the network is sparse because only a few of the 81 possible interconnections are present (17 – 21% connectivity). The term *sparse* is not strictly defined. It is generally said that brain connectivity is sparse; although a single cell may receive many thousands of connections, this is only an infinitesimal fraction of all the cells in the brain or even in a small brain region.

A recurrent network is a general network architecture since all connections are possible. In the brain, neurons that receive projections from one brain area typically also project back in the other direction. Most cortical function is probably dependent on these loops of recurrent connectivity.

In addition to recurrent networks, the other major architecture is *feedforward.* Feedforward networks are layered. All information flows in one direction from a presynaptic layer to a postsynaptic layer. Since there are no back projections in the opposite direction and no connections within either layer, there are no loops. A typical three-layer feedforward network is shown in Fig. 6.7.

In general, the first layer of a multilayer feedforward network will be referred to as an input layer, and the last layer will be considered an output layer. Any layers in between are called hidden layers — they do not interact with the environment, hence they are hidden from the environment. When we translate this network into matrix form, we see large blocks of zeros that would have entries if there were recurrence.

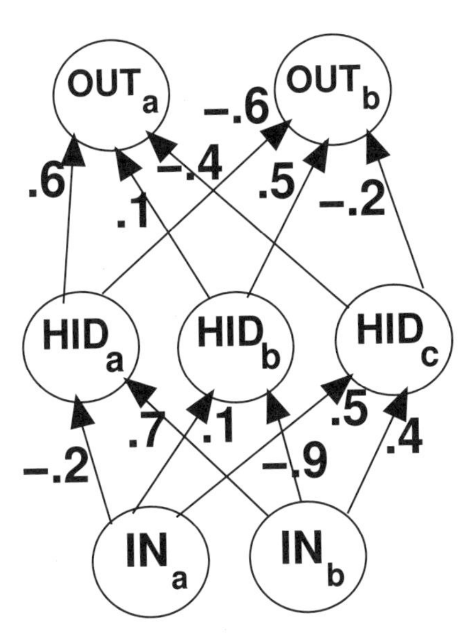

Fig. 6.7: A simple three-layer feedforward network. Activity percolates from the input layer at bottom through the "hidden" layer to the output layer at the top.

FROM ⇒ TO ⇓	IN_a	IN_b	HID_a	HID_b	HID_c	OUT_a	OUT_b
IN_a	0	0	0	0	0	0	0
IN_b	0	0	0	0	0	0	0
HID_a	**−0.2**	**0.7**	0	0	0	0	0
HID_b	**0.1**	**−0.9**	0	0	0	0	0
HID_c	**0.5**	**0.4**	0	0	0	0	0
OUT_a	0	0	**0.6**	**0.1**	**−0.4**	0	0
OUT_b	0	0	**−0.6**	**0.5**	**−0.2**	0	0

The only blocks that are filled are those at center left, which represent the projections from the input to the hidden (2nd) layer, and those at center bottom, which are the projections from the hidden to the output layer. The input units themselves have no inputs at all, and the connectivity matrix is therefore all zeros in rows a and b. If we were implementing this network on the computer, using the full connectivity matrix would be very wasteful of memory. Therefore, we would most likely just use the non-zero portions of the full weight matrix as two feedforward weight matrices. The first would map the input layer to the hidden layer, and the second would map the hidden layer to the output layer. These operations should be done sequentially, giving the following sequential update rule:

$$\begin{aligned} \text{step 1:} \qquad \vec{s}_{hid} &= \sigma\left(W_{in \to hid} \cdot \vec{s}_{in}\right) \\ \text{step 2:} \qquad \vec{s}_{out} &= \sigma\left(W_{hid \to out} \cdot \vec{s}_{hid}\right) \end{aligned}$$

$$W_{in \to hid} = \begin{pmatrix} -.2 & .1 \\ .7 & -.9 \\ .5 & .4 \end{pmatrix}$$

$$W_{hid \to out} = \begin{pmatrix} .6 & .1 & -.4 \\ -.6 & .5 & -.2 \end{pmatrix}$$

The two matrices are simply copied from the appropriate sections of the complete connectivity matrix. Note that neither of these feedforward matrices is square. $W_{in \to hid}$ is a 3×2 matrix that maps the two input units onto the three hidden units and $W_{hid \to out}$ is a 2×3 matrix that maps the three hidden units onto the two output units. There is no update rule for the input units. In fact, these don't have to be considered as units at all. They are simply state values that are presented to the system, coming from the environment.

Although the brain is certainly heavily recurrent, there are probably systems in the central nervous system that are largely feedforward. For example, in a reflex, a simple input (the tap of a knee) is quickly and directly translated into an action (the kick).

6.6 Summary and thoughts

Despite enormous progress in neuroscience, the simple units utilized in artificial neural networks are little evolved from the McCulloch-Pitts neurons and vacuum tubes of 60 years ago. These arose in a hopeful era when understanding neural processing seemed to be just around the corner. Although archaic, sum-and-squash units are not obsolete. Most of the learning theory of the past 20 years has been built upon them.

Not coincidentally, artificial neural network units fit neatly into the mathematical tools that are typically used to handle large systems. Linear algebra is a tool that provides a compact notation for describing and running a large network of sum-and-squash units. The basic artificial neural network update involves a multiplication of the weight matrix times the state vector followed by application of an activation function to the result: $\vec{s}^{\,t+1} = \sigma\left(W \cdot \vec{s}^{\,t}\right)$. In addition to providing a compact notation, the linear algebra formulation is readily translated into computer programs.

Neural networks are typically classified as either feedforward or recurrent. As we see in Chaps. 9 and 10, different learning algorithms are used for these different network architectures.

7 Neural Coding

7.1 Why learn this?

We discussed the idea of signals as codes in Chap. 4. We demonstrated that we could take the same signals and interpret them as a picture, or a sentence, or a set of commands. To interpret a one-dimensional bit string as a picture, we had to determine how to properly lay out the bits in two dimensions (Fig. 4.2). In general, both the code and the thing encoded can be multidimensional. Part of the key for code interpretation involves a redimensioning. The information to be encoded will have different defining dimensions depending on the modality (touch, smell, sight, language), but most if not all modalities will involve some sort of time-varying signal and thus include the dimension of time. Similarly, in addition to its physical dimensions, the brain also possesses time-varying signals, as indeed it must in order to respond to a time-varying world with time-varying actions.

In Chaps. 4 and 5 I speculated about brain representations by drawing analogies from standard computer-science representations. In this chapter, I present more standard ideas based more directly on the brain. We want to be able to answer the question: How does the brain store and process information? Because we can't answer this question unequivocally, this book dances through metaphors, models, speculations, relevant details, possibly relevant details, likely irrelevant details, etc. The 43rd edition of this book is tentatively scheduled for publication in 2212. It will contain a more thorough treatment.

Information, like beauty, is in the eyes of the beholder. The information content of a "message" depends on what you want to do with it. The position of planets in the solar system is generally not thought of as a message. However, an astrologer reads this as a code and an astronomer uses this information to make predictions about future celestial events. While wise men eternally wonder about the question, If a tree falls in the forest, is there a sound?, I am prepared to answer. Yes, there is a sound (a pressure wave), but is there a noise? No. A signal? No. Information? No. Similarly, patterns of light in the world can be taken as a code when they fall on a retina.

For our purposes, patterns of activity in the nervous system will generally be regarded as codes, though some of these codes may not be involved in the information processing tasks of the organism as a whole. For example, there is a large amount of information processing going on at the cellular level in order to handle the computational needs of genetic and metabolic processes. This reproductive and metabolic information processing is generally kept separate from the organism's thought and behavior. However, if you have low blood sugar (hypoglycemia), you suddenly discover that your cells' metabolic misery is quickly made your own.

We divide the question of coding strategies into three categories: 1) ensemble coding (how neurons work together to form a code), 2) neuron state encoding (what aspects of the neuron's chemical or electrical activity are relevant for coding), and 3) temporal coding (how signals are interpreted across time).

7.2 Coding in space: ensemble codes

Just as the pattern of pixels in Fig. 4.2 formed a picture, some areas of the brain are organized into maps that effectively mirror the world. The visual system has many maps, starting with the direct painting of the world by light on the retinal photoreceptors and continuing with similarly organized *retinotopic* maps in various areas of cortex. It is possible to measure the activity of cells in cortex and then re-create a blurred version of what the animal or person is looking at. One visual area projects to another in a hierarchy of processing areas. In general, as we go to higher processing areas in the visual systems, the responses of individual neurons become more specialized. For example, there are higher visual areas with specialized cells that detect motion, and other visual areas specialized for response to shapes.

In addition to greater neuron specialization from lower to higher processing centers, there is also a trend toward greater *receptive field* size. The receptive field of a visual cell is the area of visual space where suitable activity (movement, light, color, etc., depending on the processing area) will affect a response in that neuron. In higher visual processing areas, some cells have enormous receptive fields and will respond to the entire hemifield (everything off to the left, for example) or even the entire visual field. By knowing the receptive fields of cells and looking at their responses, it is possible to draw a picture in which activity in each cell represents a pixel. At the photoreceptor level in the retina, this picture will be perfect (although upside down), like the image on the film at the back of the camera. By the time you get to ganglion cells in retina it will be a little more blurred and so on up the processing pathways. In higher visual areas, the pixels would overlap enormously and the picture would be blurred beyond recognition.

Similarly, different areas of the brain have skin maps for touch and movement maps for motor activity. These maps can be superimposed on the brain to indicate areas that are more or less heavily represented in cortex. In the sensory realm, areas that are used for exploration by touch are heavily represented. In a mouse, whiskers get the heaviest representation. In a person, the hand, face, and tongue are most heavily represented. Fig. 7.1 shows a version of the humunculus, the distorted drawing of a human figure according to the location of its representation on the cortex. The tongue, face, and hand are big. The body and back of the head are small. The foot is pretty big, too. In areas with large cortical representations, there are many cells, and each cell will have a small receptive field, allowing it to produce a very fine-grained map of that area. This can be tested by having someone do two-point discrimination testing on your skin. This is done by randomly poking either with two pins placed closed together, or with a single pin. At some point, the two pins are so close together that you can't tell whether you're being touched by two pins or by one. You will find that you have much better two-point discrimination on your face or hand than you do on your back.

The motor humunculus also devotes a lot of space to hands and lips and relatively little to shoulders and foreheads. In general, left brain innervates right body and right brain innervates left body in both the motor and sensory systems (contralateral innervation). In the motor realm, areas that are not heavily represented in cortex tend to be innervated by both sides (bilateral innervation). This can be appreciated by comparing the brain's control of the forehead to its control of the mouth. Almost everyone can move one side of the mouth without moving the other side. However, most people cannot raise one eyebrow without raising the other eyebrow at least a little. Because of the bilateral innervation, when your left cortex signals a contraction of your right forehead muscles, it signals contraction of your left forehead as well. Everyone's brain is different. Some people have relatively segregated facial innervation patterns and can raise one eyebrow without difficulty. Also you can train your brain to increase the density of representation for one area at the expense of another.

There are many other maps in the brain. Hearing has both frequency and location maps. Hearing maps location by measuring differences in arrival

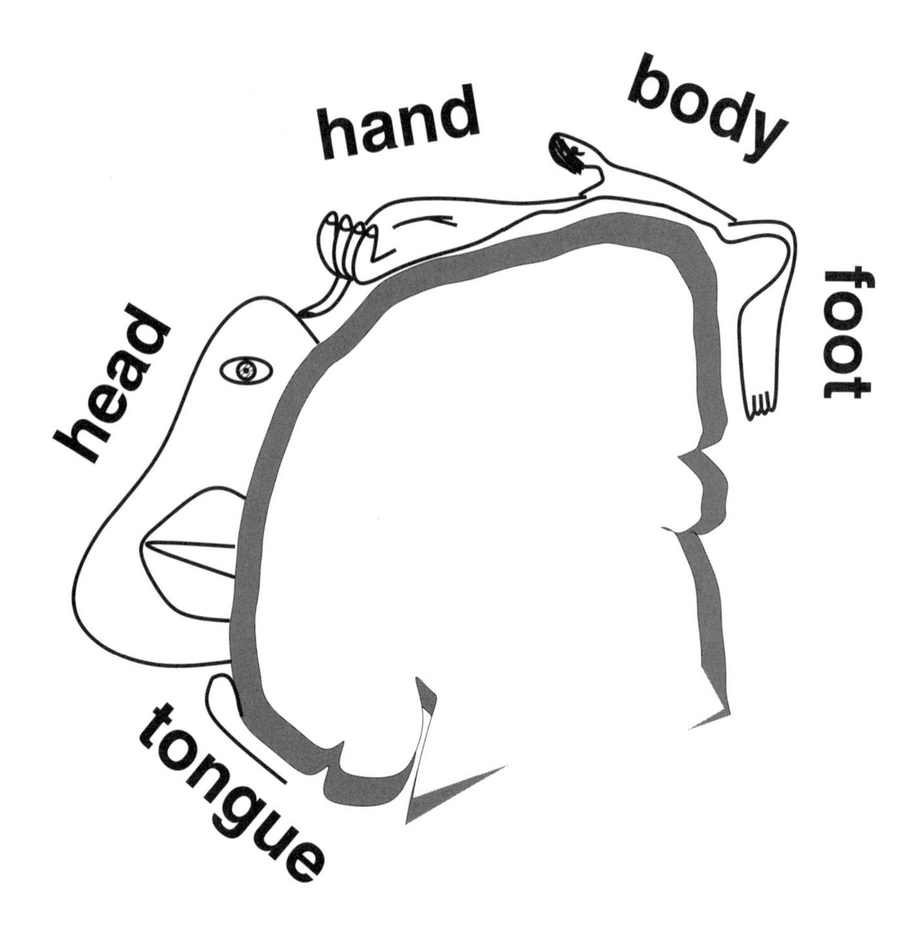

Fig. 7.1: The humunculus — right cortex viewed in coronal section.

time and intensity between the sounds received by the two ears. Bats and dolphins, using echolocation, have much more sophisticated auditory maps that involve both motor output and perception. Many aquatic animals, including electric fish, sharks, and platypuses, have electric field maps and can detect and apparently image electric fields.

Local vs. distributed ensemble coding

Ensemble codes can be classified as either local or distributed encoding schemes. A *local code* is one in which an individual neuron can be identified as representing a specific item (such an *item* could be an image, a sound, a thought, or a word) out in the world. For example, some invertebrates have command neurons that are triggered by particular stimuli and produce a stereotyped response. A local code is not necessarily one to one; there can be many neurons, all of which represent the same location (or item or thought or word). Local coding neurons are also called *grandmother cells.* This phrase implies that there is a specific neuron whose activation would indicate that grandma has been detected. If this cell was lost, one would lose all knowledge of grandma. It does not appear that local coding is used much in higher animals.

The alternative to a local code is a *distributed code.* In a distributed code a bunch of neurons need to be coactive to represent item A. In this respect, a distributed representation isn't any different from a local representation where each one of the bunch can represent A by itself. The trick in a distributed representation is that we reuse the same units when we are representing a bunch of *items:* A, B, C .... In a local representation, any given neuron represents just one of these items. If you lose all of the neurons that represent A, you no longer have any memory of A. In a distributed representation, it is the pattern of activity that counts rather than the activity in any single neuron. Different patterns, representing different items, will involve some of the same neurons.

We generally think in terms of representations involving large numbers of neurons. In a distributed encoding, the loss of a few neurons will degrade the representation of several *items.* However, no single item will be lost. As more and more neurons are lost, all of the representations will be degraded more or less equally without losing any particular representation. This property of distributed representation networks is known as graceful degradation with damage. This is a valuable property and an advantage over local representation networks, which are more prone to lose memories if damaged.

A drawback to distributed coding is that there can be significant overlap between representations of different items leading to confusion between them. Although this is bad if you are trying to engineer reliable memory systems, it is good if you are trying to model human memory, which is notorious for its capacity for confusion. Overlap is also a potential advantage where linking related items is beneficial, as in reasoning by example or metaphor.

A technological example of a distributed representation is a hologram, a three-dimensional picture made with lasers. The image in a hologram is distributed, while the image in a photograph is local. In the photograph, every item in the visual scene is represented at one spot on the photo. If you cut out a piece of the photo, you lose an item. In a hologram, the information about the scene is distributed across the image. As a result, it needs to be decoded in order to be seen. However, if you cut a piece out of a hologram, you do not lose individual items in the scene; rather, you blur all of the items. Since information about each item is distributed, you can't cut out one item without removing information about other items as well.

7.3 Coding with volts and chemicals: neural state code

In the previous chapter we introduced the notion of the state of a unit or neuron. This hypothetical state comes from a top-down computer science perspective, and represents a minimalist notion of neuron activity. To reconcile this top-down notion with the bottom-up complexities of wet biological measurements, we want to ask what we can measure in a real neuron that corresponds to the state of the neuron.

The neuron is full of chemicals. Some are measurable by currently available techniques; most are not. Nerve cells also have a large catalog of electrical responses. A neuron is big — electrical potentials or chemical concentrations in one spot will not generally correspond to their values somewhere else in the neuron. Any and many of these potentials and concentrations are likely to be involved in neural information processing. The existence of so many possible neuron states makes it hard to see how we can work within the context of the scalar-state single unit description introduced in the previous chapter.

We return to the complex internal state of neuron when we consider the neuron's electrical and chemical infrastructure in more detail in Chap. 11. For now, we simplify matters enormously by discussing only the output of the neuron. Since this output is the only way that one neuron influences another, it must represent a summary or synopsis of the complex multidimensional state of the entire neuron.

Biologically, it is important to emphasize the distinction between the neuron state, which is complex and high-dimensional, and the output state, which is relatively simple and low-dimensional. This distinction is often overlooked because there is no such difference in sum-and-squash units, where unit state and output state are one and the same. By making this state-output identity assumption for real neurons, we force the conclusion that the neuron does not do major calculations internally. A simple neuron can take only low-dimensional inputs, combine them, and produce a low-dimensional output. On the other hand, if a neuron is running complex programs on multiple internal states, it can take the input signals and do substantial processing before producing another signal that is only distantly related to the inputs. In the terms that I have previously used, the debate is between the neuron as transistor and the neuron as CPU.

Having thrown most of the neuron out of our discussion, we can now throw out the synapse as well. The synaptic mechanisms, very complex as well, will be collapsed into a single dimensional weight. As we describe later, this simplification collapses presynaptic, synaptic, and postsynaptic mechanisms. Instead of worrying about all that, we assume, as before, that the signal passes from one neuron to the next via a simple scalar weight. As with the oversimplification of the neuron state, I'm prepared to complain about this, but I'm not really ready to do anything about it.

Having thrown away the neurons and the synapses, the only thing left to discuss is the meaning of axonal spikes or action potentials. The spike is the form of electrical activity responsible for long-distance signaling to other neurons. The spike, or the pattern of spiking, can be taken to be the output state of the neuron. As the single major output channel for many neurons, this gives us a single place to measure state and a single thing to measure. The action potential is a brief (about 1 ms) positive voltage that travels as a wave down the axon to communicate via synapses with other neurons. An action potential is a sudden upward deflection of membrane voltage followed by an equally sudden return to the baseline — a brief spike in voltage. Spikes are stereotyped; every spike looks pretty much like every other. (Actually later spikes in a train tend to be smaller, but this is not believed to make any difference in their ability to trigger synapses.) This rules out two obvious possibilities for information transmission: spike shape and spike size. Therefore, spike timing is the only attribute that is thought to be important for information transmission.

7.4 Coding in time: temporal and rate codes

State in the sum-and-squash unit is instantaneous. The state value at any time is dependent only on the previous step and independent of whatever came before that. We can instead base state value on a train of spikes, making unit state a reflection of that unit's history. This is called temporal coding, a code that utilizes sequences of interspike intervals. Because utilization of temporal information bases the state values on what happened in the past, it is a basic form of memory.

Temporal information is clearly used in human communication. To understand speech, one has to process strings of phonemes (sounds) to understand a word, and then process strings of words to arrive at a complete idea. The fact that the brain is good at creating and understanding speech demonstrates that temporal coding and decoding occur in the brain, although not whether they are performed primarily at the neuron or at the network level.

In signals and systems parlance, a signal processor is called a *filter*. Interpretation of a temporal code requires that the filter store information in some kind of memory to keep track of what came before. To interpret a temporal code, it is necessary to chunk data, to group it. Consider Morse code. Like the signal on an axon, an initial analysis of the telegraph signal would suggest that it has two states: a high voltage and a low voltage. However, since we know the code, we know that it is more accurate to regard it as having at least three states: 0/dot/dash. To distinguish these three states we have to monitor activity on the line for a long enough period to distinguish the dot from the dash.

Going further, we could consider the different letters of the alphabet as distinct states (ignoring punctuation and special signs). By monitoring the line for longer periods, we can distinguish 26 different states. Interpretation of these 26 states requires that the filter store a series of high and low voltage levels. This interpretation is made more difficult by the fact that the different states are represented by signal sequences of differing durations.

If we didn't already know Morse code, it would be difficult to extract it from recordings made on a telegraph line. Usually Morse code is used to transmit natural language. If natural language is being transmitted and you know or suspect this to be the case, you could use statistical measures to infer which symbol represented what letter. This is the domain of cryptography. If the type of data being transmitted is completely unknown, other techniques must be used. This is the domain of information theory. To parse (divide up) an unknown code, information theory compares various temporal integration times, using statistical measures to look for recurring patterns. The information theoretical approach has been extensively applied to spike trains of single neurons with some encouraging results. However, it may not be possible to break the neuron code using single neuron signals, if the code is dependent on coordinated firing across multiple neurons. In that case, information theory would have to be applied to multiple spike trains simultaneously, applying spatial as well as temporal integration.

As in the case of Morse code, temporal code can generally be chunked in various ways depending on the sophistication of the filter. Longer chunking times require more sophisticated processing. To an extent, this will enhance efficiency, but at the cost of slowing any responses to be based on the results of the interpretation. Therefore, optimal chunking size will depend in part on the requirements of sensorimotor integration.

Temporal integration

Temporal integration is a much simpler filtering algorithm that neurons can readily perform. Integration is the calculus word for summation. The mathematical symbol for summation is $\sum$; integration is $\int$. Both are variations on the letter S for summation. Temporal integration means adding up all the signals that are received over a certain period. This is believed to be a major form of filtering occurring at the single neuron level. Only one number needs to be stored in memory, representing the signal sum at that moment.

If we keep integrating incoming signals indefinitely, the sum that we are storing just gets bigger and bigger. Therefore, we must define an integration time τ (tau). At any time, the value of the temporally integrated signal is the sum of all signals that have arrived during the last τ ms. Typically, we can imagine neurons as doing temporal integration by maintaining a running average of activity. As signals come in they are added to this average signal, but their influence cuts off (or more realistically wanes) after period τ has passed. Note that temporal integration would not work for Morse code. If we consider a dash to be twice as much signal as a dot and we receive dash-dot (N), we could add up $2+1 = 3$ as our temporally integrated signal. However, dot-dash (A) would also give 3, as would dot-dot-dot (S).

Individual units in an artificial neural network don't make use of temporal information. Furthermore, the artificial neural network's ensemble code only uses the state of the network at the previous instant in time (for example, $t = 4$ in Fig. 6.5) to determine network state at the next instant ($t = 5$). Despite this, artificial neural networks can do temporal processing. Events that happened in the past ($t = 2$, $t = 3$) percolate around a recurrent network, providing implicit access to historical information. In this way, artificial neural networks have been used to model trajectories that unfold in time.

Clocking

In a typical artificial neural network implementation, updating is strictly clocked: every unit is updated at the same time. This is called synchronous updating. Artificial neural networks can also be designed to use asynchronous updating: each unit updates itself on its own schedule. This does not mean that clocking has been eliminated. Doing away with a central clock requires that we delegate the clocking to the individual units. Although units update at random times, they must remain grossly synchronized in order to participate in the network. This generally requires that units all have similar time constants. A network could be designed with some units whose time constants are multiples of the time constants of other units. These units would then update regularly after several cycles.

As we discuss in Chap. 11, real neurons do have a time constant and can therefore do their own clocking. There are also various types of central clocking apparent in the brain as well. The circadian (24-hour) rhythm is the most obvious and best studied clock in the brain. The appearance of a variety of frequencies in the electroencephalogram hints at the possibility that there may be rapid clocking as well. Electroencephalogram frequencies are grouped in ranges: delta, theta, alpha, beta, gamma (these are usually written out like this, rather than abbreviated with the Greek letter). The delta rhythms are the slowest, ranging from below 1 Hz (less than once per second) to 3 Hz. These are seen mainly in deep sleep or deep coma. The theta range is 4 to 7 Hz in humans (the same frequency names are used in rats but the frequency ranges differ). Theta is mostly seen in drowsiness. The alpha rhythm, 8 to 12 Hz, is a prominent occipital rhythm with eyes closed that goes away when the eyes open. Beta, 13 to 26 Hz, and gamma, 27 Hz and above, are also seen in the awake state. High-frequency gamma rhythms don't penetrate well through skull and scalp, so they can't be seen on the electroencephalogram. These high frequencies are better recorded on the electrocorticogram, which uses electrodes placed directly on the brain.

The circadian rhythm is generated by identifiable cells in the suprachiasmatic nucleus. By contrast, faster brain waves do not appear to be projected from a central generator. They are instead an emergent property resulting from the activity of large numbers of cortical neurons. Brain waves might just be epiphenomenal, a consequence of activity but having no direct functional relevance. On the other hand, it is possible that these waves, emerging from one set of cells, are utilized by other cells as timing signals. Simultaneously occurring waves at different frequencies could then be playing different clocking roles in different brain areas.

If neurons in the brain are being clocked, neuron time constants and brain wave frequencies can tell us something about how the clocking might work. A spike lasts about 1 ms. Therefore, maximum spike rates are about 1 kHz (1000 cycles per second). This would be the maximum rate that a clock could run. Although this theoretical upper limit is probably unobtainable, fast clocking at 100 to 200 Hz is possible in small brain regions. It has been suggested that inhibitory interneurons, which have fast firing rates, might provide such synchronization. Over larger regions, or perhaps over the whole brain, gamma frequencies of 40 to 60 Hz have been suggested as possible clocking signals. Synchronization of activity at these frequencies across large areas of cortex has been found.

7.5 Frequency coding

The most widely accepted model of temporal coding in the central nervous system assumes that temporal information is utilized through temporal integration. As we saw in the Morse code example above, this means that exactly when spikes are received will not matter — everything just gets dumped into the lump sum. The only thing that matters is how many spikes have arrived during the period (τ), i.e., the rate of spike arrival. This coding scheme is referred to as *rate coding* or *frequency coding*.

Frequency coding has several advantages when it comes to translating the activity of real neurons into the nomenclature of artificial neural networks. The primary advantage is that it translates the spike train into a scalar state. A second advantage is that it gives us a way to express negative states (Fig. 7.2). Artificial neural networks that use $-1/1$ as the range of states are widely used in network memory models, as will be described in Chap. 10. Spikes are always positive voltage deflections; there are no negative spikes to use for representing -1. Using rate coding, we assume that there is a natural spontaneous firing rate that represents a zero state (Fig. 7.2 — 0). An increase from this spontaneous rate corresponds to a

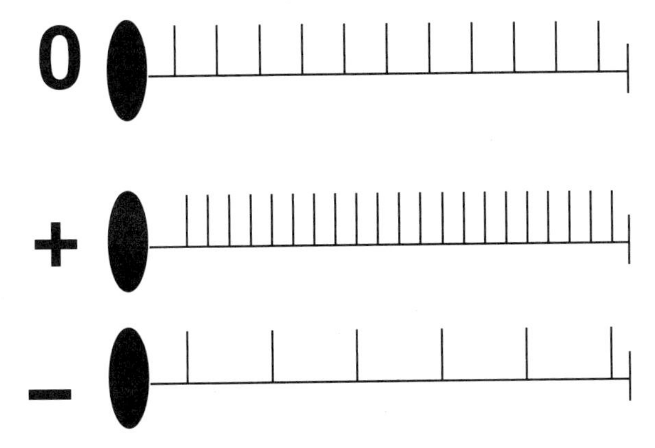

Fig. 7.2: Spontaneous rate represents a scalar state of 0. Increased rate is positive state and decreased rate is a negative state. Spikes are represented as short vertical lines atop the horizontal axons. At bottom: negative state times negative weight equals positive state.

positive state (Fig. 7.2 — +). A decrease from the spontaneous rate is a negative state (Fig. 7.2 — −).

The baseline rate is produced by a balance between ongoing excitatory inputs and ongoing inhibitory inputs. At the top of Fig. 7.3, I show excitatory and inhibitory inputs producing the spontaneous rate in a follower cell. There is a balance of excitation and inhibition, which favors excitation, driving the follower neuron. I have illustrated the excitatory synapses as little stubs and shown one inhibitory neuron (shaded). At the bottom of Fig. 7.3, the rate of firing in the inhibitory cell has decreased (a negative state). This negative state is transduced via a negative, inhibitory weight. With the reduction in inhibitory input, the follower cell shows increased firing. This demonstrates the rate coding of the following situation: negative state times negative weight equals positive state. Similarly, it is easy to see that both negative state (reduced rate) times positive weight (excitatory input) produces negative state (reduced weight), and positive state (increased rate) times negative weight (inhibitory input) produces negative state (reduced rate).

A major appeal of the rate coding model is that it gives us the simplicity and modeling tractability of a scalar state. However, there are neurobiological as well as computational reasons to believe that rate coding is important in the nervous system. There is clearly a frequency/strength relationship in the efferent (motor) peripheral nervous system. Increased rate of firing of a nerve that controls a muscle will lead to increased contraction strength in that muscle. You can electrically stimulate a nerve from outside the skin

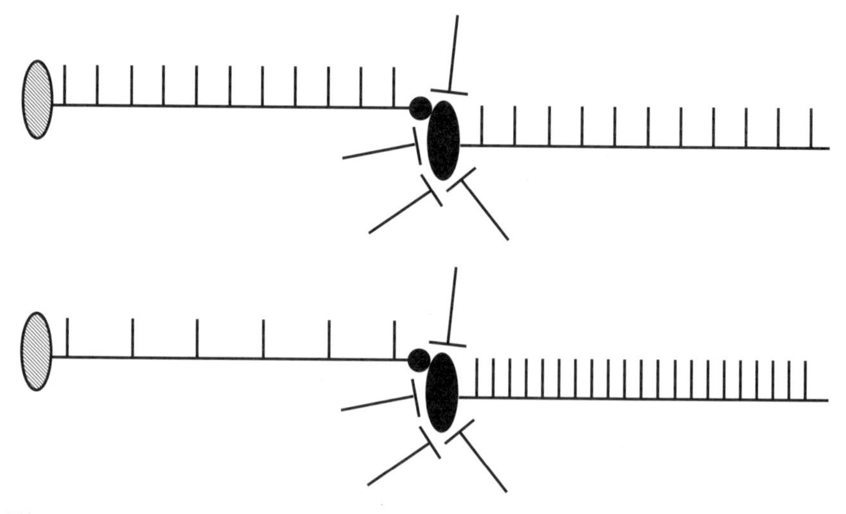

Fig. 7.3: Spontaneous rate is produced by a balance of excitatory inputs (line segment synapses) and a negative input (small circle synapse, shaded cell body). Bottom: A negative state (reduced rate) in the inhibitory neuron times a negative weight (inhibitory synapse) produces a positive state (increased rate) in the follower neuron.

and make its muscle contract harder as you increase the current, pushing the firing rate higher.

There is also substantial evidence for stimulus specific increases in firing rate in the afferent (sensory) central nervous system. In their studies of occipital cortex, David Hubel and Torsten Weisel showed large increases in the neuron firing rate in response to oriented line segments in the visual field of a cat. Recently, Newsome and colleagues performed some remarkable experiments in which they altered an animal's visual perception by injecting current into the brain. They studied area MT, a visual area with neurons that respond to motion. They identified sets of neurons that fired when most of the randomly moving spots on a display screen went from left to right. They then showed the animal a display in which most of the spots went from right to left. If they injected current so as to increase the firing rate in the previously recorded left → right cells, they could deceive the animal into thinking that the spots were moving left to right as before. Paranoid schizophrenics sometimes complain that something (the CIA, the TV networks, or a dog) is projecting thoughts into their mind. (Which of these three will be the first to get their hands on the new thought-insertion technology?)

In addition to the motor and sensory correlates, rate coding is suggested by the fact that many cells appear to have spontaneous firing, meaning that they fire even when no stimulus is present. Of course, unexplained firing

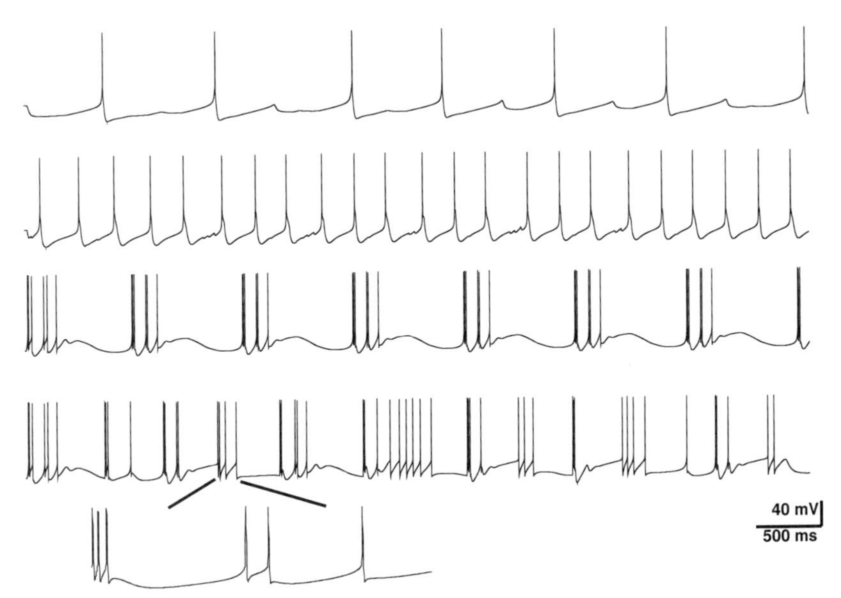

Fig. 7.4: Trains of regular firing and bursting neurons from thalamic cell simulations.

does not have to be spontaneous or meaningless. Instead, irregular firing may be coding for things that are not directly connected to stimulation, such as thought or memory access. We ask an experimental subject, or provide rewards for an animal, to just pay attention to certain stimuli and not think about anything else. I tried this in yoga; my mind didn't want to cooperate.

On the other hand, there are also neurobiological reasons to believe that rate coding may not be the whole story. Many neurons fire bursts. In that case one can define either a rate of spike firing within the burst or a rate of bursting — there is no single scalar state for rate. Fig. 7.4 shows four traces of simulated thalamic cells. The top two traces show fairly regular firing at about 1.5 and 6 Hz (spikes per second), respectively. The third trace shows bursting. The rate of bursting is about 2 Hz, but the spike rate within the burst is much higher. The bottom trace is a complex mix of bursting and regular firing with varying rates. Blowing up one of the bursts in the trace at bottom demonstrates that it is not simply a burst of spikes but is actually a burst of bursts. The bursts in the third trace are similar; they are also bursts of bursts.

The top trace of Fig. 7.4 also illustrates that even "regular" neuron firing is often somewhat irregular. In an experiment, irregular firing is typically attributed to noise. As in Fig. 7.4, however, such irregularity is also seen in a deterministic dynamical system. In both biology and simulation, apparent noise may be evidence of deterministic chaos. Chaos is a relatively new field of mathematics that has exposed the tendency of nonlinear systems to act weird. More precisely, chaotic systems appear to show random behavior despite following precise mathematical rules (deterministic system). Examination of chaotic behavior has shown that there is some slight predictability within the chaos. Chaotic systems show sensitivity to initial conditions. This means that two identical systems, started at slightly different points in state space, will show completely different behaviors. This is the source of the hypothetical "butterfly effect" — a butterfly flapping in China can change the weather a year later in South America. Additionally, a chaotic system will never revisit a point in state space; it's always going somewhere new. Despite all this wildness, chaotic systems do manage to stay on strange attractors, distinct volumes of state space. If you use a computer to graph the trajectory of a chaotic system gradually in time, the scribbles will eventually fill in an area. This is the strange attractor.

Some researchers have suggested that chaos might be used by the nervous system to store information. In this case, complex patterns of seemingly random neuron firing would fit on one of many strange attractors. These attractors would be mapped by following network state in n dimensions, where n is the number of neurons.

A simpler source of firing rate changeability is adaptation. Neurons will typically adapt to a constant input with a gradual reduction in firing rate. Because of this, a constant stimulus is not precisely associated with a single rate code. This does not necessarily contradict the rate coding model because adapting neurons could underlie the psychophysical phenomenon of habituation, the reduction in perceived stimulus strength with continued exposure. It does illustrate that coding will not be a simple matter of defining values that imply particular stimuli.

Problems in defining firing rate also arise when neurons fire rarely. Some neurons may fire only a single spike when they fire and therefore have no spontaneous rate. This kind of firing pattern is reminiscent of the all-or-none binary coding originally hypothesized by McCulloch and Pitts. Some neurons don't spike at all; clearly these neurons are not candidates for any kind of spike coding. These may be neurons that don't have the channel density needed to produce action potentials. Such neurons could communicate signals through neurotransmitter release triggered by subthreshold potentials, voltages that are too small to produce firing.

7.6 Summary and thoughts

In this chapter I have emphasized the most widely accepted coding models: local and distributed ensemble coding, and rate coding. I have also taken the opportunity to cast doubt on these models by pointing out how the many complexities of real neurons don't fit neatly into these simplified theories. In the following chapters, I try to stick with the basic theories and hold my skepticism in check.

8
Our Friend the Limulus

8.1 Why learn this?

Artificial neural networks have primarily made their mark in engineering and cognitive science. We look at one of the basic cognitive models in a later chapter. The success of neural networks in these nonbiological domains raised great hopes that artificial neural networks had captured the design of the brain. A grand notion, hard to live up to, that unfortunately led to artificial neural networks being regarded by many neurobiologists as all bluster and no substance, at least with regard to biology.

In this chapter and the next, I present two applications of artificial neural networks in biology. The example in this chapter describes the mechanisms of vision in the eye of the horseshoe crab. This work was the research of H.K. Hartline and Floyd Ratliff in the 1940s and 1950s. The horseshoe crab, often called by its scientific name, limulus, has been around for a while (about 500 million years) and has changed little over these thousands of millennia. Its visual system is of course very simple compared to our own. Despite this, there are aspects of human visual performance that can be explained through our understanding of the limulus eye.

Because this chapter represents the first real modeling in the book, I present here some general computer modeling concepts. These include simplification, scaling, sampling, linearity, parameterization, and iteration.

8.2 The biology

Sensation starts with a process of signal transduction — where a physical signal from the environment is "led through" (the meaning of transduction) a transforming step so that it can be processed by the nervous system. More generally, this is only the first of a series of transductions as the signal is repeatedly changed from one form to another. In visual transduction, for example, light energy is converted into a chemical signal. This chemical signal is in turn converted to an electrical signal, which is converted to a chemical signal, which is converted to an electrical signal, and so on through chains of synaptically connected cells. At each stage, we can speak of a physical process that is encoding the original signal.

As we saw in Chap. 7, Section 7.5, rate coding provides one way to represent a single scalar value in a neural signal. In the case of vision, an important stimulus variable to be encoded is light intensity (signal amplitude). Another major aspect of the visual scene that must be encoded is spatial pattern (signal locations). These spatial locations are directly present as locations in the retina and transferred, in higher animals, to locations in cortex. This is called *labeled-line coding* since the meaning of the message does not depend on the form or timing of the signal but rather on the identity of the transmission line where the signal is received. For example, in an old-time telegraph office in Denver, the identity of the line would tell the operator whether the message came from Albuquerque or San Francisco, even before any decoding took place.

The horseshoe crab has compound eyes — lots of little eyes work together to make one big eye. One can think of the axons projecting from these as so many labeled lines, each coming out of a separate eye segment. Each of these little eye segments has its own cornea and photoreceptor. The cornea collects light from a certain area of visual space, the photoreceptor measures the amount of a light, and a neuron transmits this information to a central visual processing station.

8.3 What we can ignore

The art of modeling consists of deciding what aspects of the object of interest have to be included and what aspects can be ignored. This transforms a real physical object into a simulacrum, which may be another physical object, a set of mathematical equations, or a computer simulation. Generally, you model something when it is too complex to understand by just thinking about it. To do this, you must first simplify, deciding what to include and what to omit. This winnowing process is the art of modeling. As pointed out in Chap. 2, Section 2.3, reference to birds during development of heavier-than-air flight proved both helpful and misleading.

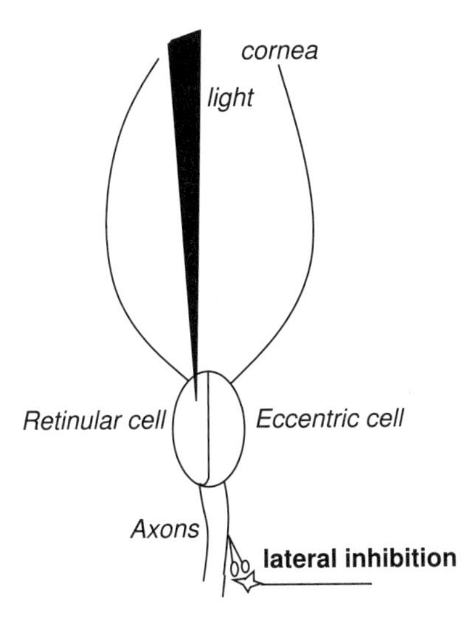

Fig. 8.1: Ommatidium, illustrating lateral inhibition plus all the stuff to be ignored.

As with most biological systems, the complexity of the limulus eye is daunting. Each little photoreceptor complex, called an ommatidium (plural ommatidia), consists of two cells (Fig. 8.1). The retinular cell contains the photoreceptor chemical, rhodopsin, which changes shape when hit by a photon of light. This chemical process, and the subsequent process of transduction that changes the chemical change into an electrical signal on the membrane of the cell, is itself complicated. The signal is then transferred to the eccentric cell and from there out the axons to central processing areas. From there, cells take the signal and transmit it to the side where the signal reduces the output from other cells. This process is known as *lateral inhibition.*

Any and all aspects of this signal handling might turn out to be critical for understanding the process of vision in the limulus. We will see that lateral inhibition is of particular interest. It will turn out that we can strip away all of this complexity and just assume that the system somehow takes light of a certain intensity and transforms it to an electrical signals coding that same intensity. Simpler still, we'll just represent light intensity (luminance) as a number, and then represent the cell signal as the same number.

8.4 Why the eye lies: the problem

Our long-distance senses use light, sound, or chemicals to transmit information from the physical world around us. Each of these has its own properties that can produce illusions. For example, when we look up to the sky, a seemingly real blue dome overhead, we are just seeing blue light from the sun highly distorted. Even after we get through these physical distortions, our information about the world passes through receptors and then through the nerves that connect to these receptors. All of these intermediaries can introduce further errors and distortions.

In an ideal world, it might be nice to have direct access to reality. It is sometimes hard to appreciate that we do not, that we are only seeing those platonic shadows on the cave wall. When Hermann von Helmholtz, the 19th century pioneer of neurophysiology and medicine, discovered that nerves conduct information at a finite rate, his father thought the idea ridiculous. It is quite obvious, his father said, that we perceive the world immediately and directly without delay. Actually, a variety of physical and neural intermediaries create both delays and distortions. Some of these distortions can be left alone, for example, the blue sky. Some distortions can be useful — delay in sound propagation through air is used for sound localization by perceiving the difference in arrival times in the two ears. Other distortions must be fixed in order for the animal to compete and survive. The limulus has been hanging around since long before the dinosaurs. It is a survivor. One of the roles of the limulus eye and retina is to correct receptor errors, sweep away illusions, and see life as it really is.

For the limulus, the world out there is an image of light and dark. In an ideal situation of minimal distortion, the activity across the limulus eye (or the back of a camera or the human retina) would accurately match the image: brightness (the perception of strong light) would exactly match luminance (the physical presence of strong light). In places where the image reflects a lot of light (high luminance), the corresponding location on the eye would have a highly active photoreceptor (appearing bright). Where the image is dark, it would have a very inactive photoreceptor. However, two physical factors get in the way (Fig. 8.2). One is scattering. As light passes through water or air from the image to the eye, it spreads out, blurring the representation of the image on the photoreceptors, just as blue light from the sun is blurred out all over the sky. In Fig. 8.2, scattering of light is seen affecting receptors a and c. Receptor a is positioned midway between light and dark bands but receives more than half the maximum light due to scatter. Receptor c is positioned over an area of darkness, but scatter causes it to register the presence of some light.

The other problem is sampling. Even in the absence of scatter, accurate detection of a sharp change in luminance requires that two receptors neatly flank the contrast boundary, as do receptors b and c. If, as in the case of a, a receptor sits in the middle, it will sample both the dark and the light side and get an intermediate amount of light. This gives the perception of an intermediate brightness, suggesting a gradual drop-off rather than a sharp contrast boundary. Additionally, photoreceptor misalignment will also add to the overlap between portions of the image received at different

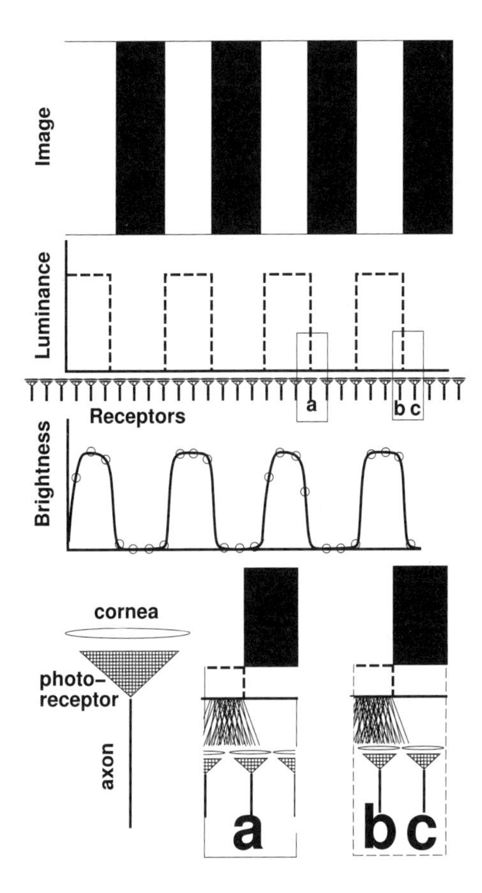

Fig. 8.2: Photoreceptor placement and light scatter turn the sharp black and white image into a blurred image. Image at top is depicted with a brightness graph, which is sampled by receptors schematically made up of cornea, photoreceptor, and axon. The resulting image can be reconstructed by drawing a smooth line between sampling points. Lower left: simplified schematic of receptor.

receptors. Although redundant sampling ensures that nothing is missed, misalignment may cause it to increase blurring still further.

Nature has designed neural processing to compensate for the blurring due to scattering and sampling. The solution is *lateral inhibition.* This solution is used not only by limulus, but also by us and by most of the in-between creatures of the great chain of being. Fuzziness is most troublesome in cases where the original image is least fuzzy — at a sharp contrast boundary from black to white, where contrast is maximal (Fig. 8.2). On one side of the contrast boundary, the image is black; little light emerges from the surface.

On the other side the image is white: maximum, saturating, light reflects from the surface. If neural activation perfectly reflected this contrast, the region facing the black part would be completely inactive, while the region facing the white part would be maximally active. A major purpose of lateral inhibition in an eye is to find this boundary location by deblurring. As we will see, lateral inhibition restores and actually enhances contrast boundaries. When a receptor detects light and transmits that information, lateral inhibition reduces the strength of signals transmitted by adjacent receptors. The amount of signal reduction on the flanks is proportional to the strength of the signal at the center. This process allows the receptors flanking the contrast boundary to reconstruct that boundary and determine boundary location as precisely as possible.

8.5 Design issues

To construct a neural network, we start with the biology and then abstract from it to make things as simple as possible. In doing so we typically make a large number of compromises in order to make the model tractable. These compromises reach into all aspects of modeling: translation from biology, calculation, understanding the model, and translation back to biology.

In this section I describe some common modeling compromises: linearity, size and dimensional reduction, wraparound, simple parameterizations, and scaling. In the present case, the biggest compromises come from the translation of biology into the model. We substitute a simple sum-and-squash unit to represent a complex signal transduction process that is actually carried out by a set of cells each utilizing multiple internal processes. The use of a sum-and-squash unit also means that we assume that signals sum linearly. This important approximation simplifies calculation and interpretation. To further simplify calculation, we use scaling and dimensional reduction for the network. To compensate for problems with scaling, we use wraparound. We also use simple images that are easily parameterized instead of real pictures of limulus predators. Finally, for the individual units, we use a scaling factor that makes it easy to translate back to biology. As you can see, we end up with interlocking compromises that make it hard to put together a single list of assumptions that can be readily assessed, and then accepted or rejected.

Making the model small — scaling

A common modeling simplification has to do with size. Neural systems are compact but pack in a lot of neurons and synapses. Certainly, the sheer number of units is a major factor in the brain's extraordinary information-processing abilities. In performing simulations, we cannot effectively emulate the large number of neurons in a given brain structure such as the retina. We generally use two simplifications to reduce the computational load: decrease the number of dimensions and reduce the absolute number of neurons (scaling).

Scaling is pretty common in modeling of all sorts, hence the term *scale model.* A sometimes overlooked aspect of the scaling process is the assumption of linearity. A linear transformation preserves scale. If the wingspan of an airplane is twice the length of its fuselage, and you create a scale model, the wingspan will still be twice the length. By contrast, a nonlinear transformation would not preserve ratios. You may have heard that, aerodynamically speaking, bumblebees can't fly. The confusion results from the fact that a large scale model of a bumblebee would not be able to fly because flight forces do not scale linearly. As pointed out in Chap. 2, flapping does you less-and-less good once you get past the size of a duck. Despite the fact that there's actually not much linearity in biology, we often make the assumption of linearity in modeling. Although it doesn't work for bumblebees, this mistaken assumption often works out well enough to get something useful out of a model.

Linearity is of course the major feature of the *linear* algebra used in neural networks. As described in Chap. 6, sum-and-squash units have both a linear step (the sum) and a nonlinear step (the squash). The linear summation step allows any number of excitatory or inhibitory inputs to be added in together without distortion. In the limulus model, linearity offers a major advantage in quickly illustrating the core issues we want to study. However, any aspects of limulus eye function that depend on its nonlinearities will not be represented in the model. This is a major compromise since sensory systems generally show a highly nonlinear response with increasing sensory stimulus strength. In many cases, this nonlinearity is logarithmic, meaning that any doubling of stimulus intensity produces the same-size increase in the perceived stimulus. This sensory response property is called the Weber-Fechner law, which explains why light/dark contrasts look the same whether viewed on a bright or cloudy day.

Making the model small — dimensional reduction

The retina is a sheet of photoreceptors and other nerve cells that lines the back of the eye in the mammalian eyeball and lies right near the front in limulus. As a sheet, it is a two-dimensional structure, which we could model as a circle or square of interacting units. However, we can make things much simpler and smaller by modeling it as a one-dimensional structure (a line segment) instead of as a two-dimensional structure. Such a reduction of dimensionality is a typical strategy in modeling that saves a lot of computer time and space; compared to modeling a square, we need far fewer units. The limulus has about 8000 ommatidia arranged in approximately a 6 by 12 mm ellipse. Modeling a line across the retina instead of a square retina

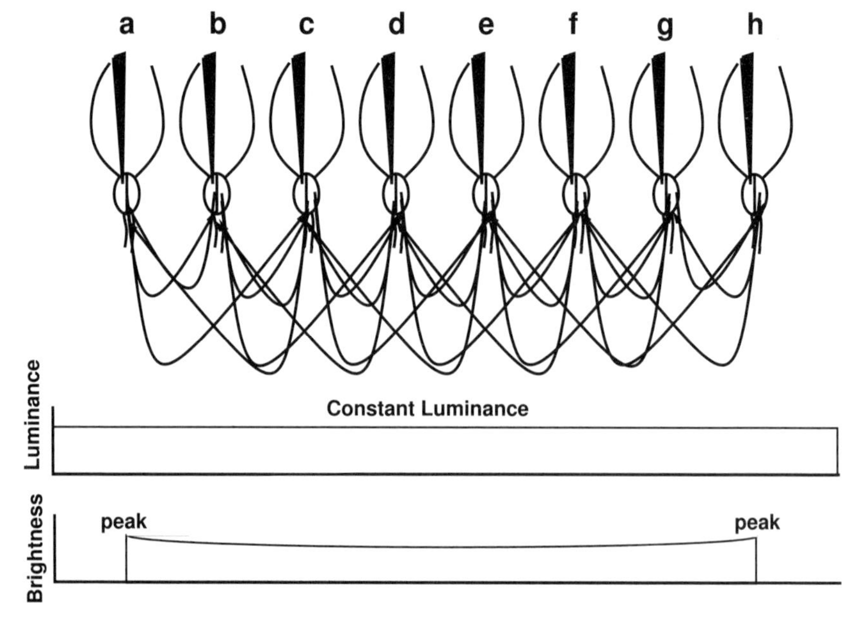

Fig. 8.3: Lateral connectivity of the limulus retinal network. Most units receive a total of four projections: one from each of two units on either side. However, unit a receives two, unit h two, unit b three, and unit g three projections. Because they are on or near the edge, these units behave differently. As a result there is an illusion of brightening at the edge as perception differs from luminance. This is an *edge effect*.

reduces the problem to $\sqrt{8000}$, which is about 90 units. Thus one main effect here is to make the problem much smaller and therefore allow it to be done with much less computer space and computer time. Working in one dimension also makes it easier to set up the connectivity matrix and easier to understand the results.

Eliminating edge effects — wraparound

In a two-dimensional network like the retina, there will be some neurons that are situated far out at the periphery, at the rim or edge. Neurons of the retina have a geometry, being connected preferentially to neighboring neurons. Edge neurons will have a different connectivity, since they have neighbors only on one side. This difference in connectivity will lead them to respond differently to a stimulus. This difference in response will in turn influence the activity of the neurons that they connect to, leading these neurons to also behave a little differently. In this way, edge effects will propagate a little way in toward the center of the network. Note that the

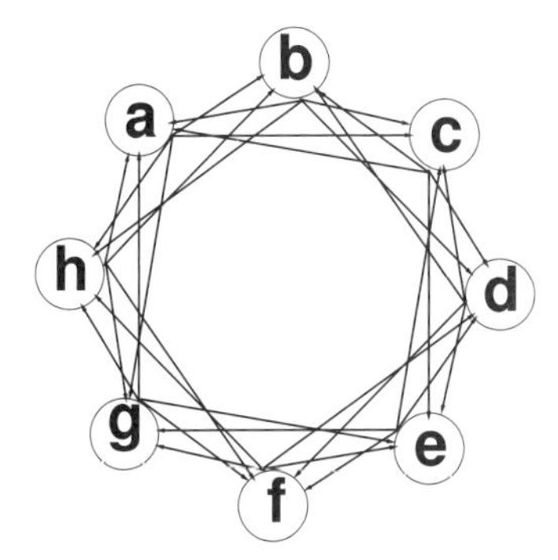

Fig. 8.4: Wrapping the network around so that one end connects to the other eliminates edges and thereby eliminates edge effects. Compared to Fig. 8.3, units a and h have been wrapped around and connected in a circle.

"edge" of edge effects refers to the edge of the retina or retina model. Above we have been discussing *contrast boundaries*, which also can be called edges, image edges between a light area and a dark area. Edge effects are probably not much of a problem in the real retina. The vast size of the network means that proportionally few cells are near the edge. However, edge effects do become a problem in the model.

In a model network, the much smaller number of neurons accentuates the edge effects. A model network will be simpler and easier to understand if it is uniformly connected, with all units having the same number and types of connections. Instead, Fig. 8.3 shows units at or near the edge with fewer connections. In the limulus model, connections are all inhibitory. If the edge units do not get as many connections, they will not get as much inhibition and therefore will be turned on more than the other units. This will give an illusory appearance of brightness at the edge of the image. The brightness curve (shown smoothed out in Fig. 8.3 bottom) does not match the square luminance curve above it. This is an example of perceptual enhancement at a boundary. Here the boundary is not a contrast boundary out in the world, but a histological boundary in the eye. Perceptually, an edge-effect illusion would look like a segment of halo around any lighted object whose image went off of the edge of the eye. If you looked at a wall or other uniform surface, it would appear to have a brightening at the edge. The brain presumably has mechanisms to get rid of edge-effect illusions.

In the model, we avoid the problem of edge effects by employing a common trick called wraparound. This eliminates the edges by simply turning the network into a circle so that unit a connects to unit h (Fig. 8.4). If we were doing the same thing with a two-dimensional model, the wraparound would give us a torus (also known as a donut). From the surface of a torus

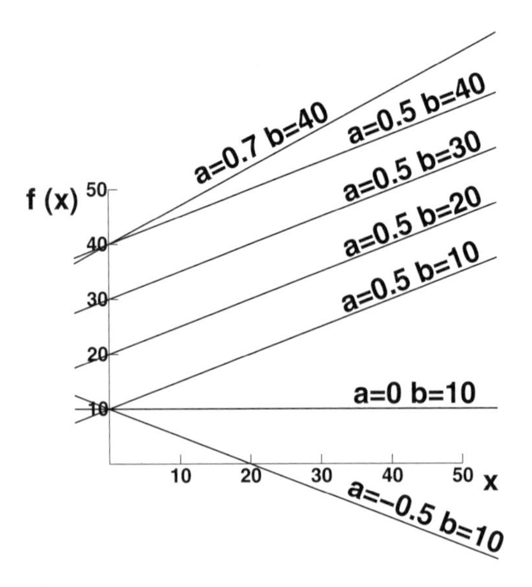

Fig. 8.5: Various parameter choices for a line: $f(x) = a \cdot x + b$.

you can go around in a circle in one of two ways: through the hole or around the entire donut.

Presenting the input — parameterization

Like scaling, parameterizing is one of the basic tasks in mathematical or computer model construction. Modeling involves translation. Different types of models require translation into different types of language: pictorial models, metaphorical models, physical models. In mathematical (or computer) modeling, we draw with equations. The art is to find the right equation and to then parameterize that equation so that it has the right form and size for the model. In general, parameterization refers to the choice of constants that give a particular class of equation a particular appearance. The form of an equation determines the general shape of the curve, while the parameters define its specific location and appearance. For example, the form of the function $f(x) = a \cdot x + b$, defines it as a line. ($f(x)$ is read "f of x" and is short for "function of x.") a and b are the parameters. a gives the slope of the line and b, the y-intercept, shifts the line up and down (Fig. 8.5).

Things that are easy to draw pictures of are not always easy to parameterize (e.g., natural scenes) and vice versa (e.g., things in four or more dimensions). An important step in using the simulator will be to draw inputs, images, to present. The image will be presented as an input vector $\vec{p}$ (p for picture) and therefore could be any arbitrary image. Nonetheless,

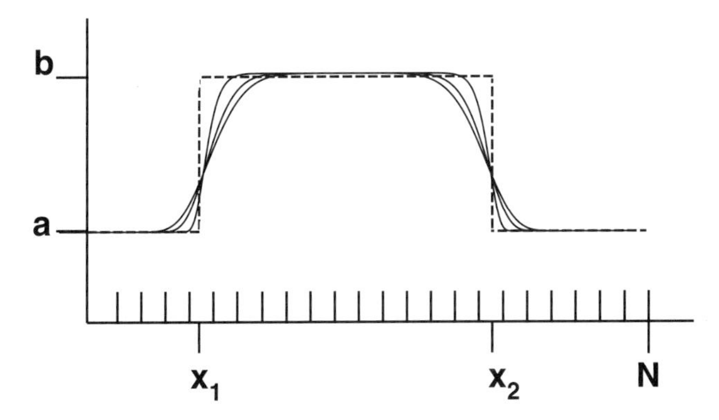

Fig. 8.6: Sample pictures ($\vec{p}$).

it will be convenient to parameterize the input in order to produce images easily without selecting individual values for tens or hundreds of inputs.

Since we are interested in how the eye handles contrasts, we want to use functions with high contrast — big jumps from high to low and low to high. The highest contrast is provided by a discontinuous function that just pops from low to high at a specified point. A discontinuous function cannot be described in a single functional form. Instead, it is made up of line segments and described as a *piece-wise linear function.* A single piece-wise linear function is written as three linear equations over three different domains: $f(x) = a \ (0 \leq x < x_1)$; $f(x) = b \ (x_1 \leq x < x_2)$; and $f(x) = a \ (x_2 \leq x < N)$ (Fig. 8.6 dashed line). The four picture parameters are x_1, x_2, a, and b. N is a parameter of the network: the number of units. This piece-wise linear function describes two contrast boundaries at locations x_1 and x_2 jumping between levels a and b.

We also want to see how the eye handles blurring. We use a blurred function that gives a flat top but sloping shoulder drop-off (Fig. 8.6, solid lines). These lumps (not a technical term) are formed by raising a Gaussian function (the familiar bell curve: $f(x) = e^{x^2}$) to an even power — the higher the power, the broader the lump and the steeper the sides. However, shifting and scaling our lump equation to get it to sit neatly on top of our little square step function requires a bunch of parameters: $f(x) = a + (b - a) \cdot e^{\{\frac{-(x-N/2)}{w}\}^{2n}}$, where a, b are again the vertical limits, $w = \frac{x_1 - x_2}{2}$ is the half-width of the lump, and n determines how flat the top is (larger n, broader top). Although both the step function and lump function are defined for the entire domain from 0 to N, we are only sampling the functions at discrete points corresponding to the location of our photoreceptors (vertical hash marks on x-axis in Fig. 8.6). The values at these points will be placed in the input vector $\vec{p}$.

Because we are using wraparound, the leftmost part of the network (unit 0) is next to the rightmost part (unit number N). Therefore, it is convenient to have the value of our functions be the same at receptor 0 and receptor N. This is why the step function in Fig. 8.6 has two steps instead of one. Alternatively, we could have just defined one step and left the other step as a discontinuity between receptors N and 0.

Parameterizing the activation function

As mentioned in Chap. 6, the activation or squashing function of a sum-and-squash unit usually uses a range of either 0 to 1 or -1 to 1. These values then need to be scaled in order to relate them to biological measurements. In the limulus model we assume rate coding and simply use the values from actual limulus firing rates to make it easy to relate our results to limulus experiments. By setting the model to suitable numbers at the onset, we don't have to do re-scaling for interpretation. Axons projecting from the limulus ommatidium spike at rates of 10 to 50 Hz. Therefore, we use a squashing function in our model that utilizes a range of 10 to 50.

We have chosen numbers for neuron state that relate to the biology. We also need to choose input values for our image — the a and b of Fig. 8.6. We are not concerned here about absolute light levels, so we won't bother with correctly defining the image intensity using units of lumens. Instead, we simplify our arithmetic by just using the same range of numbers for light intensity as we are using for neural state. This is highly artificial but reflects the fact that we are interested not in modeling the physics of light but only in modeling neural responses. Setting input range equal to state range simplifies signal transduction. The transfer from input light to neural response becomes a multiplication by 1 (identity function).

We use a piece-wise linear function for the squashing function. The simplest procedure is to just let the value alone if it is in the appropriate range and to move it to the nearest limit if it's outside of that range: $f(x) = 0 \ (x < 10)$; $f(x) = x \ (10 \leq x \leq 50)$; $f(x) = 50 \ (x > 50)$ (Fig. 8.7).

Parameterizing the weight matrix

It is also convenient to parameterize the weight matrix W so that we can easily make minor changes without going through and setting each of N^2 weights by hand. The weight matrix is going to be completely regular; the connections from each unit to its neighbors is the same since we're using wraparound. It will be easiest to just parameterize a single column of the weight matrix, a divergence weight vector for a single unit, and use this to create all the columns of the weight matrix. Each column will have the same weights shifted to correspond to the location of the postsynaptic unit. Since this is a purely inhibitory network, the numbers will all be negative. A graph of the divergence vector of a particular receptor reveals how much that receptor inhibits itself and other receptors surrounding

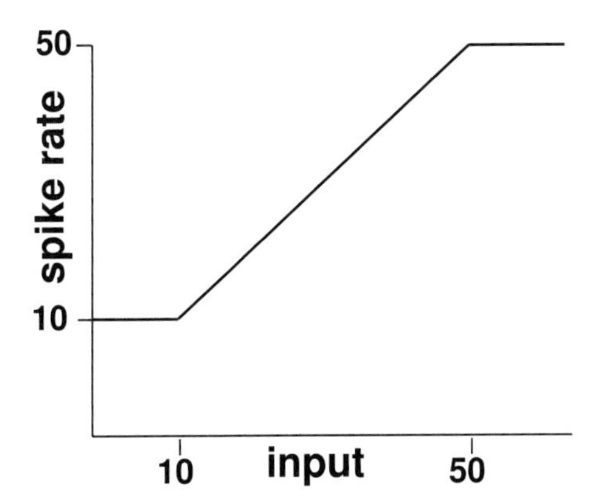

Fig. 8.7: Limulus model squashing function.

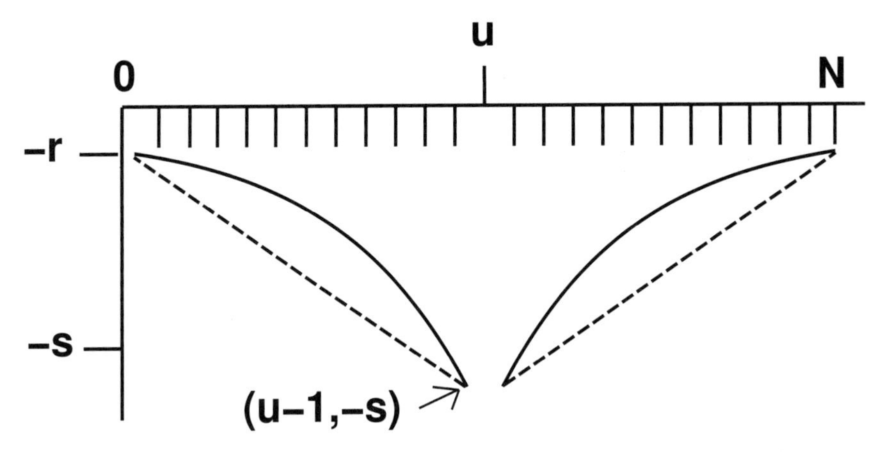

Fig. 8.8: Spatial decay of weight: exponential (solid line) and linear (dashed line) functions

it. Nearest neighbors are inhibited most, those farther away are inhibited progressively less. Two simple choices for fall-off functions are linear fall-off and exponential fall-off (Fig. 8.8).

Again, the equations are basically very simple. However, we would need to add a bunch of parameters to make it easy to scale the inhibitory drop-off for different network sizes and then to place the drop-off correctly in each row of the matrix. Rather than defining the whole kit-and-kaboodle in a series of equations, as we did with the picture functions above, we can easily do the needed mirroring and shifting algorithmically, using the computer. Referring to the weight curves shown in Fig. 8.8, we see that the fall-off is symmetrical on either side of a given unit u. Therefore, we just define one of these drop-off curves. We can easily parameterize a line segment running between the points (0,0) and $(u-1, -s)$: $w(x) = -r - \frac{s-r}{u-1} \cdot x$. We can use values from this equation to set values for the top half of the central column of the weight matrix, corresponding to the divergence from unit u at the center to units 0 through $u-1$. The following shows the resulting weight matrix for a 10-unit network with $s = 0.99$ and $r = 0.95$. Subsequent simulations were run with an 80-unit network. The 80×80 matrix used is too big to show.

$$\begin{pmatrix} 0 & -.99 & -.98 & -.97 & -.96 & -.95 & -.96 & -.97 & -.98 & -.99 \\ -.99 & 0 & -.99 & -.98 & -.97 & -.96 & -.95 & -.96 & -.97 & -.98 \\ -.98 & -.99 & 0 & -.99 & -.98 & -.97 & -.96 & -.95 & -.96 & -.97 \\ -.97 & -.98 & -.99 & 0 & -.99 & -.98 & -.97 & -.96 & -.95 & -.96 \\ -.96 & -.97 & -.98 & -.99 & 0 & -.99 & -.98 & -.97 & -.96 & -.95 \\ -.95 & -.96 & -.97 & -.98 & -.99 & 0 & -.99 & -.98 & -.97 & -.96 \\ -.96 & -.95 & -.96 & -.97 & -.98 & -.99 & 0 & -.99 & -.98 & -.97 \\ -.97 & -.96 & -.95 & -.96 & -.97 & -.98 & -.99 & 0 & -.99 & -.98 \\ -.98 & -.97 & -.96 & -.95 & -.96 & -.97 & -.98 & -.99 & 0 & -.99 \\ -.99 & -.98 & -.97 & -.96 & -.95 & -.96 & -.97 & -.98 & -.99 & 0 \end{pmatrix}$$

In each row, $W_{uu} = 0$, meaning that there is no self-connectivity. The most negative values (strongest inhibition) flank the zeros. This is the nearest-neighbor connectivity. From there, inhibition falls off linearly in both directions until the furthest unit is reached with a $-.95$ strength inhibitory weight projection. Each row (or each column) is identical to the prior one except that it is rotated by one place. Because of the wraparound, a number that falls off the right side of a row with row-to-row rotation shows up back on the left side. A number that falls off the bottom of a column with column-to-column rotation shows up back at the top.

A similar equation for exponential drop-off with distance x is $w(x) = -s \cdot e^{-\frac{x-(u-1)}{\lambda}}$. The two parameters are maximal inhibition $-s$ and a spatial length constant λ, which indicates how far the inhibition will reach. Bigger λ means that the inhibitory influence will extend further out. Here again, wraparound is most easily done algorithmically by appropriate placing, mirroring, and copying.

8.6 The limulus equation

Now that we've done all that scaling, redimensioning, and parameterizing, we can actually put together the limulus equation itself. We need to make only a small change in the basic sum-and-squash unit update equation introduced in Chap. 6. This change is needed to account for the image input. We have scaled the input, representing luminance, to match the range of values used by the units to represent firing rate. Because of this, the input can be simply added in with the interneuronal connections as a set of values all with weights set at 1. The *limulus equation* is

$$\vec{s} = \sigma\left(\vec{p} + W \cdot \vec{s}\right)$$

where $\vec{p}$ is a picture (image) vector (Fig. 8.6). As usual, $\vec{s}$ is the state vector, W is the weight matrix, and σ is the squashing function (Fig. 8.7).

Explicitly including the image vector $\vec{p}$ in the equation provides ongoing drive from the image to the system. This represents the fact that the input (light falling on the photoreceptor) is continuous during the processing. Rather than adding in the $\vec{p}$ vector explicitly, it could be handled as an external input represented as a separate layer of fixed (clamped) "units" with feedforward connectivity projecting one to one onto the processing layer.

To examine the meaning of an equation and get some insight into what it means, it's helpful to simplify the equation in various ways by setting parts of it to zero and looking at what's left. For example, if we set the weight matrix to 0 (no connection between units), then $\vec{s} = \sigma(\vec{p})$. Because σ is the identity function ($\sigma(x) = 1 \cdot x$) in the 10 to 50 range, $\sigma(\vec{p}) = \vec{p}$. This means that in the absence of lateral connectivity, the input vector is simply mapped directly onto the unit state vector. This would be the perfect solution in the absence of scattering and sampling error. If, on the other hand, we suddenly turn off the lights, setting the input ($\vec{p}$ values) to numbers ≤ 10, the negative weight matrix will drive the states back down to their minimum, 10, regardless of their starting point.

We scaled all the numbers to correspond to the range of rate coding actually seen in the limulus neurons, namely 10 to 50 Hz. By using the same values for both the input vector and the state vector, we can simply add the input vector to the state vector at every time point simulated. To start any system simulation, we need an initial condition, or starting point, for the state vector. We would typically start the system with the image already present on the retina by setting $\vec{s} = \vec{p}$. Alternatively, we could start the system in darkness by setting $\vec{s} = 0$ — the zero here is actually a vector of zeros: $\left(000 \cdots 0\right)$ — and then flip the lights on suddenly by setting $\vec{s} = \vec{p}$ at time $t = t_1$.

8.7 State calculation

The core operation of the limulus eye lateral inhibition model is the determination of the values of each unit at each discrete time interval. This is the process of simulating the limulus equation. In Fig. 8.9 we start with no image on the retina so that at time $t = 0$ activity is zero throughout. Each of the circles along the x-axis represents the state of one of the 80 photoreceptors along the line of retina. When the image is presented at

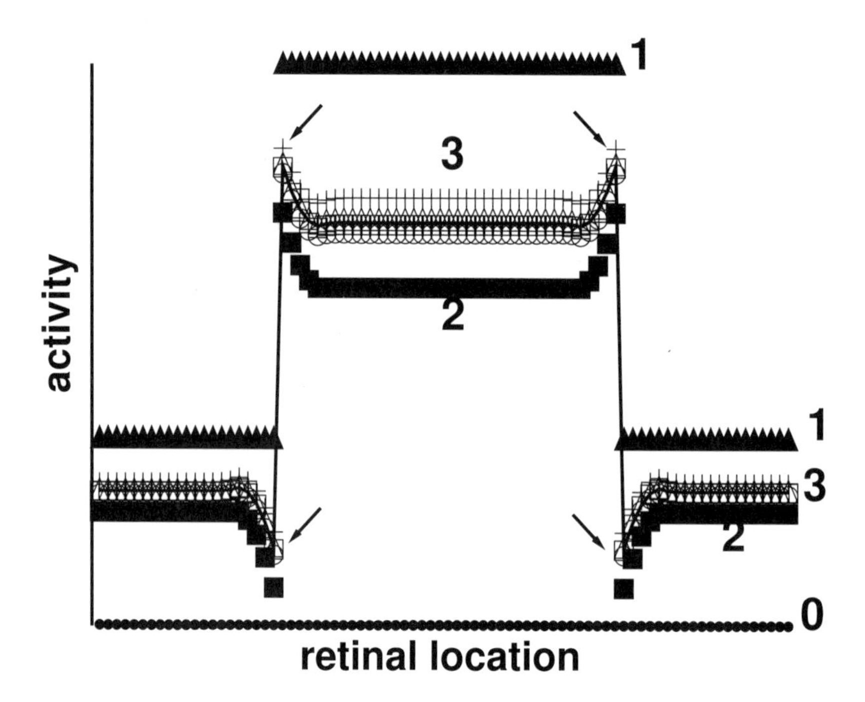

Fig. 8.9: Limulus equation simulation. Sequential time steps are shown by numbers 0–3 and by symbols: circle, triangle, square, plus, ... Square wave image presented at $t = 1$. Mach bands indicated by arrows.

$t = 1$, the states jump up to match the values of the image (triangles). Note that there is no inhibition yet since the state vector at the previous time step was uniformly zero so that all the weights were multiplied by 0 to obtain the input for $t = 1$. At $t = 2$, however, inhibition kicks in, resulting in a lowering of all values (squares) and the appearance of an exaggeration of contrast at the contrast boundaries (small arrows). The difference in activity between the receptors at the contrast boundaries is greater than the activity difference between receptors elsewhere. Contrast enhancement is a characteristic finding in all lateral inhibitory networks. In the visual domain, this enhancement is referred to as Mach bands.

As we continue the simulation for a few more steps, the values converge toward a final target (Fig. 8.9 — solid line under all of the piled-up symbols). At the steps beyond $t = 1$, the values alternate from being above or below this line, reflecting lesser or greater inhibition from the slightly lower or higher values on the previous step. Mach bands are present at all steps after $t = 1$.

In Fig. 8.10, we have started with a blurred image. The Mach bands here are less pronounced — we do not succeed in completely reversing the

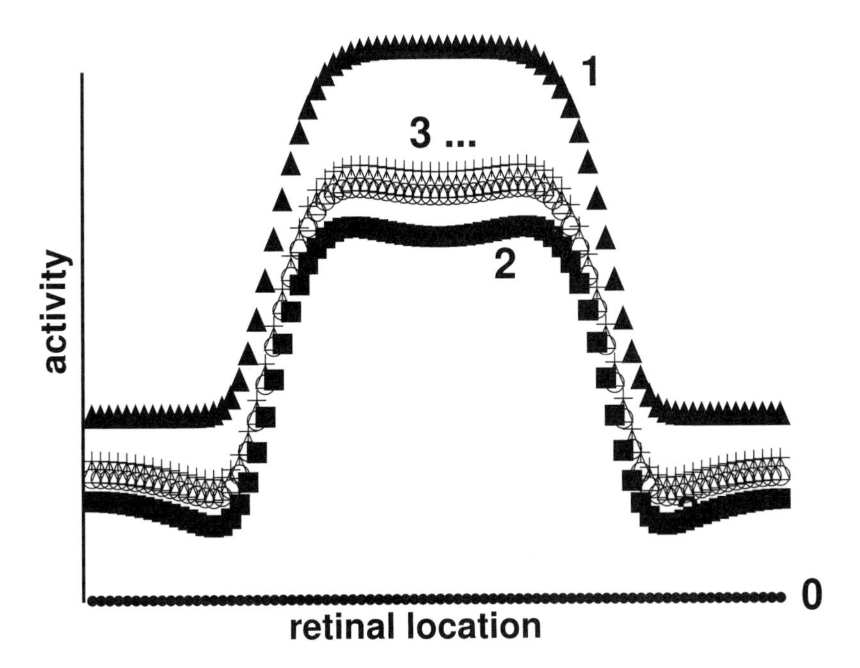

Fig. 8.10: Blurred image presented at $t = 1$. Mach bands form but are less pronounced.

blurring. If we play with the width of the inhibitory surround by changing the fall-off in the weight matrix parameterization function, we can match the width of the blurring and get improved contrast enhancement. However, such a specially designed inhibitory filter would not be optimal for greater or lesser blurring, as might occur when the limulus is in extra dirty or extra clean water. To detect how much enhancement is needed and alter the inhibition accordingly, we would need an adaptive filter. This brings us to the topic of learning algorithms, which we take up in the next chapter.

8.8 Life as a limulus

People often ask me, "What's it like to be a limulus?" This is the philosophical problem of qualia. No matter how much we explain scientifically, how can we know what experience is like for another person or another creature? Another way of stating this is as the problem of subjectivity: Do others perceive things the same way that I do?

One could argue, heck, I will argue, that if my neural responses to a particular stimulus are organized similarly to those of other persons, or animals, I will see things in the same way that they see things. Because of lateral inhibition, the limulus model demonstrates Mach bands. Inso-

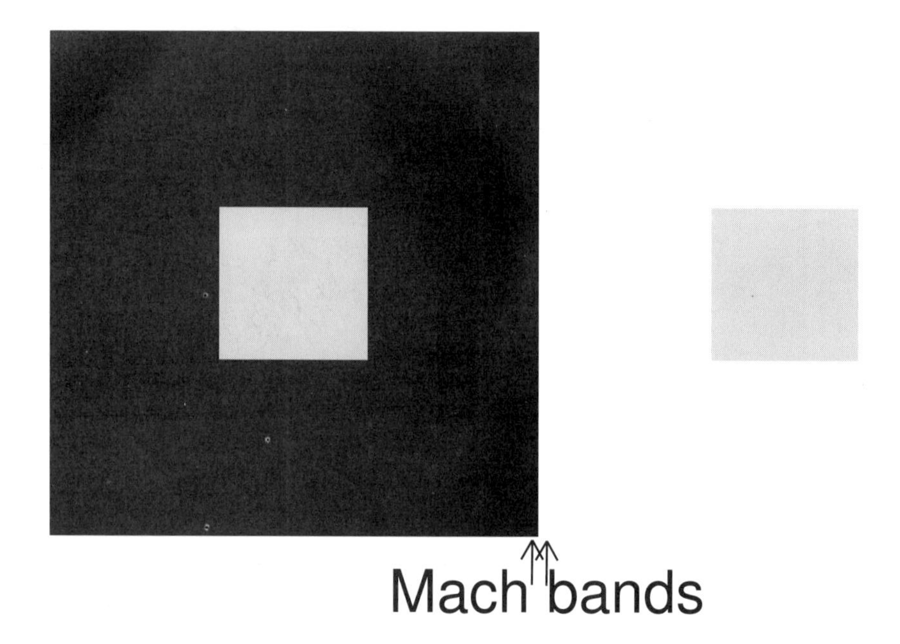

Fig. 8.11: Mach bands and contrast illusion. The Mach bands are seen at the central contrast boundary between dark and light (arrows). A related illusion produces a difference in appearance of the two identical central squares.

far as the model is right and insofar as the limulus experiences anything, we can conclude that the limulus eye tells the limulus brain about these Mach bands. Similarly, our retinas have lateral inhibition and our eyes tell our brains about Mach bands. The Mach band illusion is not restricted to limulus but is found throughout much of animalia. In this admittedly limited sphere, we can understand what it is like to see things as a horseshoe crab sees them. Fig. 8.11 shows the illusion. The central contrast between black and white is exaggerated (arrows) — the black appears blacker and the white appears whiter. Another consequence of lateral inhibition is also demonstrated in the figure and may be easier to see. The two gray squares are identical but the one surrounded by white appears darker than the one surrounded by black. This is due to inhibition of response to the square on the right in the presence of greater lateral inhibition due to increased flanking activation associated with the white surround.

Of course the question of qualia gets more thorny when we ask whether one person see things the same way that another person sees them. The typical example given is color — How can I know that the red that you see is the same as the red that I see? At the lowest level, we could measure the responses of the different cone subtypes in your eyes and my eyes (cones are the type of retinal cells that respond to color; they differ slightly from person to person). We could then adjust light frequency in order to give your cones the same amount of stimulation that my cones are receiving at the chosen "red" frequency. How about if we went up and around my cortex and discovered that red vibrates certain ensembles more than others due to my personal experience with red? We would attempt to produce a similar experiential contrast in your cortex, relative to the ensembles and wiring that you have available. This would give you an experiential illusion, evoking comparable associations. As a simple example of this, consider the condition of synesthesia, a benign condition that causes crosstalk between sensory modalities, causing a person to experience sounds while seeing shapes, or colors while listening to music. Mapping these complex sensory combinations in the cortex of a synesthetic would allow us to reproduce the same combinations in another person, giving the nonsynesthete the experience of synesthesia.

8.9 Summary and thoughts

In this chapter, we have reviewed the decisions and compromises made in designing any computer model: simplifying, scaling, parameterizing. We also considered additional common compromises, such as linearizing and wraparound. It may seem remarkable that one can learn anything after stripping away so much detail. The lesson of the limulus model is just this, however. One can contrast the limulus-model approach with its opposite: throw everything that you know into a model, set it loose, and see whether it tells you something you don't already know. This everything-and-the-kitchen-sink strategy can be a useful starting point, helping one organize information and ideas. However, such a mega-model is just a prelude to the hard work of paring back, focusing on a problem and figuring out what to take out and what to leave in.

Lateral inhibition is a common architectural feature of both natural and artifical neural networks, often seen in a context when a neural sheet is mapping some attribute of the world. In this context, the flanking neurons mean something relative to the inhibiting central neuron. As in the limulus example, this will produce sharpening, allowing a maximally activated neuron to stand out more strongly from the background. Other examples could be taken from the somatosensory (touch) system where lateral inhibition would allow more precise localization of a poke on the skin.

A mechanism similar to lateral inhibition can be used to allow a subset of neurons to emerge out of a background of many active neurons. Highly active neurons inhibit less active neurons. As the activity of the inhibited neurons decreases, they provide less inhibition onto the more active neurons, leading to a dynamic where the rich get richer and the poor poorer. In the extreme, an inhibitory network can be organized so that only a single neuron remains active in the steady state. Such a network is known as a winner-take-all network.

9

Supervised Learning: Delta Rule and Back-Propagation

9.1 Why learn this?

Learning and memory are favorite topics in both neuroscience and artificial intelligence. They have thereby become a popular focus in computational neuroscience as well. In this chapter and the next, we discuss two types of learning: supervised and unsupervised.

In childhood, we are told what things are for and what they are called. In the context of learning theory, this is considered *supervised learning*. When we learn to write, there is a specific target for the formation of each letter. As adults, we often learn things without ever clearly delineating exactly what we are learning. Thrown into a new work environment, for example, we simply absorb things and gradually sort out when to use which strategy to accomplish which objective. This absorbing and sorting is a process of *unsupervised learning*, where information is being organized in our brain without our necessarily ever being aware of exactly what is what. A still more common form of learning is *reinforcement learning*, in which the feedback is only general success or failure (or approval or disapproval) rather than an explicit target behavior to be replicated. For example, the success of a tennis serve will gradually teach improved serving, while a slightly raised eyebrow may be all of the feedback available to assist use in fitting into a new social situation.

We have previously mentioned Hebb's rule as the basis of one type of learning algorithm. Hebb's rule states that connection strength will increase between simultaneously firing neurons. In the proper context, Hebb's rule can be the basis for learning, as we illustrate in the next chapter. In its basic form, Hebb's rule is a part of unsupervised learning algorithms since it's just a rule for altering connections without regard to any goal or purpose. It can be adapted to make it part of a reinforcement learning system by only allowing the synaptic modification to take place when the positive reinforcement conditions are in effect.

In this chapter, we discuss two supervised learning algorithms: the delta rule, which is simple, and the backward-propagation algorithm, which is complicated. We use back-prop (as it is affectionately called) to explain how neurons work to move the eyes when looking at something and to stabilize the eyes when the head moves.

9.2 Supervised learning

In supervised learning, we pair an input with an output and teach a network to produce the output when it is presented with the input. Most supervised learning algorithms use a feedforward network architecture, taking the input at one end and processing it to produce the output at the other. For example, we might want the network to associate faces with names. This is difficult to do because different pictures of a face, even if all taken full frontal, will differ in shading, photographic exposure, and facial expression. Let's say we have a database of 500 digitized names with 20 frontal facial photographs for each one. Perhaps each photograph is stored as a 50 by 50 pixel array and each name is stored as 30 Ascii letters. Then we would construct a feedforward artificial neural network with 2500 input units ($50 \cdot 50$) and 240 output units (30 times 8 bits per Ascii character). We would start with random weights throughout the system. We would present the faces to the system in random order by setting the input units to the pixel values for each face and then forward propagating the values through the layers of the network to obtain output values. Initially, each of the resulting output vectors would be a meaningless bit stream. However, we would use a learning algorithm to change the weight matrix so that the system would do better next time. Over many thousands of presentations of all of the paired names and faces, the network would get progressively better at accurately producing the name when the face was presented. Finally, the network is ready for sale to the FBI.

In discussing representations (Chap. 4) and parameterizations (Chap. 8, Section 8.5), we noted that the worlds of mathematical and computer modeling presume and prescribe the use of numbers for representing everything. Therefore, when we discuss learning in this context, we are learning numbers. When we discuss supervised learning, we are getting the number wrong and correcting it to get the number right. Unlike many real-life learning situations, errors in numbers are neatly defined and quantified.

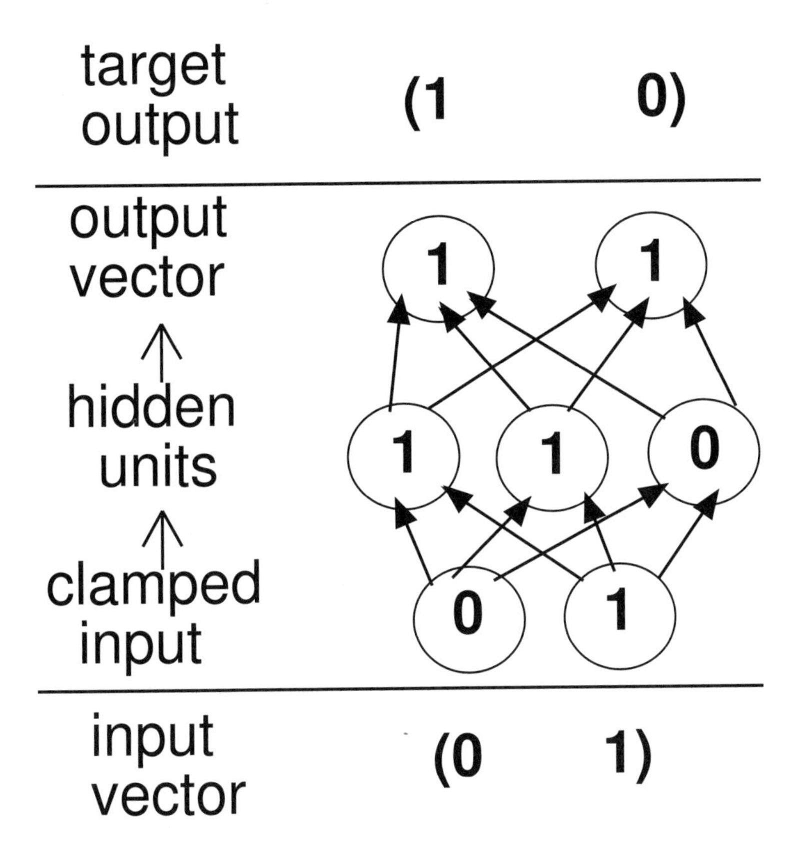

Fig. 9.1: Learning in a feedforward network. An input is clamped onto the network. Information feeds forward through a hidden layer to produce a calculated output. This output is compared to the target output. Weights are changed. Another input/target pair is tried.

Feedforward networks take inputs and provide outputs. The teaching is done by providing input/output pairs: a specific input is paired with a specific target output (Fig. 9.1). Output is calculated from each input provided. One or more processing layers between input and output are referred to as hidden layers. In a real nervous system these units are in the "black box"; they remain hidden from an experimenter who only presents inputs and measures outputs. Calculated output must be distinguished from the target output. For this reason the target output (the output of the input/output pair) is called a target ($\vec{t}$) where it must be distinguished from the calculated output ($\vec{out}$ where out_i is the state of unit i of the output layer).

On the input side, this use of an externally provided input for learning leads to another confusion. In the limulus equation, we added in the picture vector $\vec{p}$ (the input) directly as part of the update rule. In the case of feedforward learning networks, we handle the input vector as if the values were the states of a set of input units. These input units are false units; they are not processing units. It is useful to represent them as if they were processing units, because we are interested in modifying weights between these inputs and the next layer. The input units do not get updated by an update rule. Instead they are successively *clamped* (set) to the values of each input.

Similarly, in some algorithms, we associate a bias unit with each processing unit. This is another false unit. The bias unit is always set to 1. After learning is complete, the weight from the bias unit to its processing unit produces a constant bias. This bias effectively repositions the total-summed-input on the squashing curve. Changing the weight for the bias unit is equivalent to shifting the squashing function left or right. Therefore, learning the bias weight is equivalent to learning the location of the threshold for the squashing function.

In the example of Fig. 9.1, the input $\begin{pmatrix}0 & 1\end{pmatrix}$ and output $\begin{pmatrix}1 & 0\end{pmatrix}$ belong to a training set for learning to reverse the order of the two bits. The input/output pairs for the full training set are $\begin{pmatrix}0 & 1\end{pmatrix} \rightarrow \begin{pmatrix}1 & 0\end{pmatrix}, \begin{pmatrix}1 & 0\end{pmatrix} \rightarrow \begin{pmatrix}0 & 1\end{pmatrix}, \begin{pmatrix}0 & 0\end{pmatrix} \rightarrow \begin{pmatrix}0 & 0\end{pmatrix}, \begin{pmatrix}1 & 1\end{pmatrix} \rightarrow \begin{pmatrix}1 & 1\end{pmatrix}$. The clamped values in the input layer determine values in the hidden layer, which determine values in the output layer. In this case, the output is shown to be in error. The error would be $\vec{t} - \vec{out} = \begin{pmatrix}1 & 0\end{pmatrix} - \begin{pmatrix}1 & 1\end{pmatrix} = \begin{pmatrix}0 & -1\end{pmatrix}$. This error would then lead to weight changes as will be described below.

Presenting the input states as false units creates an additional nomenclature confusion. This set of false units is called the input layer. The simplest network is usually considered a two-layer network, having an input layer and an output layer. As an example, Fig. 9.2 has a two-unit input layer and a one-unit output layer. This would usually be called a two-layer network since it's drawn with two layers. This is what I'll call it. However, it only has one processing layer and could alternatively be called a one-layer network. The two inputs are just numbers that are copied from the list of input patterns and are not real units that update as sum-and-squash units. In Fig. 9.2, only the single output unit is a sum-and-squash unit.

In feedforward networks, units are typically numbered within each layer so that there is an input unit #1 and an output unit #1 in Fig. 9.2. The same numbers are then used as subscripts for a weight to indicate which units are connected by that weight. The update rule for Fig. 9.2 is $out_1 = \sigma\{w_{11} \cdot in_1 + w_{12} \cdot in_2\}$. For Fig. 9.2, σ is the thresholding function shown inside the picture of the output unit. Below, we use subscripts i and j as indices for units with s_i as the state of a postsynaptic unit (a hidden or output unit) and s_j as the state of a presynaptic unit (a hidden

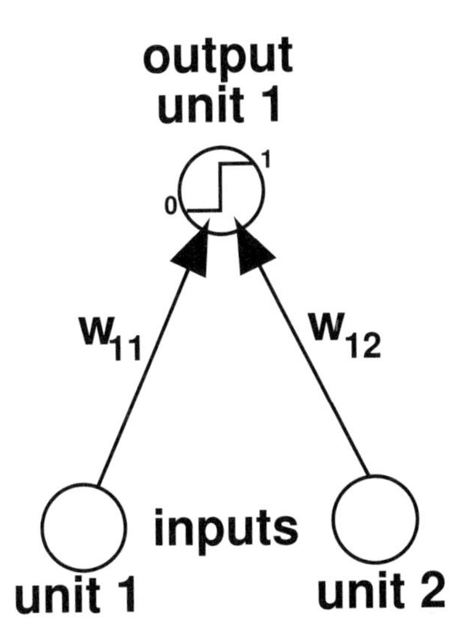

Fig. 9.2: Simple feedforward network: two inputs and one processing unit. Sharp thresholding.

or input unit). In general, total-summed-input for a postsynaptic unit is $s_i = w_{i1} \cdot s_1 + w_{i2} \cdot s_2 + \ \ldots \ + w_{iN} \cdot s_N = \sum_{j=1}^{N} w_{ij} \cdot s_j$. (The capital sigma, $\sum$, standing for "sum," says to add up all of the values obtained by setting $j = 1 \ldots N$. Here, N is the number of presynaptic units.) The order of subscripts in w_{ij} indicates the feedforward direction. This just gives us another way of writing the sum of products that is represented by the dot product: $s_i = \vec{w} \cdot \vec{s}_j$.

9.3 The delta rule

Neural network supervised learning algorithms change network weights little by little until the input patterns used in the training are reliably mapped onto their corresponding output targets. (*Mapping* is a general term for the transformation of one representation to another, just as the physical representation of a city is transformed into a graphical and textual map.) When learning is complete, we should be able to present an input pattern to the network, perform the update rules for all of the units, and have the network produce output states that match those of the paired target output. To get from a random network that will produce a random output for each input to the correct network, we need to make appropriate changes to each weight based on the errors the network makes. When an error is made, which weight should be changed and by how much? This is the *credit-assignment problem* — the basic problem that must be solved by supervised learning algorithms. Perhaps it would be better called the blame-assignment problem. We determine which weights are most to blame for the error and correct those weights most.

The Greek letter delta (capital Δ or small δ) used in math means "change." The delta rule and its variations are common supervised learning rules that are used to change weights. There are a variety of similar rules with different names that can be considered as variations of the delta rule. These include the Widrow-Hoff rule, adaline rule, perceptron rule, least mean squares (LMS) rule, and gradient descent rule. In general, the rules differ by the use of different squashing or activation functions (sigmoid, linear, or thresholding). They also use somewhat different factors to correct the weights.

Consider a single output unit out_i. As discussed in Chap. 6, the input to that unit is the dot product of presynaptic weights and presynaptic states. A learning rule should alter the weight vector in order to get the state of the unit, out_i, closer to a target state t_i. The simplest rule would be to subtract or add a small number α to all of the presynaptic weights depending on whether $t_i - out_i$ is negative (state too big so decrease weights) or positive (state too small so increase weights). Let's call this the α rule. The size of α determines how fast learning takes place. The α rule doesn't work very well. The first problem with it is that it does not account for the magnitude of the error. We can take account of this by multiplying α by the magnitude of the error $t_i - out_i$ to get a proportionate correction factor. This allows learning to go fast when the error is big and to slow down as the actual output state approaches the target.

The other problem with the α rule is that it doesn't solve the credit-assignment problem. Multiplication of each weight by $\alpha \cdot (t_i - out_i)$ changes the weight whether or not it had a role in producing the error. Unnecessary and counterproductive changes to weights can be prevented by considering the presynaptic activity that is being communicated through that weight. The delta rule includes this factor by using the value of the presynaptic state, s_j.

The delta rule is $\Delta w_{ij} = \alpha \cdot (t_i - out_i) \cdot s_j$. In the delta rule, α provides the *learning rate.* The delta rule avoids weight changes where a presynaptic state was very small or zero and therefore made no difference in the output. As the credit/blame for the error increases, the weight change increases. Using the delta rule, the update rule for weight w_{ij} is $w_{ij(new)} = w_{ij} + \Delta w_{ij}$.

The energy analogy

When we talked about the evolution of network state through time, we noted that this evolution could be mapped in a suitably high-dimensional state space. In state space, the state of the entire network can be described as a single point. Similarly, a learning network makes gradual changes in weights that can be mapped in a high-dimensional weight space.

Within weight space, the movement of a single point defines the process of learning. There will be a spot in weight space where the input/output mappings are performed as well as possible. We want a learning algorithm to be able to find this spot. Such an algorithm can be designed by defining an energy field that assigns a potential energy for each point in weight space. This potential energy is just an analogue of error. The *energy function* for the delta rule uses the *summed-square error*: potential energy = error = $\frac{1}{2}\Sigma_i(out_i - t_i)^2$, where the sum is over the number of output units. The difference is squared so that we are always summing positive numbers, regardless of whether the $(out_i - t_i)$ difference is negative or positive.

The standard derivation for the delta rule utilizes this energy function. By determining how the energy function changes with change in weight (a derivative in calculus), one arrives at the same delta rule we determined above. Using the delta rule and the energy function, learning is a process whereby network weights are altered so that the weight vector gradually flows downhill to reach the point of lowest potential energy. Big error is high energy and small error is low energy. One looks at the learning progress as a dynamical system, as if this were a bowling ball rolling downhill under the influence of gravity (Fig. 9.3). The delta rule is guaranteed to find the point attractor in the energy field that corresponds to the optimal mapping of inputs to target outputs. Error getting smaller corresponds to energy getting smaller with the reduction of potential energy down a gradient. For this reason, algorithms of this mathematical form are called gradient descent algorithms.

It is also useful to think of error in terms of this energy-minimization analogy even when using error-minimization algorithms that do not use gradient descent. For example, genetic algorithms jump around in weight space. Although they do not walk down hills, it is helpful to think of them as looking for low points on the error landscape.

The delta rule solves AND

Let's take a simple example to see how the delta rule works. Typical test examples for artificial neural networks are taken from Boolean algebra (see Chap. 16, Section 16.3). We'll look at the Boolean AND operation, which maps 0,0→0; 0,1→0; 1,0→0; 1,1→1. (Using 0 for False and 1 for True, this mapping says that statement A AND statement B are false if either A or B or both are false but true if A and B are both true.) We'll use the architecture of Fig. 9.2 with its sharp threshold activation function with a threshold of 0.5. The delta rule will be used to determine weights. Let's use $\alpha = 0.1$. We'll start with $w_{11} = 1, w_{12} = -0.6$ as our randomly chosen initial weights. Fig. 9.4 shows the evolution of the weights both in time (left graph) and in weight space (right graph).

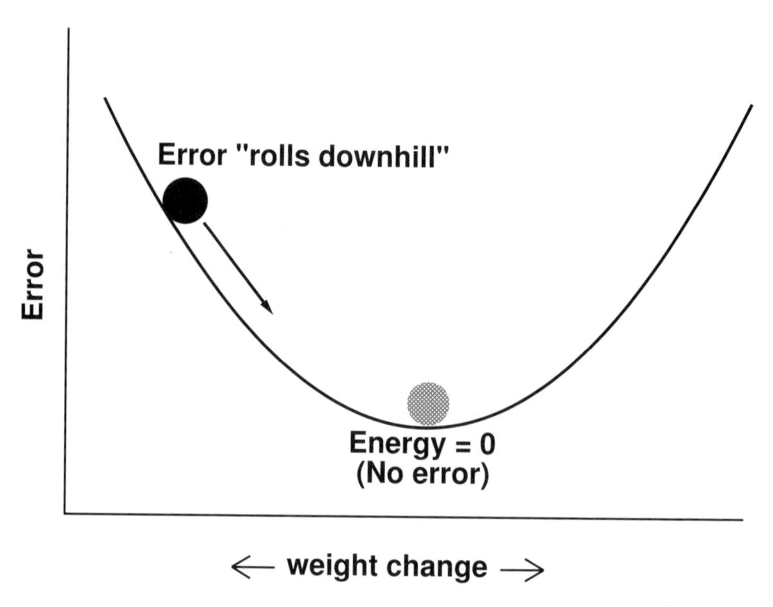

Fig. 9.3: Gradient descent. The parabola shown here: error $= \frac{1}{2}(out - t)^2$ is the one-dimensional version (one weight dimension) of the delta rule energy function.

We present each of the four patterns in the order given above. The first pattern is trivial: with the input set to $\begin{pmatrix}0 & 0\end{pmatrix}$ the values of the weights are irrelevant; the output will always be 0. The error is 0 and credit assignment tells us that neither weight would bear any blame even if there was any error. Therefore, delta is 0 for each weight and the weights remain at $w_{11} = 1$, $w_{21} = -0.6$. The next pattern presented is $\begin{pmatrix}0 & 1\end{pmatrix}$. $0 \cdot 1 + 1 \cdot -0.6 = -0.6$; $\sigma(-0.6) = 0$. Again, there is no error and no weight change. However, the third pattern $\begin{pmatrix}1 & 0\end{pmatrix}$ gives an incorrect value: $1 \cdot 1 + 0 \cdot -0.6 = 1$; $\sigma(1) = 1$. The error $t_i - out_i = 0 - 1 = -1$ is multiplied by α to give a delta of -0.1, which changes w_{11} to 0.9. w_{12} is not changed because its presynaptic state is 0, so it was not to blame. The fourth pattern $\begin{pmatrix}1 & 1\end{pmatrix}$ also gives an incorrect value: $1 \cdot 0.9 + 1 \cdot -0.6 = 0.3$; $\sigma(0.3) = 0$. Now the error is $+1$ and the delta, $+0.1$ is applied to both weights since both presynaptic states are 1. Note that w_{11}, which was corrected down from 1 to 0.9 for the $\begin{pmatrix}1 & 0\end{pmatrix}$ pattern, is now pushed back up to 1 (arrow in left graph). This type of competition in weight space is typical for neural networks as they converge toward a solution. In this case, w_{11} gets pushed down and then up repeatedly until w_{11} dips below 0.5 and w_{12} goes above 0.

Learning continues with each pattern presentation until each pattern produces the right output and the summed error is 0 over all four patterns. In weight space (right graph), we can follow the learning as it goes in the

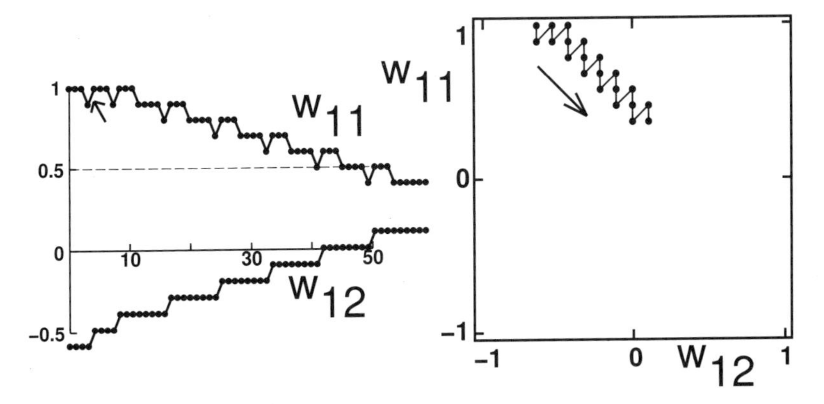

Fig. 9.4: Left graph shows sequence of weights over 56 iterations (14 presentations of four input/output pairs) during delta rule learning. Right graph shows same progression in two-dimensional weight space; arrow indicates direction with time.

direction shown by the arrow. Many pattern presentations don't cause any change in weights; therefore, the 56 points seen in the time graph can't all be seen in weight space — many lie on top of each other. The solution finally arrived at makes sense. The output state for $\begin{pmatrix}1 & 1\end{pmatrix}$ is $\sigma(1 \cdot w_{11} + 1 \cdot w_{12}) = \sigma(w_{11} + w_{12})$. To get 1 as the output ($\sigma(w_{11} + w_{12}) = 1$), the sum of the weights have to be greater than 0.5 ($w_{11}+w_{12} > 0.5$). However, each weight alone has to be less than 0.5 so that inputs with a single one, $\begin{pmatrix}1 & 0\end{pmatrix}$ and $\begin{pmatrix}0 & 1\end{pmatrix}$, will not be able to drive the output to 1.

9.4 Backward propagation

The solution to the credit-assignment problem for three-layer (and greater) networks bedeviled investigators for many years. Neural networks were popular in the early 1960s. There was much excitement about their potential for understanding how the brain works until two researchers (Marvin Minsky and Seymour Pappert) proved in 1969 that delta-rule type networks, while fine for AND, couldn't do more complicated Boolean expressions like "exclusive OR" (XOR, mapping 0,0→1; 0,1→0; 1,0→0; 1,1→1). This soured most people on neural networks for a decade or two. Additionally, Minsky and Pappert guessed (and stated in their book) that three-layer networks (two processing layers) couldn't do anything special that two-layer networks couldn't do. They were wrong about that one. David Rumelhart, Geoff Hinton, and R.J. Williams developed the back-propagation algorithm in 1986 and showed that a three-layer network could solve XOR. Neural networks rose from the grave. As with many such stories of excitement and remorse, neural networks were then again oversold as the solution to all things beautiful or bright. Once again they suffered the scorn of skeptics, as it was discovered that back-propagation probably was not happening in the brain.

The solution to the credit-assignment problem for two-layer feedforward networks is to look back at the input layer and see which units are contributing to the output. Similarly, the solution to the problem for three- and more-layer feedforward networks is to keep looking back down the convergence tree through the layers. In Fig. 9.5, the dashed lines show the convergence tree onto the shaded output. As we go backward down this tree, we assign blame to weights at each level in order to decide how much to change them. In Fig. 9.5, the error will propagate backward from the shaded unit at upper right through three weights to all of the units of the hidden layer and then through all of remaining weights between the hidden and input layer. For example, if the error at the output unit was 0.2, this value could be multiplied by each of the back weights to obtain an error value for each of the hidden units. As with the delta rule, weight changes need to be proportional to presynaptic state values to assign proper blame. In Fig. 9.5, the 0 in the left-most input means that the three weights projecting from this unit cannot be blamed for an error for this training pattern and will not be changed on this input presentation. Changing the weights from the hidden to the output layer will also take into account the presynaptic state values. In this case, the values will be the states of the hidden units obtained during the feedforward phase.

The back-prop algorithm is derived by again considering the energy analogy. As it turns out, solving the credit assignment problem is not sufficient and the algorithm will not work as described above. An additional factor based on the slope of the activation function is also required. This factor allows us to make bigger changes where they will have more effect on solving the error, that is, to make big changes in places where the postsynaptic cell will pass on a lot of that change. To make three-layer feedforward networks learn, back-prop not only solves the credit assignment problem but also solves an "opportunity assessment" problem so that changes are made in places where they make a difference. This assessment of the passage of information at one layer allows the algorithm to appropriately credit lower layers by seeing how much of their influence will pass through toward the output.

By the way, back-propagation is sometimes referred to in current neuroscience literature. However, most neuroscience references that use the term are talking about "back-propagating spikes." This has nothing to do with the back-prop algorithm. We discuss back-propagating spikes in Chap. 13.

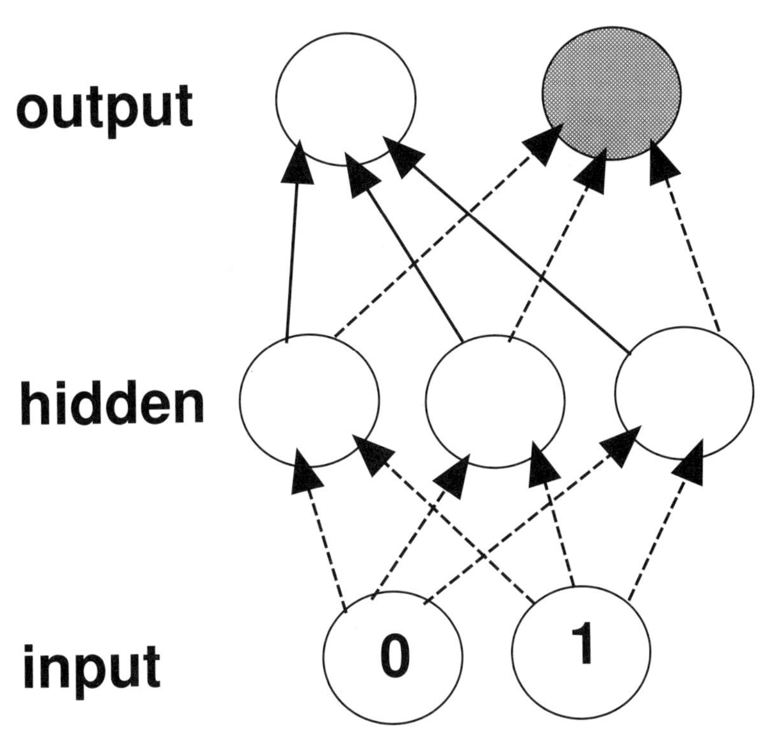

Fig. 9.5: Back-propagation: if the output unit on the right is in error (shaded), error propagates backward through the convergence tree (dashed lines). However, an input unit with state 0 (left) will not get any credit/blame for the error, and the weights from there to the hidden units will not be changed.

9.5 Distributed representations

The discovery of back-propagation made a big impact in cognitive science and a lesser but still substantial impact in neuroscience. It was also newly taken up by engineers, although ironically the same algorithm had been discovered by the engineering community some years before but had not been widely disseminated. The major conclusion from early studies with back-prop was that information processing in the brain might be parallel and distributed rather than serial and local. As has been mentioned previously (Chap. 7, Section 7.2), this represented a shift away from the prevailing artificial-intelligence paradigm to something more neural.

Study of parallel distributed processing networks showed that many features of distributed representations were reminiscent of human learning and memory. For example, the networks showed graceful degradation, falling off slowly rather than abruptly with damage. Additionally, the back-prop paradigm offered incremental learning so that new information could be added as it came in. However, this new information would tend to interfere with older information, particularly related information. Again this is a familiar attribute of human memory. In common with children, the networks tended to initially overgeneralize before getting the problem right. For example, both networks and kids first learn that "ed" makes a verb past tense and overgeneralize, making errors like "goed" before learning the correct "went."

Early successes with cognitive applications made people want to believe that back-prop was in fact how the brain worked. The only detail remaining was to find back-prop in the brain. It couldn't be found. Although back-propagating action potentials were discovered, there is no evidence that information will continue to proceed backward from the dendrite since chemical synapses are largely one-way information conduits. Although back-prop has not been found in the brain, there is evidence for distributed representations. Back-prop is useful for producing distributed representations and suggesting how they might work to solve biological problems.

9.6 Distributed representation in eye movement control

Back-propagation has been used as a tool to understand the complex responses of interneurons (hidden units) in several sensory and motor systems. Here we look at the mixture of sensory inputs that help control eye movement. There are several things that cause your eyes to move. Fast movements to look at something that has attracted your attention are called saccades. Slow movements are used to follow a moving visual target, for example a duck flying overhead. This is called visual pursuit. (Previously I've been using the word *target* to mean the goal that output states of the neural network have to match; now I'm talking about a moving visual target that you track with your eyes. To keep this straight I'll call the pursuit target a "visual target" or a "duck.") There is another type of slow movement that is used for eye stabililization. This is called the vestibulo-ocular reflex (VOR). Pursuit is based on the movement of an external visual target. The VOR is based on the viewer's own movement. Although apparent movement of the entire visual field is also used to stabilize the eyes, the major input for the VOR is an inertial sensing mechanism called the semicircular canals, located together with the sound sensing organs in the ears. The use of inertial sensing means that the VOR works in the dark. (The closeness of visual and inertial sensation also accounts for the fact that retinal slip, the movement of the entire visual field, produces the sensation of movement, as for example when a train next to yours leaves the

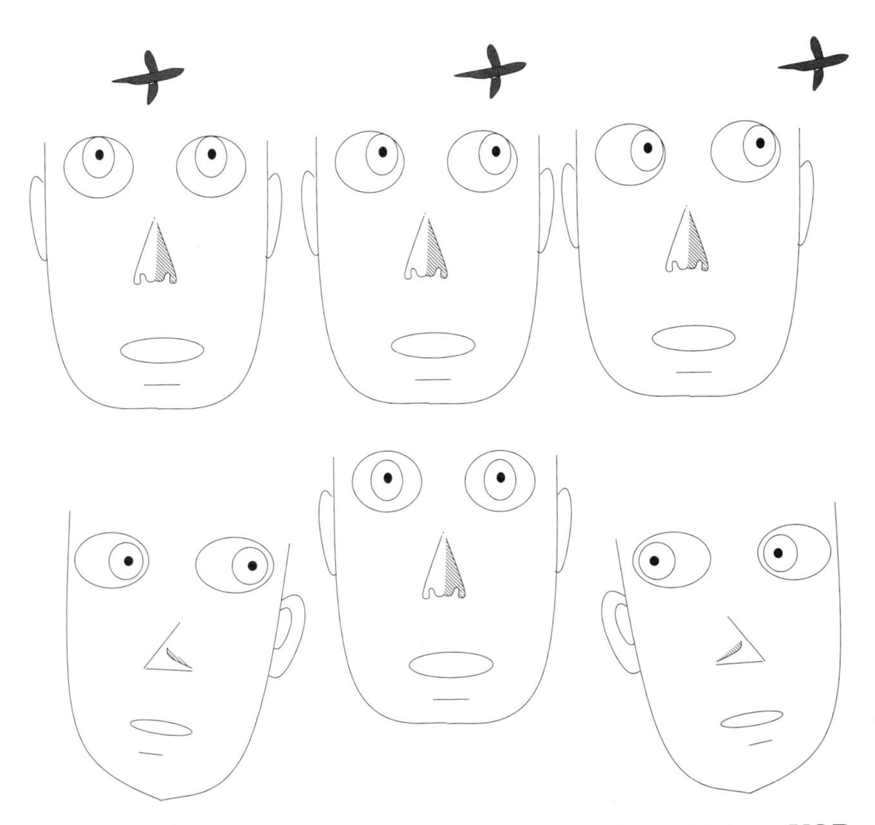

Fig. 9.6: Above, visual pursuit: guy watches duck. Below, VOR: guy turns head side to side.

station.) The VOR compensates for head movement that could otherwise cause the image to disappear off of the retina. Both the VOR and pursuit operate simultaneously, as for example when you watch a duck fly by, while sitting in a moving boat. Signals from both eye and ear (semicircular canals) influence the muscles controlling eye movement (Fig. 9.6).

The neurons that mediate eye movements are in the brainstem, a place where it is relatively hard to record neuronal activity. Before any such recordings had been made, researchers had already been constructing models of eye movement control, based largely on engineering principles. They generally assumed that the neuronal systems for pursuit and VOR systems would remain separate until they converged on the muscles of the eye. When brainstem recordings were finally made, people were surprised to find that many neurons could not be strictly defined as "pursuit cells" or "VOR cells" but had a mixed response to both visual and movement input. Even more surprising was that there were some cells where pursuit inputs were pulling in one direction and VOR inputs were pushing in the opposite direction. This is not how an engineer would usually design a control system.

Tom Anastasio and David Robinson were two researchers who looked into this by applying the then-new back-prop algorithm to the problem of how pursuit and VOR inputs might distribute the information through a bunch of interneurons (hidden units) and still be able to produce the correct effects on the eye. This represented a different approach to understanding a neural circuit. Instead of using top-down ideas to figure out how the system might work, their model was allowed to grow from the bottom up and was then analyzed by the researchers and compared to the real thing. This had some appeal since the nervous system has to develop from simpler rules that cannot take account of how the finished brain ought to look.

Design of the model

Let's look at a version of their model (Fig. 9.7). It's a feedforward network with four inputs, nine hidden units, and two output units. Unlike the binary units of Fig. 9.2, the units here use analog coding so that they can take on any value between 0 and 1 (activation function inset). Therefore, the squashing function is a continuous rather than a thresholding function.

The training input/output pairs are simple in principle, but keeping track of crossing influences pushing left and right makes it confusing. Left pursuit (visual target moving to left) makes the eyes move left, while left VOR (head move to the left) makes the eyes move right. Here are the basic input/output mappings in the context of watching a duck flying by. L stands for left and R for right.

	INPUTS					**OUTPUTS**	
	L VOR	L pursuit	R pursuit	R VOR		L eye	R eye
#1 head R	0	0.5	0.5	**1**	→	1	0
#2 head L	**1**	0.5	0.5	0	→	0	1
#3 duck L	0.5	**1**	0	0.5	→	1	0
#4 duck R	0.5	0	**1**	0.5	→	0	1

0.5 is the value of spontaneous activity for a unit; 1 means full activation; 0 means that the unit is being inhibited. Each of the training patterns represents activation of only one input pathway, indicated by the bold-faced **1**s in the table. By looking at the location of these **1**s, one can read off the meaning of each input. For example, the **1** in the first row indicates that head turning right activates the *R VOR* input.

This mapping is easiest to understand if you imagine your own head moving or your eyes tracking a duck. Pattern #1 is pure right VOR. This means that the head moves right so that the right VOR unit is activated (1) and the left VOR unit is inhibited (0). The pursuit units remain at their resting activation levels (0.5). The output is full activation (1) of the

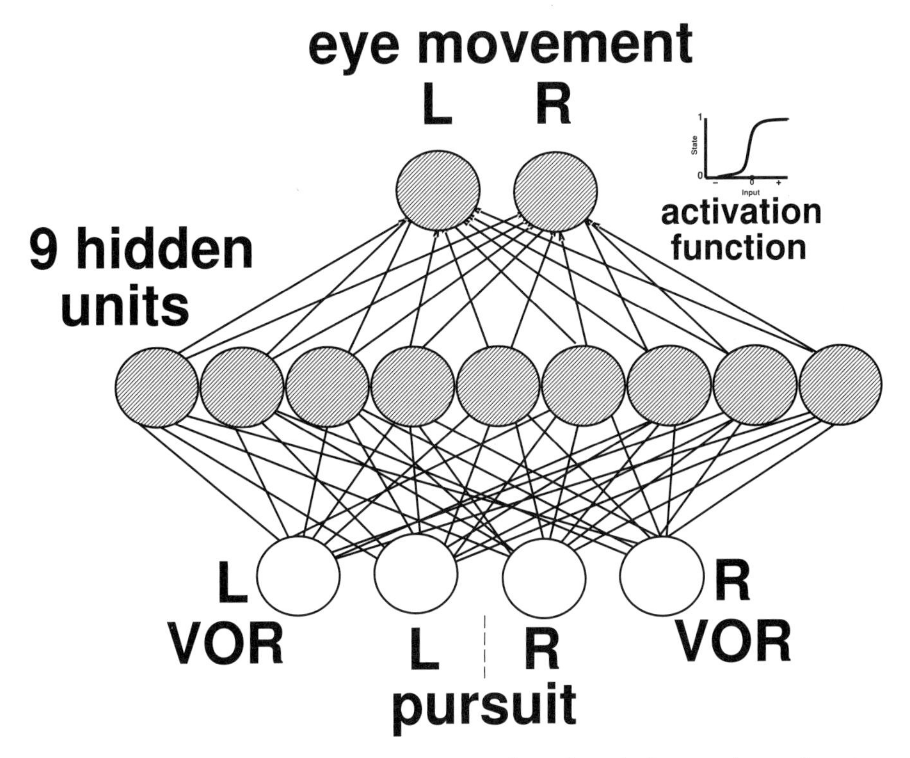

Fig. 9.7: Feedforward neural network takes information about head movement and the visual pursuit target and transforms it to eye movement. Sum-and-squash units are shaded. Inputs, which do no processing, are shown in white.

muscles moving the eyes to the left and full inhibition (0) of the muscles moving the eyes to the right. Head goes right; eyes go left. Similarly, pattern #2 is pure left VOR. Pattern #3 is pure left pursuit. The duck goes left. Left pursuit input is activated and right inhibited. VOR inputs both stay at rest. Left eye muscles are activated and right inhibited. Duck goes left, eyes go left. In pattern #4, duck goes right, eyes go right.

In this case we have an excess of hidden units and it's easy to train the system to produce these mappings. Easy is a relative term here. Training takes thousands of iterations, as we repeatedly forward propagate inputs and then back propagate error to get the total error across all patterns down to a small number. The smaller the requested total error, the more iterations required. In back-prop, the error measure for a single pattern presentation is the same *summed-square error* that was used for the delta function: $\frac{1}{2}\Sigma_i^N(out_i - t_i)^2$ (i counted from 1 to N output units). As with the delta function, this error function is used in the derivation of back-propagation. We can express the summed-square error in vector notation by defining an error vector $\vec{e} = \vec{out} - \vec{t}$. Using the error vector, summed-square error is $\frac{1}{2}(\vec{e} \cdot \vec{e})$. Given P input/target pairs, *total error* is calculated by testing each input in turn, comparing the output to the target, and summing over pattern pairs: total error $= \frac{1}{2}\Sigma_{i=1}^{P}(\vec{e}_i \cdot \vec{e}_i)$ (i counted from 1 to P patterns). *Mean error* or average error is just total error over the number of patterns: $\frac{\text{totalerror}}{P}$.

If you run the network exactly the way I've outlined it, it won't work very well. The typical squashing functions for analog units are sigmoidal with asymptotes at 0 and 1. This means they never actually reach either 0 or 1. If you use 0 and 1 as targets, the network can never eliminate the error. In trying to do so, the network will run for a long time to reduce error. It may also create enormous weights in trying to push states toward the unreachable zero or the unreachable one. To avoid this problem, it is better to use less extreme targets (e.g., 0.4 and 0.6 instead of 0 and 1) or to use a larger squashing function. In this example, I used a squashing function that ranged from -1 to 2.

As training goes along, we repeatedly assess total error. Given the proper conditions, the algorithm is likely to converge, meaning that the outputs will get closer and closer to matching the target outputs. Unlike the delta algorithm, however, the back-prop algorithm is not guaranteed to converge to the best mapping or even to a good mapping. It may find a *local minimum* of the energy function instead of the absolute minimum. In general, for cases that use binary units, like the AND problem of Fig. 9.4, error can be driven down to zero. However, with analog units, error is never completely eliminated.

If you take a bunch of networks and start them with different random weights, they will converge at different rates. After learning is stopped, the resulting networks will have different weight values and different total errors. In some cases, one of these networks will handle one set of patterns well, while another network will handle another set well. Learning rate on the VOR task varied enormously depending on the random numbers that were used as initial weights. Using eight different weight initializations for eight different networks, it took between 5800 and 61,100 iterations to achieve a mean error less than $1 \cdot 10^{-7}$. With this criterion, the worst performance for a single unit was about 1.001 instead of 1. This is error of $1 \cdot 10^{-3}$, which gives a squared error of $1 \cdot 10^{-6}$ (remember that this is the error for one unit with one presented pattern; there are two units and four patterns being averaged to give mean error). The enormous variability in the number of iterations required to converge to the answer is typical of neural networks with randomly chosen initial weights. In some cases, it is possible to choose sensible initial weights and then use the back-prop algorithm to improve the performance.

Results from the model: generalization

Fig. 9.8 shows the performance of one example network on the VOR problem. Each column of marks illustrates the response to one pattern presentation. To simplify the graph, I summarized the two output values, eye-right (output unit out_R) and eye-left (output unit out_L), as a single value representing rightward eye movement. This is calculated by taking the difference between the output states: $\text{out}_\text{R} - \text{out}_\text{L}$. In Fig. 9.8, rightward eye movement is shown as a filled circle. Similarly, rightward head movement is shown as a square and rightward duck (visual target) movement is shown as a triangle.

The patterns in Fig. 9.8 are given in the same order as in the table above. Pattern #1 is head right (square at +1) producing eyes left (circle at −1 representing output states of $\begin{pmatrix}0 & 1\end{pmatrix}$). However, while the pattern list above gives target states, Fig. 9.8 shows actual outputs after training. The training is good enough that the errors cannot be seen on the graph. The output produced for pattern #1 was $\begin{pmatrix}0.0004 & 1.0004\end{pmatrix}$ instead of $\begin{pmatrix}0 & 1\end{pmatrix}$. Therefore, the filled circle in the figure lies at −1.0008 instead of at −1.

A good test of the usefulness of a neural network is its ability to generalize. This is a fundamental aspect of human intelligence. A child learns what dogs are and then sees a St. Bernard for the first time. Even though this dog is different from all other dogs, the child can still identify it as a dog. He has generalized from experience so that new stimuli that are not identical to learned stimuli can be recognized.

In a network, we would want the system to generalize and do sensible things under conditions not in the training set. Any input will produce some output. It only makes sense to test for generalization under conditions in which the investigator (the teacher) has additional patterns that belong in the set. In the context of neural networks, a common practice is to use only some of a given input/output data set for training and reserve the rest of it for testing. For example, if we had 50 dog and 50 cat pictures available to train a network to distinguish dogs from cats, we might use 20 of each for training and save 30 of each for testing. We might also want to try out some pictures of cows and horses to see if the network was overgeneralizing, as it will commonly do for unlearned categories. Again, this is reminiscent of human learning. The city child who has never seen farm animals is likely to assume "fat doggie" when first seeing a pig.

In Fig. 9.8, I show tests for several reasonable generalizations. In pattern #5 the head is not moving and the duck is not moving. The network produces a reasonable answer — the eyes do not move. In patterns 6 and 7, the head is going one way and the duck is going the other way. This will require extra eye movement to keep up with the duck. In both cases, the eye movement is in the correct direction and is increased, although not doubled from the controls in patterns 1 and 2. In patterns 8 and 9, the

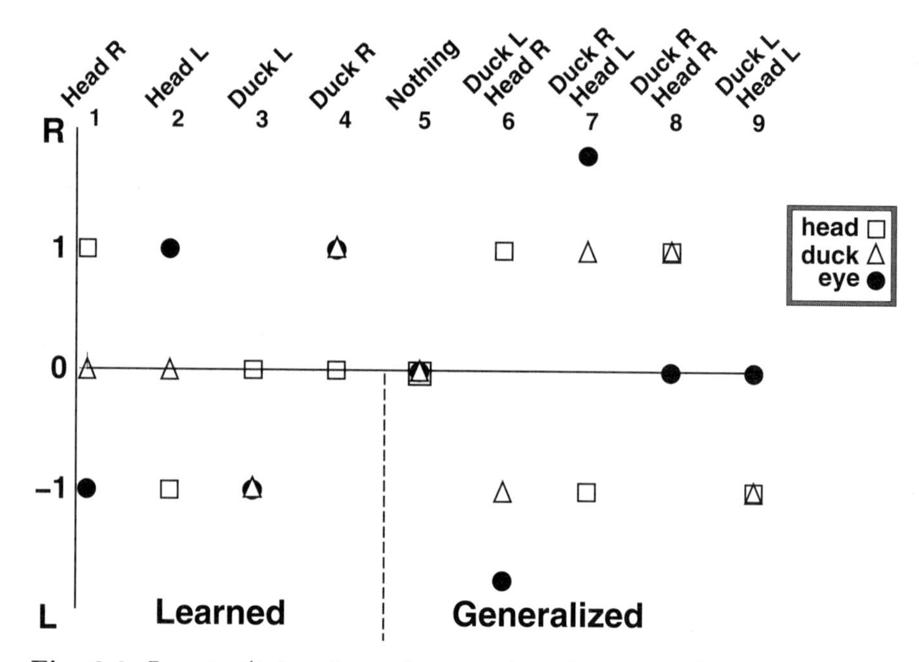

Fig. 9.8: Inputs (triangle and square) and output (circles) from neural network of Fig. 9.7. Near perfect responses are seen to four learned patterns, and reasonable generalization is demonstrated with four nonlearned patterns. (Value for each point is calculated as right-side unit state minus left-side unit state.)

entire head follows the duck so no eye movement is needed. The network again produces the correct response.

Overall, the network shows good but not perfect generalization. A limitation of the network is seen in the response to patterns 6 and 7, where the visual target is moving in one direction and the head in the opposite direction. Linear summation of the opposing head and eye direction would give doubling of the eye movement signal. Due to the nonlinearity of unit responses, it is typical that the network will not produce this linear relationship. Instead it gives somewhat less. The nonlinearity of the squashing function is the limiting factor. Output values can only be pushed as far as the sigmoid curve lets them go. For the sigmoid, the nonlinearity gets more pronounced near the ends of the range. Therefore, expanding the sigmoid's range provides more linear responses.

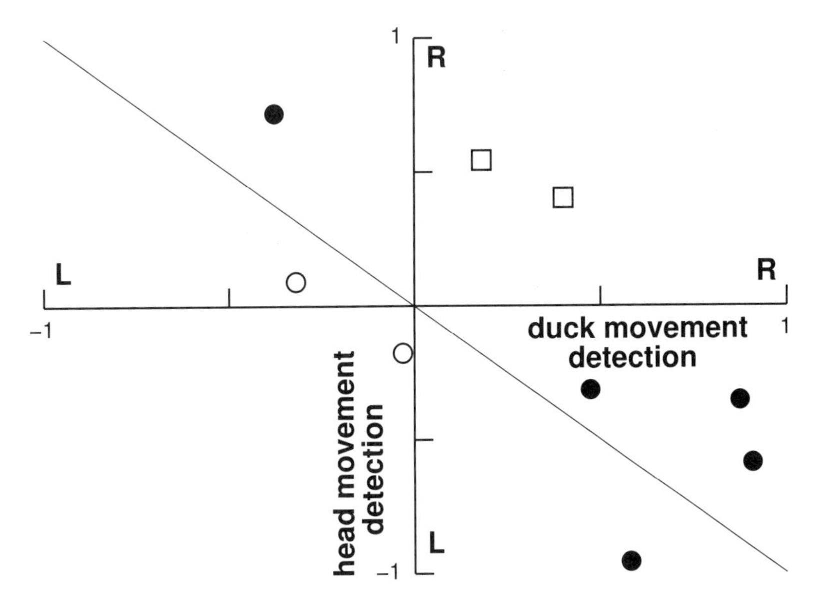

Fig. 9.9: Sensory detection profile for nine hidden units. Most units show a mixture of influences, but two units (open circles) are largely dedicated. Two units (open squares) exhibit paradoxical combinations of pursuit and VOR influence. (Value for each point represents a right weight minus a left weight.)

Exploration of the model: hidden unit analysis

Now we can look at the responses of the hidden units, shown in Fig. 9.9. This is a virtual experiment: exploring the simulation to try to understand how and why it works. The measurements here are the virtual equivalent of the biological experiment of recording from interneurons in the brainstem. If we were to do this type of recording, we would present the animal with either head movement alone or visual target movement alone, and record the response from different brainstem units. Similarly, in the virtual experiment, we measure activity in each of the hidden units under conditions of VOR activation alone or pursuit activation alone. To get everything neatly onto a two-dimensional graph, we use total rightward pursuit input and total rightward VOR input as we did in Fig. 9.8. In the network, this result can be obtained by subtracting the values of two weights that come into a hidden unit. To get the total rightward pursuit input we subtract the weight connecting the left pursuit input to that hidden unit from the weight connecting the right pursuit input to that hidden unit. We do the same for VOR and can then graph one influence versus the other (Fig. 9.9).

From an engineering perspective, we expect that the points would either lie along the axes or lie along the diagonal line from upper left to lower right. Hidden units that map to the x- or y-axis are dedicated to particular sensory input. Hidden units on the y-axis are dedicated to communicating VOR information alone. The open circle unit that lies near the y-axis in Fig. 9.9 will transmit information about the head turning to the left. Hidden units on the x-axis are dedicated to communicating pursuit information. The open circle unit near the negative x-axis in Fig. 9.9 will mostly convey information about visual targets moving left. These two units are dedicated to particular types of inputs. By contrast, each unit near the down-sloping diagonal are likely to be dedicated to a particular motor output. Units near the upper left of the graph specialize in left pursuit and right head movement and will presumably turn the eyes left. Units near the lower right of the graph specialize in right pursuit and left head movement and will presumably turn the eyes right. I say "presumably" since we are only looking at the inputs to the hidden units. We could confirm the motor influence by looking at the weights from hidden units onto output units. By again subtracting left from right, we could produce a map of motor influence from each hidden unit.

In this network, some of the hidden units are near the down-diagonal but none are on it. Most of the units have a mix of influences from pursuit and VOR. The two units at upper right (open squares) are particularly peculiar. They code for pursuit to the right and for head movement to the right. Since pursuit to the right makes the eyes move right and head movement to the right makes the eyes move left, the units represent two influences fighting to move the eye in different directions. It's like pushing on the brake and the accelerator at the same time. Units like this have been found experimentally in animals. When these were first discovered, physiologists thought they were too weird to be true and assumed that this result must be an artifact. In the neural network, one can show how units like this, although not useful in isolation, work together with other units to convey information effectively.

One advantage of the parallel distributed processing of neural network is redundancy. In this case, we are using nine units to do a job that can actually be done by one unit alone. If we train a network with only one hidden unit, the input weights always lie directly on the diagonal far from the center, either at the upper left or lower right of the graph. If we train a network with two hidden units, one of them invariably lies on this diagonal, while the other one is free to wander off. As we train with more and more units, there is less of a tendency for any single unit to sit on the diagonal. It's not clear whether this sprawling across the graph provides any advantage in terms of reliability or tolerance to noise or damage. It would be interesting to study this by comparing a system where all of the units lie on the diagonal to one with the usual sprawl. One could then compare the two systems by adding noise to the units or damage it by removing weights.

Computer modeling vs. traditional mathematical modeling

The process that we used to evaluate this model is closer to an experimental process than to the traditional mathematical modeling. Old-time mathematical models were highly simplified versions of the real thing. These old-time models were simple enough that the mathematician or physicist could hold the whole model in his mind. As the modeler became more and more familiar with his model, he could wander around in it as if he were in his own home. He would come to appreciate how various parts fit together and how the whole thing works. It's impressive to see an old-time mathematician stare for a minute at a set of equations on the blackboard and then tell stories about what the system will do.

The good news is that, with computer models, it doesn't take years of training to begin to understand them. The computer provides an interactive tool for doing the stuff the old-timer had to do in his head. The bad news is that the complexity of some computer models places them out of the reach of full mastery even after those years of training. The VOR model is fairly simple. It can be mastered intellectually. But larger, more complicated neural network models cannot be. These models are created by the computer outside of the control of the human creator (shades of cyborg). Despite being man-made artifacts, they cannot be held in the mind. Instead they provide virtual preparations to be explored experimentally by making the same measurements that one would make in the real thing. The graphs in Fig. 9.8 are the same graphs one would make after doing biological experiments.

So why bother with the computer model if you can do many of the same experiments on the real thing? After all, the real thing is a better version of itself. Well, you can do experiments on a computer model that are impossible in the real thing for either ethical or technical reasons. In the computer model, if it's there, it can be measured. In the real thing, there are lots of things that can't be measured. As importantly, it is easy to do "what-if" experiments on the computer model such as the one we proposed above: what if the brainstem evolved so that all of the pursuit/VOR interneurons lay on the down diagonal? Hey, maybe at some point animals did evolve that way and they died out because their eye-movement system wasn't good enough and they couldn't compete. If this is the case, then it might eventually be possible to use genetic engineering to re-create such a creature. However, many "what-if" situations will be biologically impossible due to both developmental and ethical constraints; genetic engineering won't put neurocomputing out of business.

9.7 Summary and thoughts

Back-propagation is an algorithm that was developed to solve the credit assignment problem. This was a landmark in the history of artificial neural networks in particular and computational neuroscience in general. Back-propagation is still used extensively in the engineering community. However, the biological community remains somewhat touchy about back-propagation because it was originally marketed as the solution to how the brain learns. It's not. However, back-prop can be used to explain some otherwise mysterious aspects of neural organization. Back-prop appears to arrive at solutions similar to those found by biology using some different, unknown, algorithm.

Back-prop solutions serve as an *existence proof*, demonstrating that it would be possible for neurons to solve complicated problems by working together in parallel, processing distributed information. The back-prop results demonstrate that, in a population of neurons working together on a task, some will be specialized, others will be doing a little of this and a little of that, and some will be perverse, seeming to specialize in both going right and going left. Alone, the action of an individual neuron may appear inadequate or senseless, but the society of neurons gets the job done.

This analogy of neural networks and societies was first pointed out by Friederich Hayek, who won the Nobel prize in economics for models based on the progress of an economic system because of, rather than despite, the contributions of individuals whose efforts are uncoordinated and undirected. Hayek also wrote a theoretical neuroscience treatise entitled *The Sensory Order*, presenting theories similar to those of Hebb. Neurons, like people, just gotta be free.

10

Associative Memory Networks

10.1 Why learn this?

Hebb's rule, an increase in synaptic strength between simultaneously active neurons, is one of the most enduring and endearing concepts in computational neuroscience. With the demonstration of long-term potentiation (LTP) of synaptic strength between real neurons, Hebb's stock rose still higher. It is now generally believed that Hebbian learning does take place in the nervous system. For Hebb's rule to result in Hebbian learning, it is necessary that the process provide the basis of some sort of information storage (memory). Hebb suggested that this would occur through the formation of cell assemblies. Cell assemblies form as cells that fire together connect together (Hebb's rule). Groups of connected cells represent memories. Since individual cells will likely be involved in more than one memory, this is a distributed memory mechanism. Various specific algorithms have been proposed that will produce Hebb assemblies using Hebb's rule. The most popular of these is the Hopfield algorithm.

In the previous chapter I showed that the delta rule and back-propagation can be used to produce neural networks that can do pattern matching. Although the back-prop algorithm is nonbiological, it's useful to create and explore distributed representations. Distributed representations do appear to be biological. The Hopfield network, in its original form, is also nonbiological. The Hopfield network is another useful tool, allowing us to better understand distributed representations and content-addressable memories.

10.2 Memories in an outer product

Hebb's rule is easy to express arithmetically using multiplication of presynaptic and postsynaptic weights: $\Delta w_{ij} = s_{post} \cdot s_{pre}$. If we use 0/1 binary vectors, this gives the pure Hebb rule:

PRODUCT

	0	1
PRE →	0	1
POST ↓ 0	0	0
1	0	1

Δ Weight

	0	1
0	none	none
1	none	⇑

This rule allows for increase in synaptic strength but there is no rule for decrease. Without the possibility of decrease, the rule leads to synaptic strengths increasing without bound. Eventually all states are pushed to their maximum value all the time. We get a more useful rule by using $-1/1$ binary values instead:

PRODUCT

PRE →	−1	1
POST ↓ −1	1	−1
1	−1	1

Δ Weight

	−1	1
−1	⇑	⇓
1	⇓	⇑

This extension to the Hebb rule sounds reasonable: if one cell fires and the other cell is silent, there is a decrease in synaptic strength. This has also been described biologically as long-term depression (LTD). However, this Hebb rule variant also states that synaptic strength will increase when two cells are both inactive. This does not make much sense and does not seem to be the case biologically.

We can bypass arithmetic and simply set up rules for synaptic changes based on biological observations. For example, a more realistic rule might be that presynaptic on (1) with postsynaptic on (1) would lead to synaptic increase, and presynaptic off (0 or −1) with postsynaptic on (1) would lead to synaptic decrease. However, the use of arithmetic-based rules has advantages. Using math allows us to utilize mathematical reasoning in order to understand what is going on. In some cases, we can even develop mathematical proofs to guarantee that a particular neural network will store and recall memories. It turns out that $-1/1$ vectors work particularly well as a basis for a content-addressable memory. They are used in one of the top-selling artificial neural networks, the Hopfield network.

Association across a single synapse

Conveniently, and not coincidentally, multiplicative associative memories with $-1/1$ binary-valued units (Hopfield networks) work because of the *associative property* of multiplication: $(a \cdot b) \cdot c = a \cdot (b \cdot c)$. If we present pre- and postsynaptic activity at a single synapse, the arithmetic form of Hebb's rule sets a weight to $s_{post} \cdot s_{pre}$. Using scalars (i.e., a two-neuron, one-weight network), we can show that this learning rule produces a weight such that a learned input, s_{in}, maps onto the corresponding target output: $s_{in} \rightarrow s_{target}$. We clamp the presynaptic unit onto the learned input: $s_{pre} = s_{in}$. We want to show that the learning and update rules produce an output on the postsynaptic cell such that $s_{post} = s_{target}$.

The update rule is $s_{post} = \sigma\{w \cdot s_{pre}\}$. σ here is a sharp thresholding at zero. The single weight, w, is determined by Hebb's rule applied to the learning of the particular input/target pair: $w = (s_{target} \cdot s_{in})$. Substituting for w in the update rule gives $s_{post} = \sigma\{(s_{target} \cdot s_{in}) \cdot s_{pre}\}$. Application of the associative property of multiplication gives $s_{post} = \sigma\{s_{target} \cdot (s_{in} \cdot s_{pre})\}$. We now present our two-unit network with the learned input, clamping $s_{pre} = s_{in}$. Now we have $s_{post} = \sigma\{s_{target} \cdot (s_{in} \cdot s_{in})\}$. In this binary network, states are either -1 or 1. This is where the choice of $-1/1$ binary-valued units becomes critical. Since $-1 \cdot -1 = 1$ and $1 \cdot 1 = 1$, $(s_{in} \cdot s_{in})$ is always equal to 1. Therefore, $s_{post} = \sigma\{s_{target} \cdot 1\} = \sigma\{s_{target}\}$. σ thresholds at zero and won't have any effect here: $\sigma\{1\} = 1$ and $\sigma\{-1\} = -1$. Therefore, the output, $s_{post} = s_{target}$, as desired. *Quod erat demonstrandum.* Latin aside, it's real simple: if the input and target of a pair are the same ($1 \rightarrow 1$ or $-1 \rightarrow -1$), then the weight is 1, which as a multiplier maps anything onto itself. If the input and target are different ($1 \rightarrow -1$ or $-1 \rightarrow 1$), then the weight is -1, which will produce an output opposite in sign from the input.

To scale this scalar associative rule up from a two-neuron network to a large network, we use vectors. We then need to use a linear algebra rule that will match up the input and target states properly when we are using vectors $\vec{s}_{in}$ and $\vec{s}_{target}$ instead of scalars. Such a rule is the *outer product.* Recall that the inner product (dot product) matches up and multiplies corresponding elements of two vectors and adds them up to give a scalar. Instead of matching up corresponding elements of two vectors, the outer product matches up every possible pair of elements. Instead of summing these products to form a scalar, the outer product places these products in a matrix. Instead of being the product of a row vector times a column vector, the outer product is the product of a column vector times a row vector. Instead of requiring that both vectors be of identical length, the outer product can multiply two vectors of different size. The outer product uses a column vector (size M) times a row vector (size N) to get a matrix of size $M \times N$.

The outer product of two vectors

To form the outer product, multiply the entire column vector in turn by each element of the row vector, placing the new column vectors side by side to make a new matrix of N columns. Alternatively, you can get the same result if you multiply the entire row vector in turn by each element of the column vector, placing the new row vectors one under the other to make a new matrix of M rows. Using linear algebra, Hebb's rule uses the outer product $W = \vec{s}_{post} \times \vec{s}_{pre}$. The update rule uses the product between a matrix and a vector: $\vec{s}_{out} = \sigma\{W \cdot \vec{s}_{in}\}$. This product is performed by forming a vector out of the ordered dot products of each row of the W matrix with $\vec{s}_{in}$. $\vec{s}_{in}$ must be oriented as a column vector. Associativity works, so the proof above is still valid. Unlike scalars, however, the outer product is not commutative: $\vec{a} \times \vec{b} \neq \vec{b} \times \vec{a}$. Therefore, the order $\vec{s}_{post} \times \vec{s}_{pre}$ is important.

The many versions of linear algebra multiplication can be hard to keep track of. Here's a table:

	dot (inner) product	outer product
symbol	$a \cdot b$	$a \times b$ or $a \otimes b$
order	row · column = scalar	column × row = matrix
	$a^T b = (\cdots) \begin{pmatrix} \vdots \end{pmatrix} = \text{number}$	$ab^T = \begin{pmatrix} \vdots \end{pmatrix} (\cdots) = \begin{pmatrix} \cdots \\ \cdots \\ \cdots \end{pmatrix}$
lengths	$N \cdot N$	any × any
process	sum of products	combined products
product	scalar	matrix

	matrix-vector product
symbol	$A \cdot b$
order	matrix · column = column
	$Ab = \begin{pmatrix} \cdots \\ \cdots \\ \cdots \end{pmatrix} \begin{pmatrix} \vdots \end{pmatrix} = \begin{pmatrix} \vdots \end{pmatrix}$
lengths	$M \times N$ matrix $\cdot N$ vector
process	matrix-row times column-vector dot products
product	vector

So far we can "remember" a single memory. At the single synapse level, this is spectacular only to the extent that you are prepared to be thrilled by the properties of multiplication. However, at the matrix or network level, we begin to enjoy some of the advantages of collective computation: the network will correct mistakes in the input. This is because the total input to a single neuron is represented mathematically by the dot product of a row of the weight matrix and the input vector. This dot product only has to have the right sign, positive or negative, and the squashing function will give the right answer. The dot product is just a sum of multiplications. Each element of $\vec{s}_x$ that's the same as an element of the original $\vec{s}_{in}$ will push this sum in the correct direction; each element that differs will push in the wrong direction. As long as the correct matches outnumber the incorrect matches, the thresholding will yield the correct output. So the input can have a few 1s where there should be −1s or −1s where there should be 1s. The dot product will still have the right sign and the output will be right.

Making hetero- and autoassociative memories

We can use the outer-product rule to form either *autoassociative* (often simply called associative) or *heteroassociative* memories. In an autoassociative memory, memory for an item is prompted by something similar but not identical to the remembered item, as if one got a glimpse of a friend through a dirty window and could nonetheless recognize him. In a heteroassociative memory, an unrelated image prompts recall; for example, seeing the friend prompts recall of his name. For associative memories, the outer-product multiplication is the learning or training operation. There is no prolonged repetitive presentation of patterns as we used in the delta rule or with back-prop. It is possible to get similar results by using gradual Hebbian changes with repeated presentations during training. In either case, I call this process "training," even when it's an immediate result of multiplication and addition.

Here's an example. Let's learn the heteroassociative mapping
$\vec{s}_{in} = \begin{pmatrix}-1 & -1 & -1 & 1 & -1 & 1\end{pmatrix}^T \rightarrow \vec{s}_{target} = \begin{pmatrix}-1 & 1 & -1 & -1 & 1\end{pmatrix}.^T$
The input and output vectors are of different sizes. The outer product gives the 5×6 weight matrix shown below. We now test an input that's a slightly messed-up version of the original learned pattern. The test vector is
$\vec{s}_x = \begin{pmatrix}-1 & 1^* & -1 & 1 & -1 & -1^*\end{pmatrix}^T,$
differing from $\vec{s}_{in}$ at the starred locations. Presentation of the test input to the memory is done by applying the update rule. This involves multiplying the memory matrix times the test input vector and squashing:

$$\sigma\{W \cdot \vec{s}_x\} = \sigma\left(\begin{pmatrix} 1 & 1 & 1 & -1 & 1 & -1 \\ -1 & -1 & -1 & 1 & -1 & 1 \\ 1 & 1 & 1 & -1 & 1 & -1 \\ 1 & 1 & 1 & -1 & 1 & -1 \\ -1 & -1 & -1 & 1 & -1 & 1 \end{pmatrix} \begin{pmatrix} -1 \\ 1 \\ -1 \\ 1 \\ -1 \\ -1 \end{pmatrix}\right)$$

$$= \sigma \begin{pmatrix} -2 \\ 2 \\ -2 \\ -2 \\ 2 \end{pmatrix} = \begin{pmatrix} -1 \\ 1 \\ -1 \\ -1 \\ 1 \end{pmatrix} = \vec{s}_{out} = \vec{s}_{target}$$

The overlap between the prompt (test) vector $\vec{s}_x$ and the $\vec{s}_{in}$ is four out of six. The product for the four matches will add up to 4 and the two nonmatch products will add up to −2 giving dot product $\vec{s}_x \cdot \vec{s}_{in} = 4 - 2 = 2$. The plus or minus sign for each of the 2s comes from the sign of the corresponding element in $\vec{s}_{out}$.

Similarly, we can form an autoassociative memory by simply using $\vec{s}_{in} \times \vec{s}_{in}$ as the associative matrix. In either the auto- or heteroassociative case, we find that these matrices have the qualities of content-addressable memories: a partial input or an imperfectly formed input can be recognized due to its similarity with previously memorized patterns. The original memory can then be reconstructed or, in the case of a heteroassociative memory, the association can be retrieved. The great thing about associative memory matrices is that they can be used to store more than one memory simply by adding up the matrices formed from the outer product of each individual memory.

As one tries to stuff more and more patterns into an associative network, adding more matrices together, the input patterns can begin to interfere with one another, producing confusion between them. This problem will be particularly severe if the learned inputs are too similar to one another (it doesn't matter if the outputs are similar or even the same). The same confusion is seen in human memory or in any content-addressable memory. If you try to remember events that are too similar, you will not be able to separate them reliably — was that fried egg Friday's breakfast or Thursday's? Unlike breakfasts, the similarity between vectors can be precisely quantified using the dot product.

Vectors that are completely nonoverlapping are called orthogonal. Two vectors are orthogonal if the dot product between them is 0. The higher the dot product, the more the overlap, up to a maximum of n for two identical n-length $-1/1$ binary vectors. If the input patterns for an associative memory are orthogonal, then there is no risk of confusion between them. Unfortunately, there are only a limited number of orthogonal vectors of a given size. One can use nonorthogonal vectors to form associative memories, realizing that if the dot product between memories is too high there will be confusion between patterns. In general, one wants to train the network with input patterns that have low overlap (small pair-wise dot products). After the training phase is complete, the network will work best when presented with unknown input patterns that have high overlap (large dot-product) with one of the remembered patterns.

For example, the following three 4-bit vectors are pair-wise orthogonal:

$$\vec{a}^T = \begin{pmatrix}1 & 1 & 1 & 1\end{pmatrix}, \ \vec{b}^T = \begin{pmatrix}-1 & 1 & 1 & -1\end{pmatrix}, \ \vec{c}^T = \begin{pmatrix}-1 & -1 & 1 & 1\end{pmatrix}.$$

This is because $\vec{a} \cdot \vec{b} = 0$, $\vec{a} \cdot \vec{c} = 0$, and $\vec{b} \cdot \vec{c} = 0$. We can add up the three outer product matrices to get a full associative memory matrix:

$$\begin{pmatrix} 1 & 1 & 1 & 1 \\ 1 & 1 & 1 & 1 \\ 1 & 1 & 1 & 1 \\ 1 & 1 & 1 & 1 \end{pmatrix} + \begin{pmatrix} 1 & -1 & -1 & 1 \\ -1 & 1 & 1 & -1 \\ -1 & 1 & 1 & -1 \\ 1 & -1 & -1 & 1 \end{pmatrix} + \begin{pmatrix} 1 & 1 & -1 & -1 \\ 1 & 1 & -1 & -1 \\ -1 & -1 & 1 & 1 \\ -1 & -1 & 1 & 1 \end{pmatrix}$$
$$= \begin{pmatrix} 3 & 1 & -1 & 1 \\ 1 & 3 & 1 & -1 \\ -1 & 1 & 3 & 1 \\ 1 & -1 & 1 & 3 \end{pmatrix}$$

Typically, we would normalize the network by dividing the final matrix by the number of memories stored. In this case, we would divide by three and get: $\begin{pmatrix} 1 & \frac{1}{3} & -\frac{1}{3} & \frac{1}{3} \\ \frac{1}{3} & 1 & \frac{1}{3} & -\frac{1}{3} \\ -\frac{1}{3} & \frac{1}{3} & 1 & \frac{1}{3} \\ \frac{1}{3} & -\frac{1}{3} & \frac{1}{3} & 1 \end{pmatrix}$

This is not a critical step but is useful in the case of networks with large numbers of remembered patterns that would otherwise give very large weights.

This autoassociative memory will map each of the input vectors onto itself, as it is guaranteed to do. Also it does a fairly good job of mapping incomplete vectors, correctly mapping 100% of vectors with one bit set to zero and 67% of vectors with two bits set to zero. (In the setting of $-1/1$ vectors, a zero can be viewed as an indeterminate value.) This is of course a very small network with limited storage capacity and completion ability.

Since this is such a small network it is easy to illustrate in the stick-and-ball form. Fig. 10.1 shows this network with the weights from the original sum of matrices, without dividing by three. If you set the value of each unit to a value and then update the other units step-by-step (as was done in Fig. 6.5), it will converge on the closest remembered pattern. This is about the smallest associative memory network that can do anything interesting and about the biggest where you can draw a stick-and-ball picture.

Fig. 10.1 is an example of an autoassociative network with full connectivity using $-1/1$ binary units. Networks with these features are called Hopfield networks after John Hopfield, who explained why they work as memories. Remember that state space is the multidimensional space in which you map the entire state vector as a single point (Fig. 6.6). In this case, the state space is four-dimensional, so it can't be drawn. An autoassociative mapping maps a pattern onto itself and therefore defines a fixed point in state space — once you arrive there, you stay there. In the case of an associative memory network, every fixed point is a point attractor. This means that points near the fixed point will be sucked into it as into a gravitational body. The set of points that will be sucked into a particular

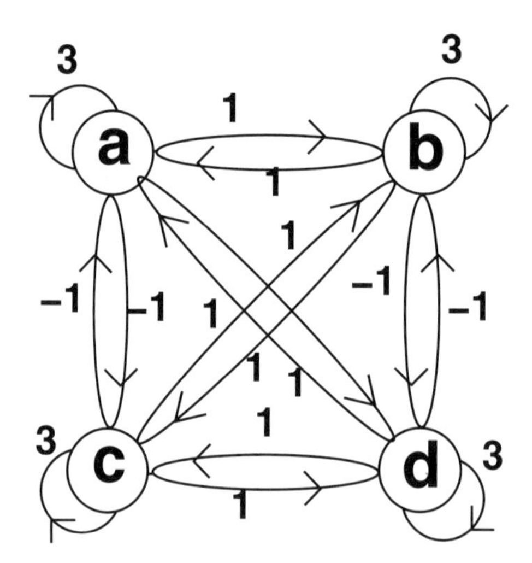

Fig. 10.1: Stick and ball diagram of a four-unit neural network from the summed matrix above. In its weights, this network carries the bittersweet memories of three orthogonal vectors.

attractor is the basin of attraction for that attractor. From any starting point in the basin the system will evolve to end up at the point attractors.

Hopfield's influential contribution, and the reason that he got the network named after him, was to prove that the fixed points (the memories) of an autoassociative memories are point attractors. This guarantees that inputs similar to remembered vectors will move toward these memories. The process of memory recall in autoassociative memories could then be thought of as gradient descent in state space. In Fig. 10.1, we have a four-dimensional state space to map 4-bit vectors. The fixed points formed by the three learned patterns lie at three of the 16 corners of the hypercube. Each will have a corresponding basin of attraction. In addition to these three desirable attractors, the matrix addition will also produce three spurious attractors. Each of these "false memories" is the negative of one of the trained memories. In Fig. 10.1, $\begin{pmatrix}-1 & 1 & 1 & -1\end{pmatrix}$ is an attractor and $\begin{pmatrix}1 & -1 & -1 & 1\end{pmatrix}$ is a spurious attractor. These attractors are also called mirror attractors. It is characteristic of Hopfield networks that the negatives of the memories (think in terms of a photographic negative for a large bitmap image) will also be attractors.

These attractors and basins of attraction bring us back to the energy metaphor. In Chap. 9, Section 9.3, we were looking at gradient descent in weight space (Fig. 9.3). Now, as in Chap. 6, Section 6.4, we are back to looking at gradient descent in state space. A Hopfield network with many memories is a landscape defined by the Hopfield energy function: $E = -\frac{1}{2}\Sigma_i \Sigma_j w_{ij} \cdot s_i \cdot s_j$, where the s are states, and both i and j count over all of the units. Although the two Σ's make this equation look intimidating, it is easily explained. We are talking about energy in state space after learning, so all the weights are fixed. For a given connection w_{ab} between units a and b, the product $w_{ab} \cdot s_a \cdot s_b$ will be large if the weight w_{ab} is large and if units a and b are both active. Notice that these are the same conditions that were required to make w_{ab} large in the first place. Since heavily connected coactive units will make the product large, state vectors with lots of heavily connected coactive units will make the entire summation large. Heavily connected coactive units are Hebbian assemblies. The energy equation is just a way of associating a large number with each Hebbian assembly. There is a negative sign at the front of the equation that turns this large positive number into a large negative number, allowing us to do gradient descent, climbing downhill to more negative values of energy.

Following pathways of decreasing energy the state vector will creep gradually downward, descending through valleys into the depths of memory (Fig. 10.2). Depending on the starting point (the initial conditions), the network will progress to a different resting place. These resting places are point attractors that correspond to each of the memories.

In the four-dimensional space of Fig. 10.1, an energy can be calculated for each location in the space (each state vector). Since a five-dimensional graph (four dimensions for the vector and one for the energy) can't be drawn, I illustrate the concept at a lower dimension in Fig. 10.2. The x, y values are the states of a two-dimensional vector. The z value gives the energy for each possible vector. If one starts with a vector near an energy peak, the update rule will lead down to an energy nadir that is a memory (arrow). This nadir or energy well is a fixed point attractor. The whole system works because of two effects: 1) the learning algorithm (Hebbian learning) produces point attractors, giving the system somewhere to go; and 2) the standard update rule gradually reduces energy so that the network will get there.

Limit cycles

Autoassociative memories map an input onto itself. By definition they utilize fixed points. These fixed points must be attractors if the memory is going to be useful. Heteroassociative memories map one pattern onto another. Therefore, they do not use fixed points. In general, a recurrent network can show all kinds of dynamics, including, for analog-unit networks, chaotic dynamics. A chaotic system will never visit the same point in state space twice. A binary-unit network cannot be chaotic since there are a fixed number of states available to the system and it will eventually have to revisit places it's already been.

Binary networks can settle into oscillations where the same set of states are revisited repeatedly. In dynamical system nomenclature, these oscillations are called limit cycles. At a fixed point the system stays in one place.

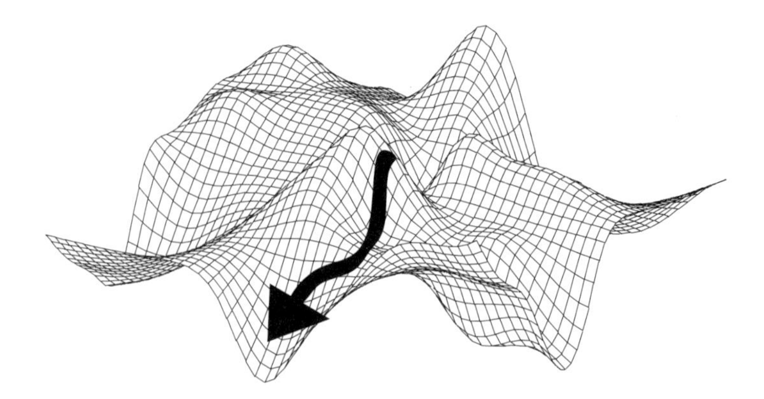

Fig. 10.2: A made-up memory landscape for a two-dimensional system. Starting on top of a hill, the update rule will produce a trajectory moving downhill to recover a memory (arrow). Compare with one-dimensional landscape of Fig. 9.3.

On a limit cycle it "orbits" in state space, visiting two or more points in the space. This notion of an orbit is of course abstracted from the concept of a physical orbit, which visits all points over the course of a revolution.

For example, a simple associative network can be set up by using the outer product rule to memorize the heteroassociative pair $\begin{pmatrix}-1 & -1\end{pmatrix} \rightarrow \begin{pmatrix}1 & 1\end{pmatrix}$ and the autoassociative pair $\begin{pmatrix}-1 & 1\end{pmatrix} \rightarrow \begin{pmatrix}-1 & 1\end{pmatrix}$. The resulting memory matrix, after normalizing by 2, the number of memories, is:

$$\begin{pmatrix} 0 & -1 \\ -1 & 0 \end{pmatrix}$$

Fig. 10.3 shows the simple stick and ball. Since the system is only two-dimensional, we can graph state space. Because of the heteroassociative mapping, the network flops back and forth in an oscillation between $\begin{pmatrix}-1 & -1\end{pmatrix}$ and $\begin{pmatrix}1 & 1\end{pmatrix}$. This is the limit cycle. The autoassociative mapping maps $\begin{pmatrix}-1 & 1\end{pmatrix}$ onto itself and produces a fixed point attractor. This network has three basins of attraction. The first is the basin of attraction for the remembered vector. This basin is the upper-left quadrant of state space. The limit cycle has a basin of attraction consisting of the upper-right and lower-left quadrants of state space. Then there is the spurious attractor at $\begin{pmatrix}1 & -1\end{pmatrix}$ with its basin in the lower-right quadrant.

Since we restrict ourselves to binary inputs to the network, these basins don't have much meaning. The only binary point in each basin is the attractor itself. We can present the network with analog vectors lying elsewhere in the basin and they will go to the appropriate attractor. This is partly a trivial consequence of the squashing function, which will take any analog vector and map it into the binary vector in the same quadrant. However,

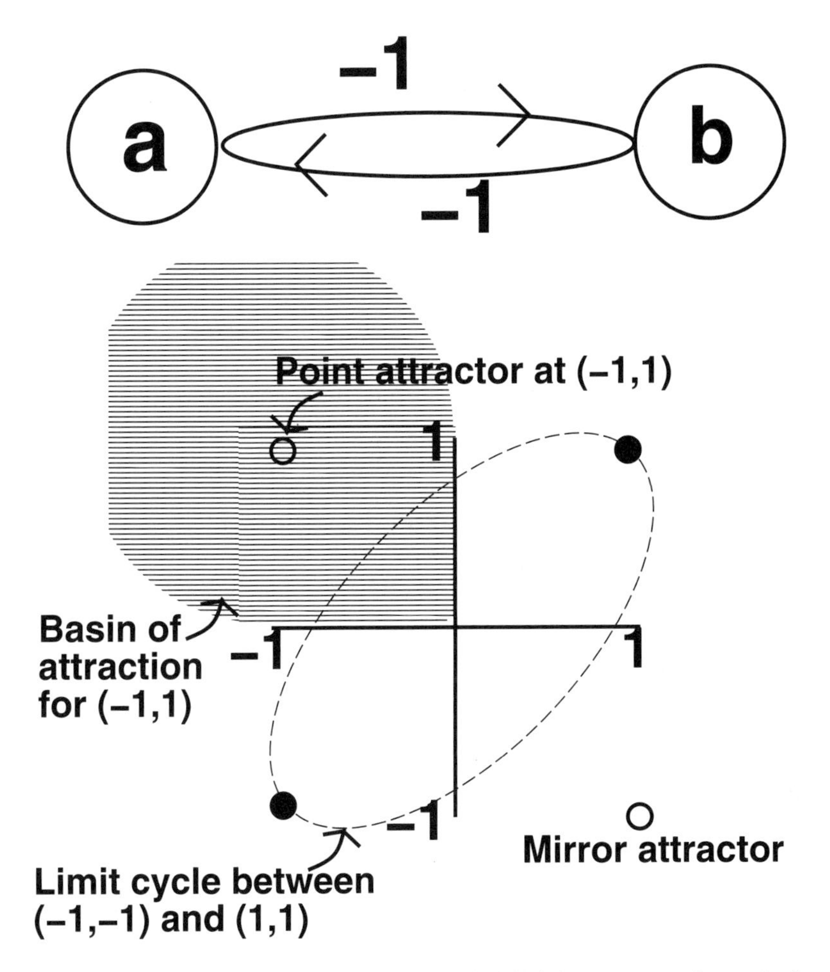

Fig. 10.3: A simple two-unit mutual inhibition network and the corresponding state space diagram. The point attractor represents the autoassociative memory. Half of the limit cycle is the heteroassociative memory. The network also has two spurious memories: a mirror of the point attractor at $(1,-1)$, and the other half of the limit cycle, which maps $\begin{pmatrix}1 & 1\end{pmatrix} \rightarrow \begin{pmatrix}-1 & -1\end{pmatrix}$. Only one of the three basins of attraction is shown, for simplicity.

if we take a point on the basin of attraction for the limit cycle, it will get sent into the opposite quadrant before returning. For example, the vector $\begin{pmatrix}0.5 & 0.5\end{pmatrix}$ will jump to $\begin{pmatrix}-1 & -1\end{pmatrix}$ on the first update.

Instantaneous vs. gradual learning and recall

With outer-product summation for learning and synchronous updating for recall, both learning and recall occur instantaneously and globally. If we had real neurons or real electronic components doing this, we would require a central clock that told all the synapses to change at the same time during learning and all of the units to update at the same time during recall. This is probably unrealistic for the brain. Hopfield proposed that the units in an associative network could operate independently. Instead of all of the units being updated at the same time (synchronous updating), individual units could update themselves at different times according to some probability function (asynchronous updating). Alternatively, one can get gradual recall by using continuous units and changing them slowly (either synchronously or asynchronously) depending on the results of the matrix multiplication. Gradual recall occurs slowly as network states gradually head "downhill" toward one of the stored memories. Additionally, one can do gradual learning by adjusting weights little by little as individual patterns are presented.

The use of a Hopfield network for complex images is shown in Fig. 10.4. This network is constructed using the same outer-product rule used in Figs. 10.1 and 10.3. However, instead of four or two units, this network has 46,592 units, which are fully interconnected. Hence it has $46,592^2$, which is $\sim$2 billion synapses. The three stored memories are seen at the bottom of the figure: Beardsley, bamboo, and Mozart. Each image is $182 \times 256 = 46,592$ units, where each unit is shown as either white (-1) or black ($+1$).

With only three memories in this big network, the attractors are very attractive. Simple matrix multiplication of any of the images at the top will recover the memory in one step. To give a better picture of the gradual downhill slide to the attractors, we have run the system so that it gradually converges on the result over 10 iterations, of which three are shown. Such gradual convergence would occur naturally if the attractor landscape were more crowded, as units will initially be pulled toward different attractors as they wander downhill. To show the gradual descent here, we had the associative memory matrix gradually push continuous units (by changing them a little bit on each iteration) instead of simply setting binary units. Another way to achieve this gradual settling or relaxation of the network to each attractor would be to use asynchronous updating.

Fig. 10.4 illustrates two aspects of the behavior of the content-addressable memories in general: pattern completion and pattern cleanup. The center image is incomplete and the network gradually fills in the missing region. The images on either side are complete but are obscured by noise. The network removes the noise and restores the pristine image. Both of these properties are present in the brain. The brain does pattern completion in vision, filling in the information that is missing from the visual field due to the blind spot. You are unaware of the blind spot even if you

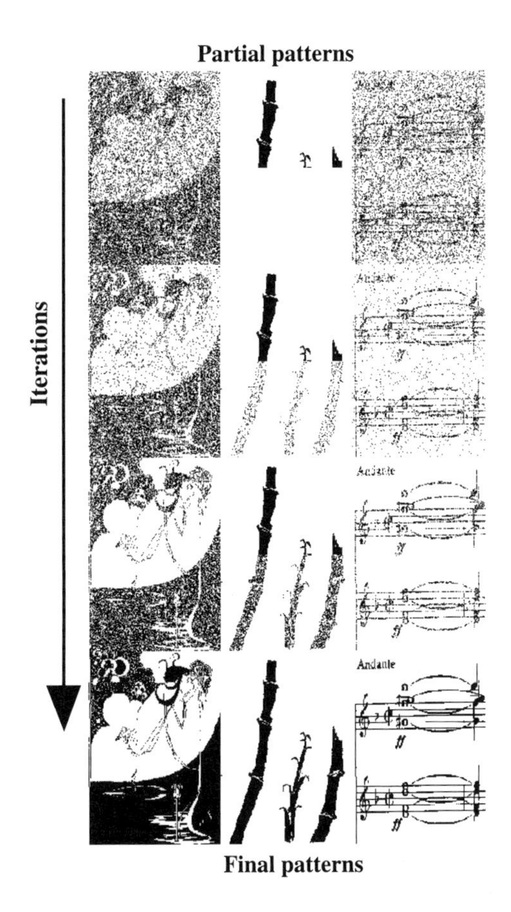

Fig. 10.4: Gradual settling of a Hopfield network into each of three attractors (bottom). "Time" (number of iterations) progresses from top to bottom.

close one eye, and you must test carefully to demonstrate that it is there at all. An example of the brain's extracting a signal from noise is the cocktail party effect: your brain can filter out all the extraneous noise at a party in order to focus on the relatively weak signal coming from the person you're talking to.

10.3 Critique of the Hopfield network

There are several problems that occur when we try to apply the Hopfield network literally to the brain. The one that bothers me the most is that point attractors are points where activity gets stuck and remains the same. The electroencephalogram (brain wave recording) demonstrates that the brain is oscillating continuously. Intuitively, our thoughts seem to flit easily from one topic to another, sometimes having trouble staying focused long enough to finish an exam or, later in life, an IRS form. On the other hand, some people have obsessive-compulsive disorder and get stuck all the time.

In the Hopfield network, we can deal with the problem of getting stuck by saying that after the attractor is reached (or just approached), the brain heats up (produces more noise) and kicks the network out of the attractor. This then allows the network to descend toward another attractor. (Heat, energy, and noise are all closely related both mathematically and physically.) Maybe attention-deficit disorder is a brain that's too hot (too noisy) and obsessive-compulsive disorder is a brain that's too cold. Alternatively, one can get away from the point attractor notion entirely and build similar networks that depend on convergence onto limit cycles or even onto strange attractors.

Another problem with the Hopfield network is that it works best with $-1/1$ vectors. This means we need negative state values. In the context of spiking neurons this is interpreted as firing rates below the spontaneous rates of regular firing neurons. As we noted in Chap. 6, rate coding does not seem like a valid model for neurons that fire rarely or fire in bursts. One can build Hopfield networks using $0/1$ vectors, but they have much lower storage capacity due to crosstalk, the tendency of patterns to bleed into one another as more memories are stuffed into a network. This, in turn, can be solved with additional mathematical jiggering. However, at some point all the mathematical manipulations begin to overwhelm the model.

In software marketing, when customers complain about a bug in the program, the marketers check with the software engineers. The engineers explain that though it seems like a bug, it's really a feature, something good that they put there purposely. Similarly, in neural networks, attributes that seem like a problem from one perspective may have benefits from another. For example, crosstalk is generally taken to be a bug. You stuff more and more memories into an associative neural network and they start to get mixed up. An image that should go to one attractor instead goes to another or to a spurious attractor. On the other hand, crosstalk has some appeal as the basis of mental association. One thing reminds you of another, or becomes a metaphor for something. Such crosstalk linking would be heteroassociative recall without explicit heteroassociative learning. So crosstalk, as a feature, is a candidate for a form of cognition. It's not clear how to put together a neural network so as to get fruitful crosstalk without fruitless confusion.

In real life, memories are formed through experience, forward in time, one on top of another. Another source of crosstalk arises when we attempt to replace the unbiological instantaneous outer-product matrix summation with Hebbian rules applied to incoming memories. Basically, a new memory that has any overlap with an old memory will tend to activate the old memory. The old memory will then be dragged into the learning and a hodge-podge of old and new will be learned. Various algorithms have been developed to prevent this mixed learning. The simplest is to suppress all spread of activity during learning. Then activity spread has to be turned on again for recall. There is some evidence that something like this might actually occur in the brain. These issues will be explored further in Chap. 14.

10.4 Summary and thoughts

Learning can be defined as the registering of associations. Memory can then be defined as registering the same association after a delay. By utilizing the convenient fact that the number 1 maps onto itself (i.e., $1 \cdot 1 = 1$), arithmetic associativity provides a way of restoring binary vectors from either partial versions of themselves (autoassociativity) or other vectors (heteroassociativity). This yields what is currently the standard paradigm for memory storage — storage is in the synapses.

This viewpoint is supported by the fact that a variant of Hebb's rule can be demonstrated biologically. Long-term potentiation (LTP) is a biological process defined by a sustained synaptic strength increase following coactivation of a presynaptic cell and a postsynaptic cell. In the original Hebb's rule, an increase in synaptic strength occurred only if the postsynaptic cell fired. This is not a requirement in LTP and may not even be sufficient to produce LTP in many cells. Additionally, there are many variants of LTP that have somewhat different properties.

The cornerstone of the Hebb hypothesis is the Hebb assembly rather than the Hebb rule. Cell assemblies cannot be readily demonstrated biologically due to technical limitations. We have no way to measure activity in thousands or tens of thousands of neurons in a behaving, learning animal in order to show that certain subsets fire together in association with particular stimuli or while thinking about those stimuli later. Evidence for some small cell assemblies have been found in the early stages of smell processing (olfaction), however.

An alternative to classical Hebbian assemblies that has been popular lately is the synchronized-firing assembly. The classical Hebbian assembly depends on a chain reaction where cells recruit other cells in a repetitive firing mode until a large group of cells are all firing. A synchronized cell assembly would involve a group of cells that fire a single spike at the same time. In that case, there would be no time for one cell to activate another in the assembly. It is not clear what role Hebb's rule would have in permitting synchronized assemblies. Such synchronization has been demonstrated in the visual and motor systems.

Although synchronized-firing assemblies are a popular alternative, it's probably fair to say that Hebb's rule and Hebb assemblies remain the dominant paradigm at present, partly because they can be so well worked out mathematically. However, there remains a huge divide between the elegant simplicity of outer products and energy fields and the elegant complexity of idiosyncratic dendritic trees, myriad neurotransmitters and second messengers, and all that other stuff that creatures actually think with (before it's ground into slop to feed to other creatures). That is why I now finish up with the nice neat reasonable stuff and dive into the slop in the next chapter.

Part IV

Brains

An electrode is placed inside a neuron. It goes through a membrane to enter the salt solution inside. On the outside of the cell the electrode showed the same potential as the grounded reference electrode. Once inside, the voltage drops suddenly to about −75 mV. If we record for a while we may see signals: little 10-mV bumps going up or down and big, but brief, up-going 100-mV spikes. We also see a lot of higher frequency noisy stuff. Some of this is an artifact of our recording equipment: it is easy to accidently pick up nearby radio stations on an electrode. Some of it is real, showing the opening and closing of small pores in the membrane or reflecting signals from far-off neurons. An electrode placed outside of the neuron will pick up more of these far-off units, eventually providing a mixed-up signal of many neurons combined.

Imagine yourself as the prototypical green alien visitor guy. After a visit to area 51 and an evening spent performing unspeakable procedures on sleeping humans, you will naturally want to monitor electrical activity on earth. You will assume, correctly, that much of this electrical activity has to do with information processing and communication. But you will have a heck of a hard time pulling apart any of the signals so as to interpret them. One big problem is that there are multiple protocols in use, both the electronic languages used to encode and decode signals and the underlying human symbolic languages. Apparently simple signals may actually be complex multiplexed signals, meaning that there are many different signals mixed up inside what appears to be a single signal. Also on top of all of this you have loads of noise, a mix of distant signals, and the background radiation that comes from all electronic devices (and from outer space). It's hard being an alien.

From concept to data

In the foregoing chapters, I presented a view of the nervous system as a well-built, rational, comprehensible machine. The theme of the following chapters is that the nervous system is a hodge-podge, a hack, and a puzzle. This perspective shift takes us from concept-based theory to data-based theory, from what might be there to what is there. This shift is disconcerting since it's natural to feel that these reasonable theories of the prior chapters really ought to be there, because they make so much sense. In this way, the theories tend to take on a virtual life of their own, influencing what is observed. To some extent, this is right and good since it is hopeless to go mucking around in the brain without any plan or guidance, perhaps counting red blood cells instead of counting neurons. When taken too far, however, this reliance on theory leads to what has been called "the heartbreak of premature definition." If taken too seriously, the theory provides blinders rather than guidance.

Current theories of neural information processing are probably inadequate. We all await the grand paradigm shift, when we throw over the old theories and enjoy the revolution. In some paradigm shifts, for example the Copernican revolution (the paradigmatic paradigm shift), old theories are completely discredited and discarded. In others, for example, the Bohr atom to quantum mechanics, initial adolescent iconoclasm gives way to an appreciation that the old theory maybe wasn't so bad after all. This is meant to be reassuring; maybe you didn't entirely waste your time reading the first half of the book. On the other hand, maybe you did. Sorry, I should have mentioned that earlier.

Leaving high-level theorizing behind, we look at the guts and volts of the brain. Unfortunately, most of these details can't be understood at the big-picture level. We replace the top-down conceptual world with the messy bottom-up world. There are still concepts, but they are limited in scope and don't pretend to answer major questions. We give up the comfortable heft of neatly encapsulated concepts well understood. We trade this for the confidence of knowing about real things.

Isaiah Berlin, in a humanities context, described a corresponding contrast in perspectives by quoting a Greek poet (Archilochus) who wrote: "The fox knows many things, but the hedgehog knows one big thing." In his famous essay entitled "The Hedgehog and the Fox," Berlin compares "those, on one side, who relate everything to a single central vision" to "those who pursue many ends, often unrelated and even contradictory." The first half of this book was meant to appeal to hedgehogs (and physicists), describing coherent ideas. The second half is for foxes (and biologists), presenting lots of little ideas that are intriguing in themselves but don't really add up.

Neurons are cells

A central example of this discrepancy between happy fancy and hard reality is found at the level of the single cell. The happy-go-lucky single cell we met in neural networks has a primitive appeal reminiscent of some ancient cave painting (Fig. 11.1). The real guy whom we are about to meet looks hopelessly organic — tree rather than I-beam. The artificial guy is made of numbers, while the real guy is soap, water, and salt.

A neuron is a cell, too. Before it can do any seeing or signaling or thinking, a neuron has to take care of itself. This means it has the same enormous machinery for energy metabolism, protection against toxins, and communication both intracellular and extracellular as other cells in the body with more prosaic job descriptions. When we find out about some remarkable signaling mechanism available to neurons, it is generally the case that kidney and liver cells have their own versions of the same mechanism.

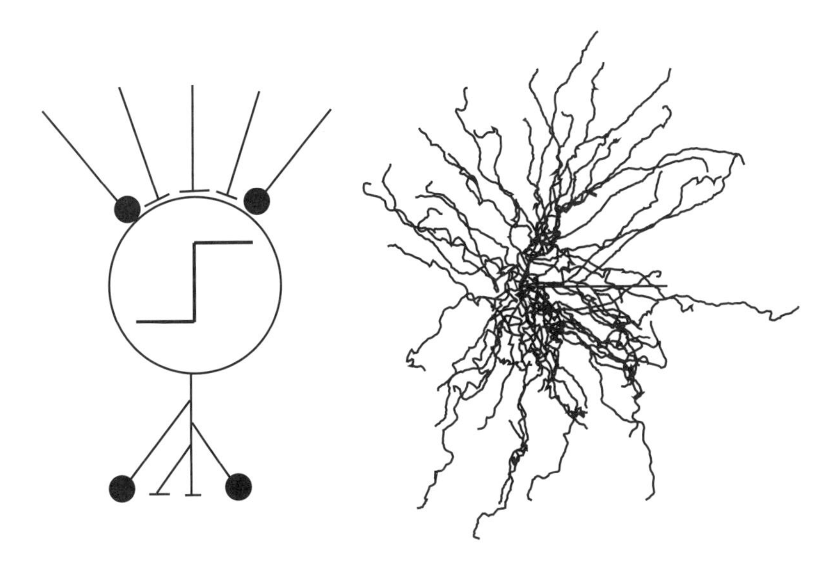

Fig. 11.1: The mythical neuron meets its counterpart.

All of the machinery of the cell is entwined with neural information processing technology. Chemicals passed around by the cell to obtain energy during metabolism are reused for communication. Three of the basic food groups for cells — glucose, adenosine triphosphate (ATP), and acetyl-coenzyme A (acetyl-CoA) — are either used directly or have close congeners that are used for information processing. The major neurotransmitters, glutamate and GABA (gamma-aminobutyric acid), are spin-offs from the tricarboxylic acid cycle (Krebs cycle), the main sugar digestion route. Similarly, ACh (acetylcholine) is half brother to acetyl-CoA, the main product of the Krebs cycle. Glucose is processed in order to produce ATP, the main energy storage medium. ATP is also a neurotransmitter. Various ATP by-products — adenosine diphosphate (ADP), adenosine monophosphate (AMP), cyclic AMP (cAMP) — are also used in signaling. Teleologically, one can imagine that shortly after the simple single-cell creatures of long ago learned to smell food, they began to use the same receptors to commune among themselves. This would gradually evolve into the sophisticated food-based data processing technology of our own brains. Are you thinking about food or is your food thinking about you?

This sharing of resources is not confined to information processing. The same compounds or their close relatives are also amino acids, the construction materials of the cell and the body. Both glutamate and GABA are close relatives of amino acids. Glycine, another neurotransmitter, is an amino acid. Glycine is the major constituent of collagen, making it the building block for skin and bones. Similarly, the nucleotides of DNA, the blueprint of the cell whose code is used to construct proteins, is also shared with both neurotransmission and metabolism (e.g., ATP). Cell maintenance also requires additional intracellular communication between and among various organelles (subcellular organs), as well as extracellular communication with various supporting cells. This type of communication cannot always be cleanly separated from classic neurotransmission. For example, during development neurotransmitters and second messengers are being used to grow neurons and coordinate their wiring and relations with various supporting cells. While a child is using his transmitters to grow a brain, he also has to think using the same transmitters.

Because of this close enmeshment of functions, it may never be possible to cleanly separate fancy neural information processing from the boring information processing of housekeeping chores. Even the electrical charge at the cell membrane, the key attribute that permits action potential signaling between cells, has generic cell maintenance tasks as well. Similar electrical potentials are present in all body cells and in yeast and bacteria as well. Information transmission (and related encoding and decoding) has to be done in every organ. Bone cells communicate in order to adapt to changes in stress patterns when you learn to rollerblade. The liver and endocrine systems all have complex nonneural communication protocols in place. The immune system has a remarkable interplay of cell types that chat with one another and with other cells in the body, all of which have to continually remind white blood cells that they belong there and should not be eaten.

What is the neuron state?

In previous chapters, a single number (a scalar) was used to define the state of a neuron. An implicit assumption of the scalar model is that we are dealing with the functional equivalent of a point neuron — a neuron that has no geometry or spatial extent. Of course, real neurons have a complex three-dimensional structure. Some of them are big enough to be seen with the naked eye. Signals come in at dendrites that may extend as much as a millimeter from the central soma (cell body). Signals then go out through axon terminals that may be more than a meter away from the cell body (e.g., the axon that goes to your big toe). However, from a signal-processing standpoint, what is important is not the physical size of a neuron but its electrical size — how far across the cell can a signal spread. Given the many different morphologies and different electrical properties of dendritic trees, some will turn out to be electrically large and others electrically compact, concording with the role they have to play. Axons, on the other hand, are all effectively compact; the action potential ensures that a signal that starts at one end will get to the other.

In addition to being three-dimensional in shape, neurons are also multi-dimensional in terms of the many different kinds of signals that are used.

Many types of electrical signals and a variety of chemicals participate in neural signaling. These include the neurotransmitters that transmit information across synapses as well as a variety of second (and third, and fourth) messengers that transmit the signal further on inside the neuron. Many of these chemical signals are likely to be important in information processing. Thus, rather than think of neural state as a scalar, it might make more sense to consider a vector of voltages and chemical concentrations to describe the cell.

Although the chemical and electrical complexity of neurons makes it clear that there is no single scalar state, it may be that multiple states occur in series as a temporal chain. In this case, each state determines the next, and any of the states could be used to represent the information processing state. Serial states are assumed in rate-coding theory: presynaptic firing rate determines synaptic potential determines soma potential determines soma firing rate determines axon firing rate. We can treat these different states as if they were the states of independent units and work out the signal transduction (weights) between them. In the next two chapters, we do this analysis; working out how synaptic inputs determine summed membrane potential and how summed membrane potential determines spike rate.

On the other hand, if multiple states are operating in parallel, then the neuron is more like a computer central processing unit (CPU) than like a transistor. In that case, analysis of single-neuron information processing becomes far more difficult. Many neurons have a single major output pathway, the axon. Even if such cells are processing multiple information streams in parallel, all of the information has to come through the axon. We can then take this axon output to be neuron state. However, this output state would just be the answer to whatever computational problem the neuron was calculating. We would have missed all the information processing.

Still more complex are neurons that have multiple inputs and multiple outputs. Thalamic cells are a prominent example of this in mammalian brains. The thalamic cell is likely not only multiprocessing but also multiplexing. Multiple signals come into the thalamic cell, multiple signals are spit out. Some of these inputs combine, others remain separate. As long as they are separate, the states can be understood separately. When the inputs combine into a single measurable state, this is multiplexing. The causes and results of such a signal can be hard to disentangle. If single neuron processing turns out to be this complex, then techniques for analyzing activity in the single neuron will be similar to the techniques that we currently use to analyze a neural network.

No scalar state; how about a scalar weight?

In artificial neural networks, scalars were used to define neuron states. Scalars were also used to define the weights between any two neurons. The use of a scalar weight simplified learning theory, since learning was

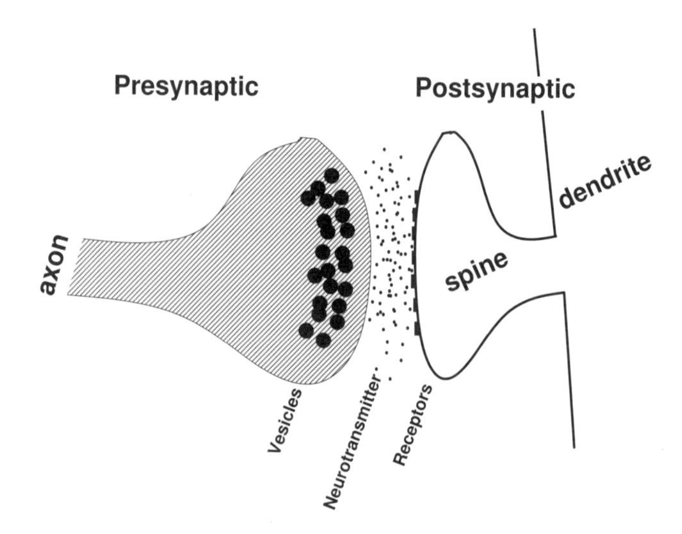

Fig. 11.2: Basic synapse design: presynaptic vesicles (big dots) release neurotransmitters (little dots) into the synaptic cleft. They float across the cleft to activate postsynaptic receptors (fat lines).

expressed as an increase or decrease in this single number. As you might guess, the scalar-weight concept will also need some revision. Basic synapse design looks pretty simple (Fig. 11.2). An electrical signal, an action potential, invades the presynaptic axon terminal. This causes vesicles to release neurotransmitter into the synaptic cleft between the neurons. The neurotransmitter molecules float across and bind to receptors on the postsynaptic neuron. These receptors are linked to channels that open up and create an electrical signal, a postsynaptic potential, on the membrane of a spine or dendrite of the postsynaptic neuron.

Although there are multiple processes in this description, the synapse can still be assigned a single weight as long as these processes are occurring in series. Just as with our description of neural state, we would then define a scalar value for each process in sequence and determine transduction between them. Again things gain in complexity if there is multiplexing going on at the synapse. Just as there is a wide variety of neuron types, there are also many different types of synapses, differing in their complexity.

There are various sources of synaptic complexity. Many synapses have more than one kind of receptor postsynaptically. Some synapses release more than one neurotransmitter. Some synapses are electrical rather than chemical connections. Some synapses are not strictly one-way but have retrograde transmission or are involved in perverse three-ways (thalamic synaptic triads). Some axons synapse onto postsynaptic spines, as shown in Fig. 11.2. The function of spines is a mystery; it seems likely that they don't provide a straight conduit to the main dendrite. Some researchers suspect that significant chemical information processing may be happening inside these spines. Many neurotransmitters don't just head straight across the synaptic cleft but instead spread far across neural tissue. This is called volume transmission. Some nonclassical neurotransmitters are gases. Gases can pass straight through neural tissue, expanding in a cloud from their release site.

Different modeling tools

Another distinction between the first and second half of the book is the set of tools used. The tools of the top-down approach were primarily those of discrete mathematics. "Discrete" here refers to the use of distinct digital numeric values instead of continuous analog values. Discrete math is the math of computer science and digital electronics: binary numbers, Boolean algebra, transistor-transistor logic.

By contrast, in the following chapters we mostly will be dealing with continuous analog phenomenon. For continuous mathematics, calculus is the primary tool. We discuss electronics and explain the mathematical descriptions of capacitors, resistors, and batteries. As it happens, the math of electronics is identical to the math of dynamics, the study of movement. This is the original realm of calculus, beginning with Isaac and the proverbial falling apple.

Whoops

The foregoing was meant to be an introduction to the realistic neural modeling that will be discussed in the rest of the book. However, I have to backpedal a bit here. I emphasized the importance of chemical signals. I will now proceed to ignore them. The details of chemical signaling are not well understood, and there hasn't been very much modeling done on this. Therefore, the emphasis in what follows will be on electrical modeling.

I also spent some of the foregoing maligning scalar state and scalar weight theories. I take that back, too. These remain the standard theories because they are still the best theories. Therefore, rather than abandoning these theories, I will explore their implications.

11

From Soap to Volts

11.1 Why learn this?

In this chapter, I introduce the hardware of the brain. I develop some of the concepts of neural membrane modeling. I show how currents, capacitors, and resistors arise from the interactions of salt water with membranes. To demonstrate this, I have to address the relatively hard concept of capacitance, which must be handled with calculus. I explain some basic calculus algebraically by using numerical, rather than analytic, methods. Using numerical calculus as a simulation tool, I then explore a couple of fundamental concepts of neural signal processing: temporal summation and slow potential theory.

Moving quickly from soap to volt to signal processing requires that I skip over some key concepts and issues that will be addressed in later chapters. As an alternative approach, I could have introduced all of the underlying material first and then moved on to discuss specific models and what they mean. I chose the get-there-quick approach for two reasons. First, delayed gratification is hard: the technical and algebraic details can get pretty dull. Applications and ideas brighten this chapter up a bit. Second, making the right simplifications is a major part of modeling. The use of a stripped-down model in this chapter demonstrates how ignoring some details can help focus attention on those details that are essential for the issue at hand. Specifically, I demonstrate why time constants are critical for understanding neural signal processing.

In the following chapters, I fill in the details. In this chapter, the circuit only has one resistor and one capacitor. In the following chapters, I put in batteries and a few more resistors. In this chapter, I only use injected currents as signals. In the following chapters, I show how biological signals are largely conductance changes that cause current to flow secondarily.

In this chapter, I have also tried to avoid stumbling over units, the volts, amperes, farads, siemens, hertz, and other famous guys immortalized as stuff. Units are unloved and underappreciated (see Chap. 16, Section 16.2). Sizes, durations, and magnitudes offer insights about the limits, capacities, and capabilities of the hardware. Unfortunately, focusing on the units now will entangle us in a nest of confusing conversions. I have largely avoided this distraction.

11.2 Basic cell design

Cell design is based on the separating of inside from outside by means of membranes. Prokaryotes, bacteria, are single cells with one big compartment. In eukaryotes, like us, each cell has many separate subcompartments doing separate tasks. Most of the body is salt water. Water and salt can't pass through fat since oil and water don't mix. Soap is the compound that connects oil and water, thereby allowing showers to wash oily stuff off your skin. Soap works by having a fatty part that connects with the dirt and a hydrophilic (water loving) part. Fat is hydrophobic (water fearing). Soap thus provides a link that allows water to drag oil away.

Soap bubbles form with the fatty part of soap pointing inside and outside, away from the water that stays on the interior of the bubble membrane (Fig. 11.3). Cell membranes are called lipid bilayers. They are configured in the opposite way: the fatty part is in the interior of the membrane, and hydrophilic heads point both in toward the cytoplasm and out toward extracellular space. Just as soap bubbles form spontaneously on the surface of soapy water, lipid bilayers form spontaneously from a mixture of phospholipids (biological soap) and water. The *inside of the cell* and the *interior of the membrane* are not the same thing. The cell is filled with a volume of intracellular solution (cytoplasm) that the membrane separates from the extracellular solution outside of the cell. Although the membrane is only a thin layer at the surface of the cell, there is enough room in the interior of the membrane for chemicals to float around and react with one another. Only hydrophobic compounds can exist in the interior of the membrane.

Ions and many proteins are hydrophilic. Ions are the charged versions of certain elements, such as sodium, potassium, chloride, and calcium. They are charged because they have either lost or gained an electron, which carries a negative charge. The little plus or minus signs tell you how many electrons were lost or gained: Na^+, K^+, Cl^-, Ca^{++}. Positive and negative

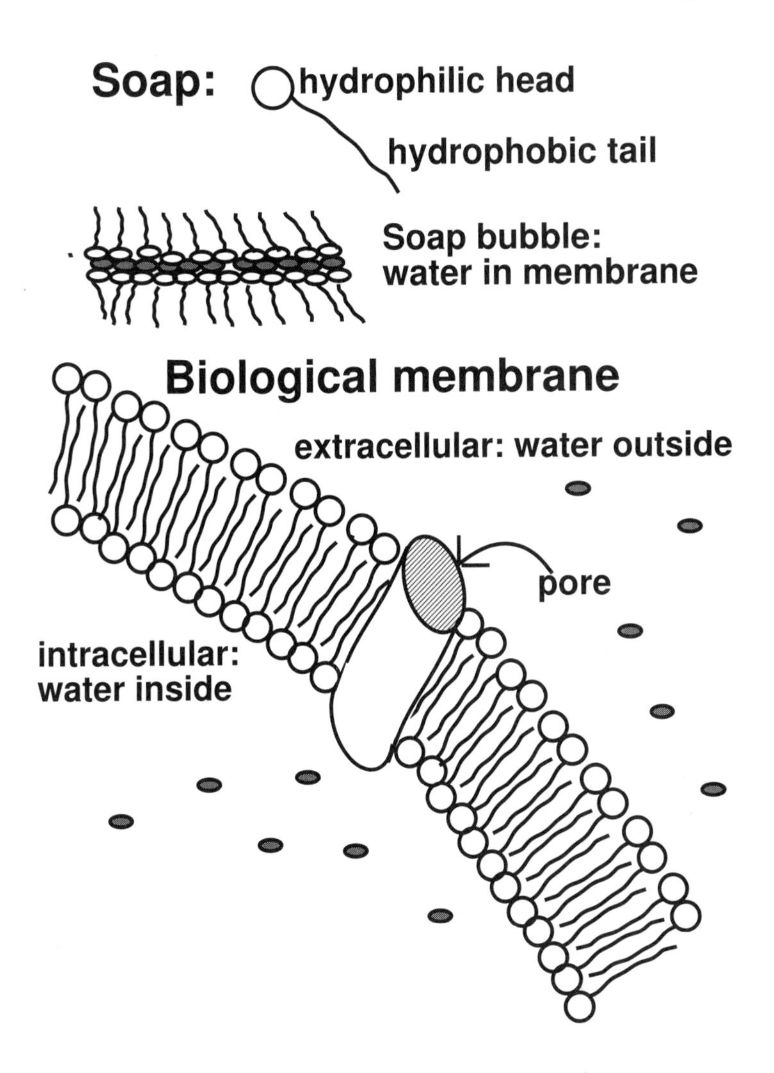

Fig. 11.3: Soap bubbles have fatty tails pointing in and out and water (shaded) interior. Biological membranes keep the fatty tails interior with hydrophilic heads sticking in and out. Proteins orient themselves using both hydrophilic and hydrophobic amino acids. They can form pores that allow ions to move between intracellular and extracellular space.

ions stick together to make a salt; NaCl is table salt. Many proteins are charged as well. Charge makes a compound hydrophilic since the hydrogen of H_2O will stick comfortably to a negative charge and the oxygen will stick comfortably to a positive charge. Fat will stick only to uncharged molecules. Therefore, ions and charged proteins can move around freely in either extracellular or intracellular space but can't pass through the fatty part of the membrane. Some proteins and various other compounds are hydrophobic and can float around the interior of the membrane but not move into the water on either side. There are also information-carrying compounds that are gases. These don't care whether they're in fat or water. They can pass through anything. The most well known of these is nitric oxide (NO), a former molecule-of-the-year winner and the functional basis of Viagra.

Proteins are the active components of cells, involved in metabolism, cell reproduction, and practically every other function. Transporters and pores are transmembrane proteins (e.g., the pore in Fig. 11.3). Although ions cannot move directly across the fatty membrane, they can flow through these pores. Ions will move passively from a high concentration to low concentration, a process of diffusion down a concentration gradient. Other proteins can provide ion pumps or transporters that push ions up against a concentration gradient. Additionally, proteins provide transport and structure within both the extracellular and intracellular spaces. Although the extracellular and intracellular space are usually thought of as being salt water, the large amount of protein gives it structure, making it more like jello than seawater.

11.3 Morphing soap and salt to batteries and resistors

Descriptions like the above are both model and metaphor: the extracellular space is jello, the membrane is an inside-out soap bubble, the proteins form holes. By switching to another representation, another language, we can describe the same thing in different terms. A particular model/metaphor will be better than another for a particular purpose. In this case, translation of the model into the language of electronics (batteries, resistors, capacitors) will allow us to describe neural signaling much more easily than we would be able to do just chewing the salt and the fat.

Morphing from a membrane and salt representation to an electrical engineering representation obscures the biology somewhat. Any representation offers trade-offs and compromises. The advantage of the electrical representation is that we can now use the many mathematical tools developed for electrical circuit analysis. Ions moving through water carry charge. The movement of charge is electrical current. Current can flow through the cytoplasm and extracellular fluid freely. The pores in the membrane provide conduits (conductors) for current to go through the membrane as well.

Protein pores of a particular type provide conductors through the membrane that are selective for a particular ion. These pores are parallel conductors providing parallel paths for current. A conductor is defined by its tendency to permit the flow of current. A resistor is defined by its tendency to resist the flow of current. A resistor and a conductor are the same thing. It's just one of those pessimist-optimist things: half-empty or half-full; resisting or conducting. Resistance is expressed in ohms (Ω) and is represented by R. Conductance is measured in siemens (S) and is represented by g. Mathematically, conductance is the inverse of resistance, $g = \frac{1}{R}$. Because of this, siemens, the unit for conductance, is also called "mho," which is ohm spelled backward (seriously).

In the circuit diagram of Fig. 11.4, all of the pores of a single type are lumped together as a single big conductor (resistor). This ubiquitous conductance is known as the leak conductance (g_{leak} or R_{leak}). Later, when we add different types of pores to the circuit, we will have different conductors in parallel with this one.

Made of fat, the membrane acts as an insulator. Current that does not flow through one of the pores will just sit next to the membrane. Charge acts at a distance to attract unlike charge or repel like charge. Charge that is sitting on one side of the membrane will cause equal but opposite charge to sit on the other side of the membrane. This phenomenon is known as capacitance. In electronics, a capacitor is built by placing an insulating material between two parallel metal plates that are attached to the wire leads of the capacitor. Since these plates do not touch, electricity cannot pass directly through the capacitor. However, electricity can flow indirectly as one plate induces electrical flow in the other plate. The two parallel lines in the standard symbol for the capacitor (Fig. 11.4) represent these two plates. In the biological situation, the two plates are just thin accumulations of ions in the water adjacent to either side of the membrane.

Capacitance has this name because it is a measure of the capacity of the plates to hold charge. This capacity will have to do with the size of the plates, the distance separating them, and the nature of the material between them (its dielectric constant). A capacitor, or membrane, with high capacitance has the ability to hold a lot of charge with only a small voltage difference between the plates. Voltage times charge is a measure of energy. A high capacitance means that charge can be stored easily, i.e., without requiring as much energy. As you put more voltage across a capacitor, it will hold more charge and more energy. A capacitor "expands" with voltage, as a balloon expands with pressure. Capacitance tells how much the capacitor can hold at a given voltage, just as a capacity measure for a balloon would tell how big the balloon would be at a given pressure.

A resistor (R) and capacitor (C) placed next to each other (in parallel) is called an RC circuit (Fig. 11.4). In electrophysiology, we insert an electrode through the membrane and then inject current into the inside of the cell (arrow in Fig. 11.4). This current will pass out of the cell through the resistor or the capacitor (through the pores or via the capacitance). Once the current reaches the outside of the cell it will disperse to the rest of the body and the surrounding world. In the diagram this is represented by the

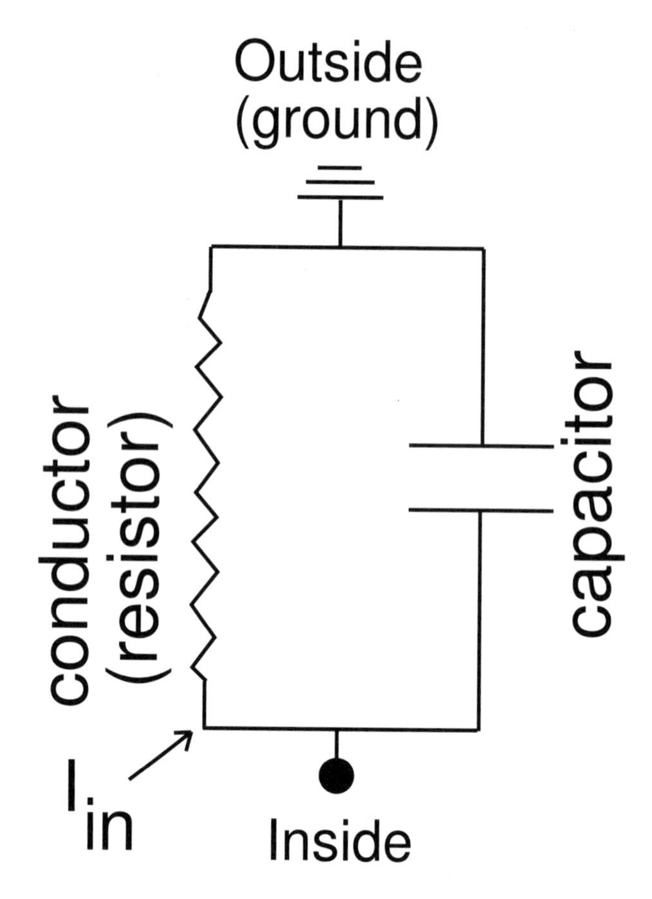

Fig. 11.4: The membrane is represented as an RC (resistor-capacitor) circuit. Current can be injected inside the cell (arrow) by using a hollow glass electrode.

ground symbol. The human body is big enough and the currents are small enough that the body can easily sink (conduct away) the current that cells produce. When we put a cell in a dish and inject current, we place a wire in the dish to conduct the current out to ground. This wire is attached to something metal that leads to the physical ground. This is the idea that Ben Franklin came up with for lightning — provide an electrode to lead current harmlessly to suburbia.

11.4 Converting the RC circuit into an equation

Modeling the neuron as an electrical circuit involves predicting voltage changes based on current flow. One can also model an electrical circuit the other way around: given the voltages, predict the currents. In the case of neurons we typically are injecting currents, or thinking about the effects of extra pores that cause extra current to flow into the cells. Here are three major electrical laws needed for neural modeling:

- Kirchhoff's law: Conservation of charge. Charge does not disappear or appear out of nowhere.
- Ohm's law: $V = I \cdot R$. The voltage (V) drop across a resistor (R) is proportional to the current (I).
- Capacitance: $Q = C \cdot V$. The charge (Q) on a capacitor (C) is proportional to voltage (V).

In electronics, Ohm's law is usually expressed in terms of resistance, while in biology it is more often expressed in terms of conductance. Since $g = \frac{1}{R}$ (not a law, just a definition), Ohm's law can be expressed as $V = I/g$. Turning this around gives $I = g \cdot V$. This is the preferred formula for biology. When we talk about pores opening up, it is most natural to think of them as conductors of ions rather than as resistors to ion flow. The ability to conduct is given by the magnitude of g. Ohm's law states that a bigger conductor will allow more current to flow down a given voltage gradient. g will be a constant for a simple resistor but will vary for a rheostat, a variable resistor.

We can't relate these three laws directly in the forms given above. Ohm's law is expressed in terms of current, while the other two laws are expressed in terms of charge. Current is the movement of charge. We can use this fact to rewrite the latter two laws in terms of current. This is easy to understand for Kirchhoff's law: charge that is moving has to go somewhere. Therefore, any current that comes out of one wire must go into some other wire. Where one wire feeds two other wires, the sum of currents in the two will equal the current that came out of the one. Conservation of charge translates into conservation of current. This is the translation for Kirchhoff's law. Translating the capacitance law from charge into current is a little harder, requiring that we use calculus.

Capacitance and current

In Chap. 16, Section 16.5, I introduce several notations commonly used in calculus. For present purposes, I stick with a single numerical notation, suited to numerical analysis by computer. The capacitance law given above is $Q = CV$. Current, I, is the change in charge, Q, with time. Let's look at the charge on a capacitor plate at two times, time t_a and a later time t_b, separated by a small duration. At time t_a, the charge on the plate is Q_a. At time t_b, it is Q_b. The amount of charge that has moved off of the plate during this time is $Q_b - Q_a$. The current, I, is $\frac{Q_b - Q_a}{t_b - t_a}$, the change in charge divided by the change in time. In mathematics, the Greek letter delta (Δ)

is a short-hand notation used to mean "change." $\Delta Q = Q_b - Q_a$ means the change in charge on the capacitor. $\Delta t = t_b - t_a$ is the time interval. So we can write the expression for current as $I = \frac{\Delta Q}{\Delta t}$. ΔQ divided by Δt is the rate of change of charge — how fast charge changes.

Now we can consider the capacitance equation: $Q = C \cdot V$. At time t_a, $Q_a = C \cdot V_a$ and at time t_b, $Q_b = C \cdot V_b$. If we subtract these two equations we get $Q_b - Q_a = C \cdot (V_b - V_a)$. In between these two measurement points, $t_b - t_a$ of time has passed. We divide both sides by the time duration to get $\frac{Q_b - Q_a}{t_b - t_a} = C \cdot \frac{V_b - V_a}{t_b - t_a}$. Using the Δ notation, this is $\frac{\Delta Q}{\Delta t} = C \cdot \frac{\Delta V}{\Delta t}$. Since $I = \frac{\Delta Q}{\Delta t}$, the law of capacitance can be expressed in terms of current as $I = C \cdot \frac{\Delta V}{\Delta t}$.

Current through a capacitor equals capacitance times the rate of change of voltage. What does this equation mean? If voltage is changing fast, then a lot of current is passing through the capacitor (high I). When the voltage stops changing, the capacitor will stop carrying current. A capacitor can be used to store charge that can then be released abruptly as a large current. An example is the capacitor used for the flash bulb in modern cameras. This capacitor has high capacitance. A little camera battery with its little voltage can place a lot of charge on this capacitor. Capacitative charging is accompanied by a humming noise, as the two plates vibrate slightly. Once the capacitor reaches the battery's voltage, the charge will sit on the capacitor until needed. Pushing the shutter button suddenly connects the capacitor to ground, meaning that voltage changes abruptly from battery voltage to zero volts. This is a large ΔV so it causes a much larger current than could be generated by the battery directly. This large current allows the bulb filament to release a large amount of energy ($Q^2 \cdot R/s$ where R is the resistance of the filament) suddenly as a bright light.

Adding up the currents

We can now use our three laws to calculate what will happen in an RC circuit when the current (I_{in}) is injected. Kirchhoff's law says that charge is conserved so the injected current has to go somewhere. It will go either through the capacitor or though the resistor to get to ground. From there, the current runs away to the rest of the body and the world beyond. We can express this conservation of current arithmetically as an addition $I_{in} = I_C + I_R$. I_{in} is a known number, let's say 1 μA (microamp). I_C and I_R can be re-expressed using the current-voltage equations for capacitance and resistance. Therefore:

$$I_{in} = C \cdot \frac{\Delta V}{\Delta t} + g \cdot V$$

With Δt set to an infinitesimal (infinitely small) value, this equation is called a differential equation. Differential equations of various sorts are one of the main topics in calculus. If you are familiar with calculus, you may recognize this as a standard differential equation that can be solved *analytically* in order to get the precise relationship between V and I. We solve it using a numerical integration, getting very close to the analytic solution by making Δt very small. By using finite numbers and finite time steps, we create a discrete approximation to the differential equation.

We set $\Delta t = 1\ \mu\text{s} = 0.001\ \text{ms} = 1 \cdot 10^{-6}$ s. We can expand out ΔV as a difference: $V^+ - V$. The negative sign in V^+ is not an exponential but is used to denote voltage in the future. Plain V means voltage in the present. V^- would mean voltage in the past. The reason that we use a superscript here instead of a subscript is that the subscript is typically used to indicate the location in a neuron where the voltage is measured.

Substituting for ΔV leaves us with:

$$I_{in} = C \cdot \frac{V^+ - V}{\Delta t} + g \cdot V$$

Notice that the voltage multiplied by g is also V; future voltage will be based on present voltage. This is called the explicit Euler integration. There is also an implicit Euler equation, which is described in Chap. 16, Section 16.5. We can now gather up the present voltage terms, V, separating them from the future voltage V^+.

$$I_{in} = V^+ \cdot \frac{C}{\Delta t} + V \cdot \frac{C}{\Delta t} + g \cdot V$$

and solve for current voltage in terms of past voltage:

$$V^+ = (1 - \Delta t \cdot \frac{g}{C}) \cdot V + \Delta t \cdot \frac{I_{in}}{C}$$

This is an update equation, similar to the ones we discussed in the context of neural networks. The voltage (V^+) is updated at each time step based on the previous voltage V . At the same time, time, t, is updated by simply adding Δt: $t^+ = t + \Delta t$. The new voltage then becomes the V and the new time the t for the next update step. We typically store all of the Vs as we calculate them. We can then graph voltage against time where time increases by Δt at each update.

Let's look at the numbers. Units and unit analysis are a big part of solving these equations. I will use standard units, but avoid worrying about them by making sure everything divides out neatly. The horror and beauty of unit conversion is discussed in Chap. 16, Section 16.2. As is standard for neuron simulation, I use units of millivolts (mV) for potential and milliseconds (ms) for time. We set $g = 1 \frac{mS}{cm^2}$ and $C = 1 \frac{\mu F}{cm^2}$. (The common use of square centimeter for these units contrasted with the usual use of microns for neuron size is an example of why units are such a nuisance. There are 100,000,000 square microns in a square centimeter.) Using the chosen values, $\frac{g}{C} = 1/ms$. Similarly, we will set the injected current to $1 \frac{\mu A}{cm^2}$ so that $\frac{I_{in}}{C} = 1\ mV/ms$. We'll set $\Delta t = 0.001\ ms$. With these convenient choices, the update equation is

$$V^+ = (1 - 0.001) \cdot V + 0.001$$

At every step, V^+ will be updated according to this rule and t^+ will be updated by Δt. So the update for the full simulation requires equations for updating both t and V:

$$\begin{aligned} t &= t + 0.001 \\ V &= 0.999 \cdot V + 0.001 \end{aligned}$$

Notice that I have now left out the superscripts for V and t. This is common practice since this is the form the update rules will take when entered into a computer program.

We now just need a starting point, called the initial condition, and we can simulate. We'll start with $V = 0$ at $t = 0$. Then at $t = 0.001$, $V = 0.001$ and at $t = 0.002$, $V = 0.999 \cdot 0.001 + 0.001 = 0.001999$. Values during the first 10 μs are

t (ms)	V (mV)
0	0
0.001	0.001
0.002	0.001999
0.003	0.002997
0.004	0.003994
0.005	0.00499001
0.006	0.00598502
0.007	0.00697903
0.008	0.00797206
0.009	0.00896408
0.010	0.00995512

as shown graphically over 5 ms of simulation in Fig. 11.5.

11.5 Parameter dependence

We are interested in how the properties of a neuron provide its information processing capabilities. In Fig. 11.5, a continuously injected current provides an input signal. The response of the membrane (the "system" in signals and systems parlance) depends on fundamental properties: membrane resistance and membrane capacitance. As we move to more complicated models, we find that membrane conductance (g) can change with different inputs or in response to voltage change. In a network of neurons, input signals received by a particular neuron will also vary, as activity moves around the network. For now, we use the highly simplified model of Fig. 11.5 to see how varying conductance and injected current alters

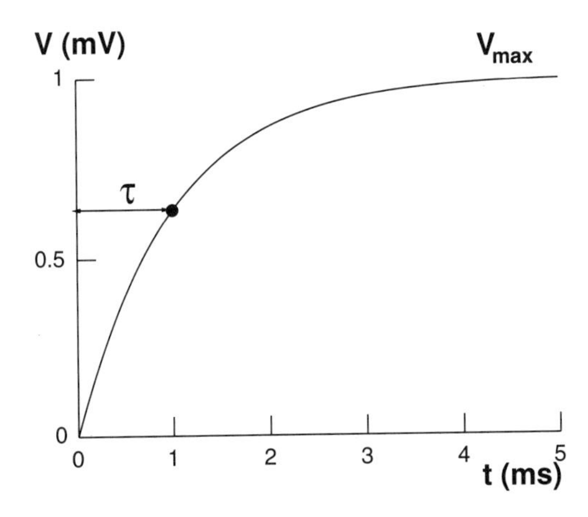

Fig. 11.5: Membrane charging curve simulation with C = 1 μF/cm^2, g = 1 mS/cm^2 , I$_{in}$ = 1 μA/cm^2 This gives time constant $\tau_{memb} = \frac{C}{g}$ = 1 ms and V$_{max}$ = $\frac{I_{in}}{g}$ = 1 mV.

neuron response. Because of the simplicity of the differential equation in this example, we could use the techniques of analytic calculus to solve the differential equation. This would precisely define how the g and I_{in} parameters determine voltage response. We instead use a computational numerical analysis. This type of numerical technique will also be applicable to more complex situations where analytic calculus cannot be used.

For starters, we need to understand how the update rule, $V = 0.999 \cdot V + 0.001$, generates the curve in Fig. 11.5. The value of the first coefficient, 0.999, reflects a tendency of V to stay put — its stickiness. The addend, 0.001, meanwhile, is the drive that pushes voltage up. The stickiness coefficient is just a little less than 1. The drive addend is just a little more than 0. If the stickiness coefficient was 1 and the drive addend was 0, then the voltage wouldn't move at all. If we had used an update time step of 0.0001, the stickiness coefficient would be 0.9999 and the drive addend would be 0.0001. This would give us a more precise approximation of the solution to the differential equation. The curve in Fig. 11.5 would look identical since in this example, the improvements would be tiny: i.e., V =0.001099 instead of 0.001 after 1 μs.

In Fig. 11.5, we start with an initial condition of $V = 0$. At the beginning of the curve, with V small, the membrane will charge quickly. The $0.999 \cdot V$ stickiness is relatively small and the 0.001 drive dominates the update rule. As V gets bigger, the stickiness begins to dominate and the rate of rise gets slower until the curve reaches one. When $V = 1$, the update rule will give $V = 0.999 \cdot 1 + 0.001 = 1$, so V will stop increasing. In general, one can find the maximum value (V_{max}) by taking the update rule as a regular equation and solving for V to get the steady state: $V_{max} - 0.999 \cdot V_{max} =$

0.001, $0.001 \cdot V_{max} = 0.001$, $V_{max} = 1$ mV. We can similarly solve for V_{max} in the parameterized equation to show that in general $V_{max} \sim \frac{I_{in}}{g}$. We need to say approximately equal ($\sim$) rather than equal because the numerical solution is just an approximation, as will be described further below.

Notice that $V_{max} = \frac{I_{in}}{g} = I_{in} \cdot R$. This is Ohm's law. The value of this maximum reflects the fact that at the end, the voltage is not increasing. Therefore, $\frac{\Delta V}{\Delta t} = 0$ and there is no capacitative current. All the current flows through the resistor and the voltage is determined entirely by Ohm's law.

We referred above to the "stickiness" of the update equation, dependent on how close the coefficient for voltage is to 1. This in turn depends on the ratio of Δt to $\frac{C}{g}$. A bigger $\frac{C}{g}$ (bigger $R \cdot C$) will make the coefficient closer to 1 and make the integration more sticky, causing the voltage to rise slower. The measure of this stickiness is τ_{memb}, the time constant: $\tau_{memb} = \frac{g}{C} = R \cdot C$. Using analytic calculus, τ_{memb} can be shown to be the time required for the voltage to reach about 63% (or more precisely $1 - e^{-1}$) of its final value (see Chap. 16, Section 16.5). In Fig. 11.5, $\tau_{memb} = 1\ ms$.

In real life, we would get our parameters from measurements made in neurons. Capacitance and membrane conductance cannot be easily measured. However, both V_{max} and τ_{memb} can be measured using an electrode that injects current into a cell and measures voltage (current clamp). V_{max} is the maximum voltage deviation reached during a prolonged current injection. τ_{memb} is the time required for the voltage to reach $\sim$ 63% of V_{max}. Although g_{memb} and C_{memb} are the basic parameters used in the simulations, these are only estimates based on gross neuronal properties that can be directly measured.

Advantages and disadvantages of numerical integration

The differential equation of membrane charging is usually handled analytically, as shown in Chap. 16, Section 16.5. This allows us to directly calculate voltage as a function of time. In our specific example, the precise solution is $v(t) = 1 - e^{-t}$. When $t = 0$, $e^{-t} = 1$ and $1 - e^{-t} = 0$. As t gets bigger and bigger, e^{-t} gets closer and closer to 0, and $1 - e^{-t} = 0$ gets closer and closer to 1. The precise solution, the real solution, never reaches V_{max}. It just approaches it asymptotically, meaning it will just get closer and closer to V_{max} without ever reaching it.

A numerical integration is just an approximation to the real solution. In the present example, the numerical solution reaches V_{max}, while the real solution never does. Since the charging curve can be solved analytically, it is easy to check the numerical solution against the correct solution. However, most neural simulations can't be solved analytically. The true, precise solution of the equations is unknowable. This shouldn't bother us too much since the equations are themselves just an approximation of the physical reality. Approximating the approximation doesn't take us too much further away from the reality.

In the usual case, without an analytic solution to compare the numerical solution to, one has to do computational controls to be confident about the adequacy of the approximation. Usually this is done by rerunning the simulation with smaller Δt's to ensure that very similar curves result with increasingly accurate approximations. Additionally, one can redo the simulation with a different integration technique. Many such techniques are available. One other integration method, the implicit Euler technique, is given in Chap. 16, Section 16.5. Numerical integrations can and do fail. In our example, if we choose $\Delta t = 2$, bigger than the time constant, then the update equation is $V = -V + 2$, which gives voltages that just bounce back and forth between 0 and 2. This is an example of an unstable numerical solution that doesn't converge onto the correct solution. Instability often occurs when Δt is too big. Generally, a good Δt is a small fraction (e.g., 10%) of the time constant of the fastest process in a simulation. The choice can be tested by demonstrating that further reduction in Δt produces little change in the solution.

11.6 Time constant and temporal summations

In the first part of the book, we studied the simple signal transduction properties of artificial neural network sum-and-squash units (Chap. 6): step 1) multiplication of weight times state; step 2) summation; step 3) squashing to final output. All of this is much more complicated in real neurons. We now start to look at signal summation (step 2) in the context of a more realistic model. In artificial neural networks, everything took place in discrete time cycles so an input coming in at one time didn't add to an input coming in at another time. In real neurons, running in real time, a later input can sum with an earlier one. This is called temporal summation.

Real synaptic signals are usually conductance changes that lead to current flow. For the time being, we use current injections rather than conductance changes in order to simplify the model and make it easier to understand. We look at the summation of two brief current injection signals to see how parameter changes alter temporal summation.

The slow charging and discharging of the membrane is a major factor in temporal summation. The other major factor is the duration of the signal itself. Long-duration signals are more likely to add together with other signals. When the signal stops, the membrane does not suddenly drop back to zero but instead discharges with an exponential discharging curve similar to the charging curve, but upside down (Fig. 11.6). The time constant, which determines charging time, also determines discharging time. In Fig. 11.5, the current injection (the signal) lasted for the full 5 ms of the simulation.

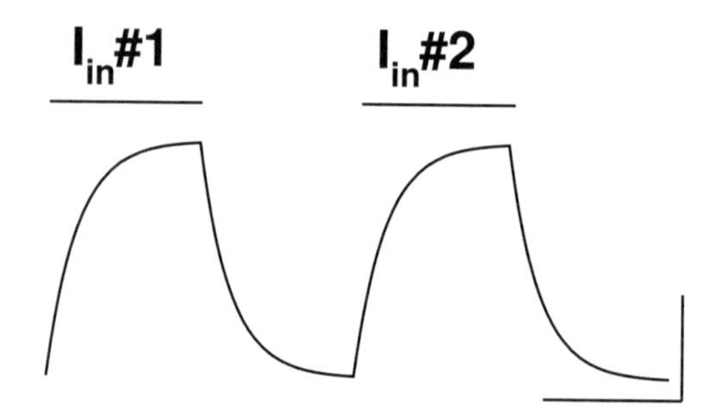

Fig. 11.6: Two brief current injections in a row. Membrane charges exponentially and then discharges exponentially. 5 ms, 1 μ A/cm^2 current injections start at $t = 0$ ms and $t = 10$ ms. Scale bar: 5 ms, 0.5 mV.

In Fig. 11.6, the current stops at $t = 5$ ms. At this time $I_{in} = 0$. The update equation becomes $V = 0.999 \cdot V$, with an initial condition of $V = V_{max} = 1$ mV. This gives an exponential discharging curve that has the same time constant as the charging curve and drops to 0 mV (asymptotic to 0 for the analytic solution).

In Fig. 11.6, the two current injections are separated in time so that the first doesn't appreciably affect the second. As in Fig. 11.5, the time constant, τ_{memb}, is 1 ms and V_{max} is 1 mV. After 5 ms ($5 \cdot \tau_{memb}$) the membrane is pretty much fully charged: $V = 0.9933$ mV. Then it discharges for 5 ms, ending up almost fully discharged: 0.0067 mV. At $t = 10$ ms, the second current injection starts up, sending the membrane potential on the same trajectory again. The simulation ends at $t = 20$ ms.

Each of the current injections in Fig. 11.6 is a signal. In this example, there is no appreciable summation. Bringing the two signals closer together may result in their adding up to a bigger signal. If we bring the two signals closer together in time, this is temporal summation. If we bring them closer together in space, this is spatial summation. Since this model (Fig. 11.4) is a point neuron, a single compartment with no spatial extent, there is no spatial summation. Using multicompartment models, we can get spatial summation as two signals arrive at different locations in the neuron (Chap. 13).

A variety of different measures can be considered as evidence of signal summation. The simplest is to look at the peak voltage. The signals are positively summating if the two signals together give a higher voltage than either one alone. Another measure is the average voltage or the integral of voltage with time (adding up the voltages at every time point). For now, we just use peak voltage, since it's easy to assess by eye, and is a better predictor of whether a neuron will spike.

As a first parameter exploration, I started the second signal at earlier and earlier times (Fig. 11.7). Because the second signal now starts before

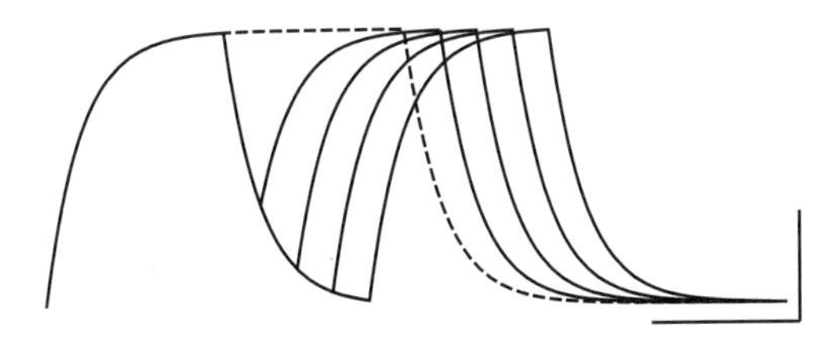

Fig. 11.7: Starting second signal during fall-off of first signal results in no temporal summation by peak criterion. Scale bar: 5 ms, 0.5 mV.

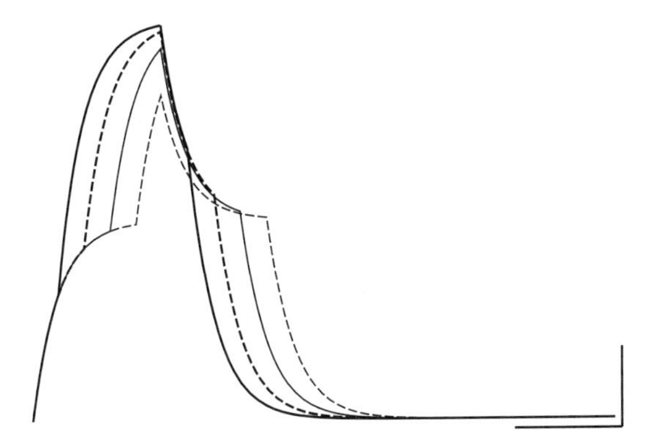

Fig. 11.8: Starting second signal during the first signal results in temporal summation. Different line types are used to tell the lines apart. Scale bar: 5 ms, 0.5 mV.

the first signal ends, the second signal takes off from a higher voltage. Despite this, it doesn't end up going any higher than otherwise. This lack of temporal summation occurs because the first signal is dropping off at the same rate as the second signal is increasing. When the second signal starts at $t = 5$ ms (dashed line), there is no drop-off between the signals. This is not surprising since the two signals are now just one continuous signal lasting 10 ms. The membrane charges over 10 ms instead of over 5 ms.

For a second exploration, I started the second signal still earlier so that there is overlap between the two signals (Fig. 11.8). During the period of overlap, the total current is twice what it would otherwise be and the membrane is charging toward twice the single-signal V_{max} ($V_{max} = 2 \cdot I_{in}/g$). The greater the duration of signal overlap, the greater the membrane depolarization. Since the membrane time constant is unchanged, the membrane will fully charge over 4 to 5 ms as before. Temporal summation is maximal for these two signals when the currents are coincident for that duration or longer.

Fig. 11.9: Reduction in conductance increases V_{max} and temporal summation. Second current injection starts at $t = 6$ ms; $g = 1$, 0.5, 0.2 mS/cm^2. Scale bar: 5 ms, 0.5 mV.

Third, I explored the effect of conductance change (Fig. 11.9). By decreasing g, we simultaneously increase the time constant and maximal response. Lower conductance (higher resistance) gives a bigger, slower response. For Fig. 11.9, I decreased g from 1 mS/cm^2 (lower trace) to values of 0.5 (middle) and 0.2 mS/cm^2 (upper trace). The second signal starts at $t = 6$ ms. The lower trace, identical to one of the simulations of Fig. 11.7, shows no temporal summation. By decreasing g, we prolong the time constant so that the membrane is not fully charged by the end of the first signal. Now when the second signal kicks in, there is room to grow and temporal summation occurs. The increased temporal summation is a direct consequence of the increase in τ_{memb}. We can demonstrate this by playing with g and C so as to independently manipulate V_{max} and τ_{memb}. This is the type of artificial exploration that cannot be done in real life, since there is no way to independently control membrane capacitance experimentally. A decrease in g with a proportional decrease in C will lead to a larger voltage response (increased V_{max}) but no increased temporal summation between the two signals (same τ_{memb}). An increase in C alone will increase τ_{memb} and lead to increased temporal summation, although the final voltage will be lower due to less complete charging during the time of current injection.

In Fig. 11.10, I used a positive followed by a negative current injection. Here the second signal was a current of -1 mA/cm^2, which by itself would give a down-going charging curve that would asymptote at -1 mV. This current injection was started at $t = 2$ ms, at which time the trajectory turns around and starts downward. The trajectory goes down still faster after $t = 5$ ms when the first current injection ends and the second is working unopposed. At $t = 7$ ms, the second current injection ends and the

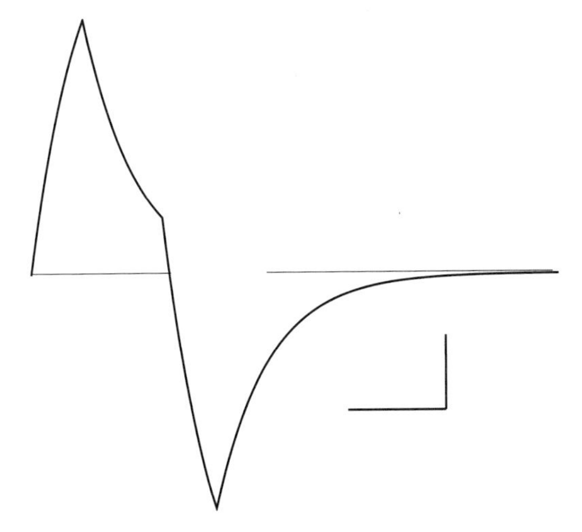

Fig. 11.10: Temporal summation of a positive and a negative signal. Negative signal is -1 mA/cm^2 and starts at 2 ms. Dashed line shows resting membrane potential. Scale bar: 5 ms, 0.5 mV.

negative membrane potential discharges back toward 0. This summation of positive and negative currents is analogous to certain types of inhibition, where a negative signal aborts or reduces the excitatory effect of a positive signal.

Figures 11.5 through 11.10 demonstrate several critical properties of temporal integration in passive neurons. The time constant determines how fast the membrane will charge (Fig. 11.5). The membrane will discharge at the same rate (Fig. 11.6). Even through the membrane has not fully discharged, onset of a second signal will not lead to a higher final voltage because the response to the first signal will fall off as fast as the response to the second signal comes on (Fig. 11.7). If the second signal comes on while the membrane is still charging, temporal summation will be seen (Fig. 11.8). The contrast between Fig. 11.7 and Fig. 11.8 demonstrates that current injection summation depends on the duration and overlap of the stimuli, not the duration of the responses. Decreasing g (Fig. 11.9) will increase both V_{max} and τ_{memb}, allowing inward currents to produce larger and longer depolarizations. Only the time-constant effect will increase the degree of temporal summation. Finally, hyperpolarizations can sum with depolarizations and cancel them out (Fig. 11.10).

11.7 Slow potential theory

In Chap. 7 we discussed rate coding, also called frequency coding. The rate-coding hypothesis states that the neural state can be measured at the axon as frequency of spiking. Rate coding uses the firing frequency of a presynaptic neuron as that neuron's output state. According to the basic artificial neural network model, this output state will be multiplied by a weight and added to the other *state·weight* products to produce the *total-summed-input* for the postsynaptic unit's squashing function (Chap. 6). To look at the artificial neural network model in the context of neural membrane theory, we need to find membrane-model parameters that correspond to the weight in the artificial neural network model. This weighting process will transduce presynaptic firing frequency to a postsynaptic value corresponding to the *state·weight* product. The obvious postsynaptic value for this purpose is membrane voltage. The voltage due to one input can add to voltages due to other inputs to give the *total-summed-input* needed for the squashing function. As we see in the next chapter, membrane voltage can also be transduced into firing rate (postsynaptic neuron state) through the dynamics of the Hodgkin-Huxley equation

To determine a firing frequency, one has to wait until at least two spikes have arrived. If we measure the time between two spikes and invert the interval between them, this gives us the instantaneous frequency. For example, if two spikes occur 2 ms apart, then the instantaneous frequency is $1/(0.002\ s) = 500$ Hz (hertz, the unit of frequency, is equal to events per second). In the context of rate-coding theory, instantaneous frequency is not a very useful measure because instantaneous frequency values tend to jump around a lot. Biological spike trains look noisy. Noise interferes with frequency estimation and makes it necessary to smooth out the noise by signal averaging.

Is this spiking irregularity really noise? What looks like noise may actually be some kind of uninterpreted signal that is being used by neurons in their calculations. This speculation is the province of an entirely different class of models (e.g., synchronization models) and won't be pursued further here. Rate-coding theory postulates that the apparent noise really is noise. So, for present purposes, noise is noise and that's that.

Noise means that instantaneous frequencies vary and do not consistently reflect the underlying average frequency. Frequency can only be reliably assessed after the arrival of several spikes. Spikes are counted over a period of time to determine average rate. This counting and adding up of spikes is signal integration. Slow potential theory describes how signal integration is performed as spikes trigger postsynaptic potentials (PSPs), which add up to produce a voltage that estimates the average presynaptic frequency. These PSPs are the slow potentials. They have to be relatively slow, long in duration, in order to give a reliable estimate. Temporal summation occurs as more spikes come in during the time while previous PSPs are still active (e.g., Fig. 11.8).

From listening to the radio, you know that amplitude modulation (AM) and frequency modulation (FM) are two methods of transmitting information using an oscillatory signal. Since biological spikes do not vary

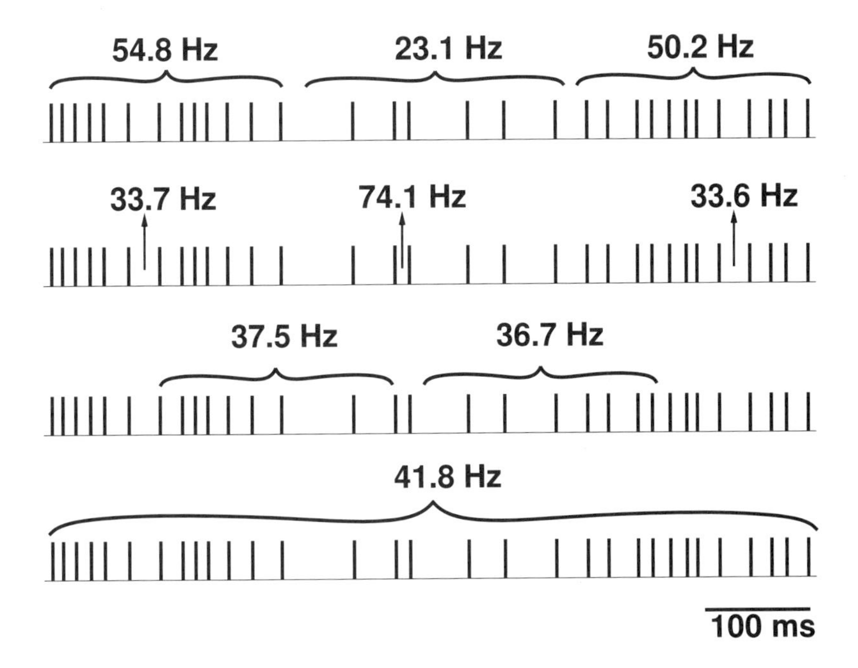

Fig. 11.11: Different frequency estimates from a noisy FM spike train. Top trace: estimates are fairly accurate reflections of the underlying 50→25→50 frequencies. Second trace: Measurements of instantaneous frequency can be highly inaccurate. Third trace: Averaging with correct duration but wrong phase (timing of start of average) also gives bad estimates. Fourth trace: averaging for too long a duration misses the modulation entirely.

meaningfully in height (Chap. 3), the system is not using AM. Rate coding theory assumes that the system is using FM. To describe an FM signal, we need to discuss two different frequencies: the carrier frequency of the spikes and the frequency at which the rate changes. In Fig. 11.11, I show a noisy spike train. The carrier frequency varies between 25 and 50 Hz. This frequency is modulated at a rate of 4 Hz, meaning that we have a shift in frequency every quarter second (250 ms — the length of the brackets at the top of Fig. 11.11). Clearly, the frequency of modulation must be considerably lower than the carrier frequencies.

In Fig. 11.11, rates for a single artificial noisy frequency modulated spike train are estimated correctly, and then incorrectly in several ways. Biologically, it is not possible to be confident of whether a particular signal estimation is correct or incorrect. In Fig. 11.11, I made the spike train, so I know what was signal and what was noise. This spike train was produced by a noisy frequency generator set to 50 Hz for 250 ms, followed by 25 Hz for 250 ms, followed by 50 Hz again for the last 250 ms. The top trace is the most accurate estimate of the underlying signal that can be obtained. Averaging is started and stopped at exactly the times of the generator's frequency shifts. However, instead of 50 Hz $\rightarrow$ 25 Hz $\rightarrow$ 50 Hz, the frequency estimates are 54.8 Hz $\rightarrow$ 23.1 Hz $\rightarrow$ 50.2 Hz. This discrepancy is due to the underlying noise and limited duration of sampling. If longer samples were available, the estimates would be more accurate. In the second trace, three instantaneous frequencies are chosen that give particularly inaccurate estimates of average frequency. Long intervals are measured during the fast firing at beginning and end. A short interval corresponding to 74.1 Hz is measured during the slow firing in the middle. In the third trace, reasonable averaging intervals are used but averaging is started and stopped at exactly the wrong time (180 degrees out of phase with the frequency modulation). As a result, the estimates miss the frequency shift and average across two different frequencies to give an intermediate frequency of about 37 Hz. In the fourth trace, the estimate also misses the frequency modulation by using too long an averaging time.

Averaging by adding PSPs

As you can see from the above example, accurate frequency interpretation depends on averaging over a reasonable duration and starting the averaging at the right time (right phase) to capture frequency shifts. Finding the right averaging duration or time constant depends both on the range of frequencies to be communicated and on the amount of noise that obscures them. The ideal time constant can be calculated using communication theory. Here, I'll just make note of some straightforward conclusions: 1) signals with low carrier frequencies require long averaging periods; 2) signals with a lot of noise require long averaging periods; 3) signals with rapid frequency modulation require short averaging periods. Noise can easily overwhelm signal in cases where a low carrier frequency is modulated rapidly.

In Figs. 11.6 to 11.10, I showed temporal summation of square waves and noted factors that would increase temporal summation. Artificial current injections from electrodes are square. Real biological signals are more curvy. The alpha function (Fig. 11.12) is a popular model for a PSP. The alpha function is parameterized as

$$I_{in} = I_{max} \cdot \frac{t}{\tau_\alpha} \cdot e^{(-(t-\tau_\alpha)/\tau_\alpha)}$$

This is a complicated equation that produces a simple curve. It rises quickly for τ_α time and then falls slowly over about 5 τ_α (a little slower than exponentially). Note that we now have a couple of time constants to discuss. We call the alpha function time constant τ_α and the membrane time constant τ_{memb}.

The time constant of the slow potential must be chosen so that the PSPs are long enough to average a reasonable number of spikes. Choice of too

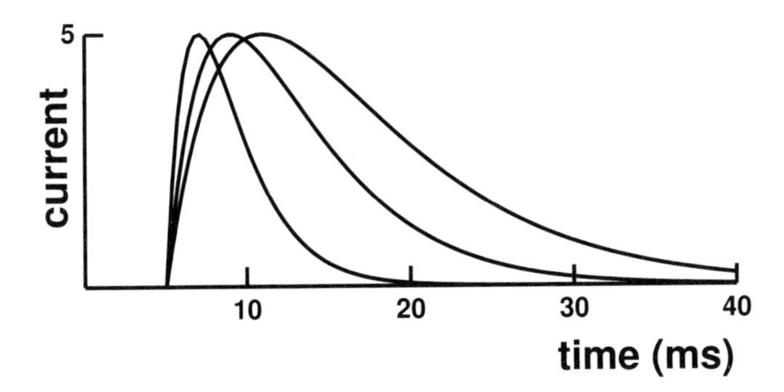

Fig. 11.12: Alpha functions with onset at 5 ms and amplitude of 5 μA/cm^2. Three curves with τ_α=2, 4, 6 ms. Note that τ_α equals the time from onset to peak. τ_α also determines decay time. This is used as a model of a postsynaptic potential (PSP).

short a τ_α will not allow integration to occur — the PSP will end after only one or two impulses so that only instantaneous frequency can be measured (e.g., Fig. 11.11, trace 2). An overly long τ_α will not allow enough PSP decay during the arrival of many spikes, and will therefore average across the frequency modulation (e.g., Fig. 11.11, trace 4).

For Fig. 11.13, I produced an FM signal with spike-train frequencies varied randomly between 20 and 100 Hz. The frequency changed every second (1 Hz modulation frequency) for 10 seconds. The carrier frequencies are shown graphically in the top trace with the values given in Hertz. I didn't show all the spikes since at this scale they would just scrunch together into an indistinguishable blob. However, at the bottom of the figure I expanded an 800-ms period to show the spikes and membrane response together.

With carrier frequency between 20 and 100 Hz, the interspike intervals (ISIs) ranged from 50 to 10 ms (period is inverse of frequency: ISI $= \frac{1}{f}$). A good choice of τ_α would be somewhere in this range; I used 30 ms. Membrane time constant, τ_{memb}, is 1 ms, which allows the membrane to follow the PSP without substantial lag and without adding any additional delay. Another way to do this would be to use a shorter τ_α but prolong the signal with a long τ_{memb}. It is more difficult to get this model to work well, since the longer τ_{memb} delays signal onset as well as signal offset.

Fig. 11.13 shows membrane potential in response to two spike trains with the same underlying frequencies but without and with noise. There is an initial charging period (arrow) when the membrane rises from resting membrane potential by about 3 mV. Then the potential plateaus in response to the constant-frequency input. When the input frequency shifts upward from 47 to 81 Hz there is another charging period as the membrane rises another 2 mV and plateaus. The correspondence between shifts

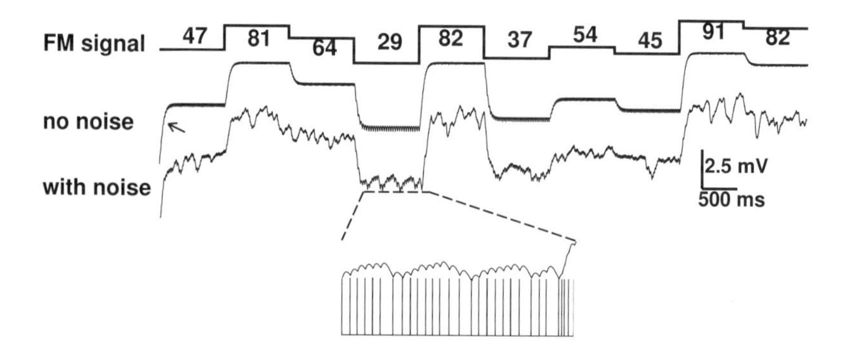

Fig. 11.13: Slow PSP response to FM square wave (carrier frequencies shown in Hz). PSP parameters: τ_α = 30 ms; τ_{memb} = 1 ms.

in frequency and shifts in membrane voltage can be easily appreciated. The scaling is arbitrary: the change in millivolts as a number is not the same as the change in frequency as a number. However, the relationship is linear: a doubling of input frequency leads to a doubling of potential. During the plateau there is a low-amplitude oscillation from the waveforms of the constituent alpha functions. In the absence of noise, membrane potential closely reflects frequency modulation, even reflecting small frequency shifts such as the shift at right from 91 to 82 Hz.

Once noise is added to the FM signal (Fig. 11.13, with noise), frequency estimation suffers. It is no longer possible to reliably identify frequency shifts of under 10 Hz. Even the shift from 81 to 64 Hz is hard to see. The expanded trace below shows the alpha function responses to individual spikes up to the shift from 29 to 82 Hz.

If we increase τ_α to 100 ms (Fig. 11.14), averaging occurs over a greater number of spikes and most of the noise is filtered out. As well as averaging over a greater period, the longer τ_α also produces longer charging delays. This means a longer wait for the voltage to stabilize on an estimate of the incoming signal (arrows). With a charging delay of nearly half a second and frequency modulation of 1 Hz (period of 1 second), the frequency estimate barely registers as a plateau before the frequency shifts again.

Notice that slow-potential averaging solves the problem of phase choice. As shown in Fig. 11.11, trace 3, one will obtain the wrong frequency average if averaging starts and stops at the wrong times. With slow potentials, averaging is going on continuously with activity in the remoter past gradually wearing out. Instead of starting and stopping at certain times, the sum of slow potentials is always most strongly influenced by the preceding

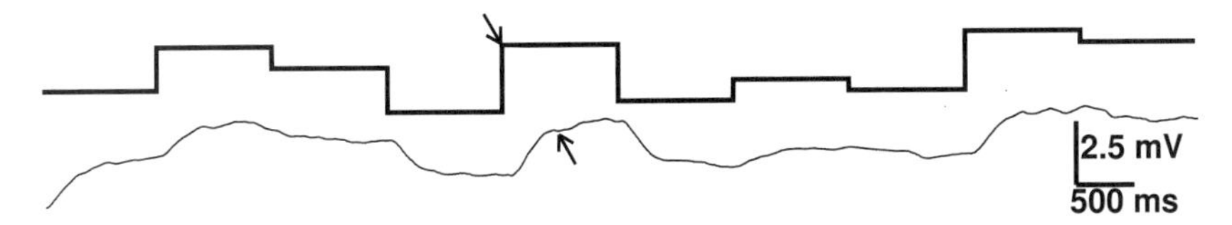

Fig. 11.14: Slower PSP response to noisy FM signal of Fig. 11.13. τ_α = 100 ms.

instantaneous frequency, less strongly by the one prior to that, and so on backward in spike history for the duration of the slow potential.

However, this figure also illustrates a problem with slow potential theory — it's too damn slow. Depending on the degree of noise, which will vary across different cell types in the brain, reliable frequency estimation can take upward of 100 to 500 ms. The route between perception and action in the spinal cord, thalamus, and brain encompasses many synapses. If each of these synapses requires prolonged slow-potential processing, it could take many seconds to react. This is plenty of time for a tiger to eat you (unless you ask that your tiger use rate coding as well).

11.8 Summary and thoughts

My recurring theme is that hardware determines software. In this chapter, I started with salt, water, and soap, the basic ingredients of the brain. Out of these, the body builds capacitors and resistors. In the next chapter we'll see that it builds batteries as well. The physical limitations of these building blocks makes neurons very slow compared to transistors. In particular, the relatively large capacitance translates into slow signaling. Rather than be dismayed by this slowness, I presented models that use it to advantage. The slowness allows the membrane to hold onto a signal, permitting temporal summation. The amount of summation has to do with the length of the membrane time constant. Slow potential theory is a model that makes slowness a feature, using long time constants to do signal averaging that blurs out noise.

This chapter showed how the parameters of the membrane and of the signal itself determine the influence a signal has on the neuron. These interactions explained signal transduction from presynaptic spike rate to postsynaptic membrane potential. They also explained the signal summation required to arrive at a total-summed-input. The next step of an artificial neural network update rule is signal transduction from total-summed-input to output state, in this case from membrane potential to spike rate. In the next chapter, we explore how the size of the membrane potential will determine the likelihood and frequency of neuron firing.

Although rate coding and slow potential theory are about the best we can do right now, I find them unsatisfying. As mentioned above, they are too slow. Also as I'm sitting and thinking, it bugs me to think that most of my substantial metabolic effort is just producing noise. Worse yet, I have to wait around just so I can ignore most of what my brain is doing. Anyway, I'm always complaining about other people using intuition to understand the brain, and here I am doing it.

12
Hodgkin-Huxley Model

12.1 Why learn this?

In the 1950s Alan Hodgkin and Andrew Huxley worked out the ionic basis of the action potential and developed a mathematical model that successfully predicted the speed of spike propagation. Their work can be regarded in retrospect as the beginning of computational neuroscience. It remains the touchstone for much neural modeling today. The Hodgkin-Huxley model demonstrates how computer models can reveal biological properties that cannot be examined directly.

Hodgkin and Huxley described two ion channels. Since then hundreds have been described and some of the basic parameterization has been updated. Despite this, the modeling techniques that Hodgkin and Huxley developed are still used and remain the standard model today.

12.2 From passive to active

The simulations in Chap. 11 used a simple RC circuit. This is a passive membrane model because the conductance remains constant. In this chapter, we use the same passive components (the R and the C), and add active components as well. Active components are conductors that change their conductance in response to changes in membrane voltage or due to activation by a chemical. Voltage-sensitive channels are responsible for the action potential. Chemical-sensitive channels are responsible for synaptic activation in response to the arrival of a ligand at a synapse. In general, voltage or chemical-sensitive channels are considered active channels because they are activated in response to some signal. Channels that remain at a fixed conductance are called passive channels.

The resting membrane potential is about −70 mV

The simulations in the previous chapter were simplified by starting at a resting potential of 0 mV. In real life, the membrane rests at a negative potential of about −70 mV (different cells differ). This is the resting membrane potential (RMP) of the cell. The resting potential is set up by pumps that separate charge across the membrane, producing various fixed potentials associated with different ions. This charge separation will be represented by batteries in the electrical circuit diagram. They are denoted the sodium battery, the potassium battery, etc. The polarity of the individual batteries depends both on the polarity of the ion involved and on the direction of the inside-outside concentration gradient for that ion. Each battery will only affect membrane voltage when the ion channels open for that particular ion. In the circuit diagram, this is represented by connecting each battery through a variable conductor.

The membrane is insulator, capacitor, and battery

The addition of batteries and variable conductances to the circuit diagram demonstrates several other roles played by the cell membrane and by membrane proteins. As part of being an insulator and a capacitor, the membrane also allows the charge separation that sets up the batteries, each of which is associated with a different ion. Protein transporters pump the ions to charge each of these batteries. Active protein channels form the variable conductors (rheostats). These proteins will be sensitive to changes in membrane voltage or to chemicals or both.

The resting (inactive) potential on the membrane is negative. Hence, both negative-going inhibitory signals and many positive-going excitatory signals will be negative relative to ground. There is a standard nomenclature to describe voltage deviations from rest (Fig. 12.1). Negative deviations, which make the membrane even more negative that at rest, are called hyperpolarizing (*hyper* means more). Hyperpolarizing inputs are generally inhibitory. Positive deviations, which make the membrane less negative than it is at rest, reducing its polarization, are called depolarizing. Depolarizing signals move the membrane potential toward or past 0 mV. Depolarizing inputs are generally excitatory.

Hyperpolarization and depolarization are not symmetrical. The membrane can be naturally depolarized by about 120 mV. 120 mV above rest is +50 mV relative to ground ($-70 + 120$), approximately the value of the sodium battery. Natural activity will only hyperpolarize the cell by about

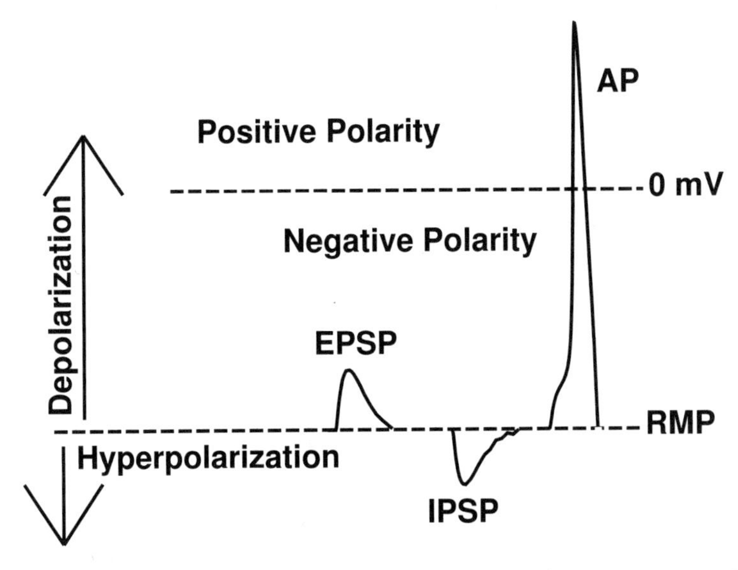

Fig. 12.1: Resting membrane potential (RMP) is typically about −70 mV (inside negative). Membrane can be depolarized as much as 120 mV or hyperpolarized as much as 30 mV from rest. Excitatory postsynaptic potentials (EPSPs) depolarize; many inhibitory postsynaptic potentials (IPSPs) hyperpolarize. Action potentials (APs) are depolarizations that can overshoot 0 mV, temporarily reversing membrane polarity.

20 to 30 mV. 20 to 30 mV below rest is about −90 to −100 mV relative to ground ($-70 - 30 = -100$). This is about the range of values for the potassium battery, which may differ somewhat among different cell types. Experimentally, one can hyperpolarize the membrane beyond −100 mV by injecting negative current through an electrode inside of the cell. Artificial depolarization with injected current is limited by the tendency of prolonged depolarization to kill the cell.

Synaptic inputs aren't current injections

In the simulations in Chap. 11, I started by using square wave current injections as signals. These were similar to the current injections used by physiologists. Then I moved on to the alpha function as a more natural signal. I was still using current injections. Most synapses in the nervous system are chemical synapses. The associated synaptic potentials, the natural input signals for neurons, generally arise as conductance changes rather than current injections. The current that flows is secondary to the conductance change. (There are also electrical synapses where current flows through a channel connecting two neurons.) As shown in Fig. 11.2, a chemical ligand (neurotransmitter) is released presynaptically. It floats across the synaptic cleft and binds to a receptor on the postsynaptic membrane. This receptor is connected either directly or indirectly to one or more ion channels that conduct current across the membrane. Depending on which ions the channels allow through, the synaptic current can be inward (depolarizing) or outward (hyperpolarizing). In our electric circuit diagrams, this will correspond to connecting up a particular battery by activating a switch or a controllable conductance (a rheostat).

12.3 History of the action potential

The story of the action potential starts with Luigi Galvani's 1791 discovery that electrical signals from lightning or primitive batteries could cause contraction of the leg of an otherwise dead frog leg. This was an inspiration to mad scientists, and researchers started performing similar experiments on the heads of hanged criminals, attempting to bring back the dead with electricity at around the time that *Frankenstein* was written. The National Institutes of Health do not encourage this line of research nowadays. Perhaps these demonstrations had some value in that they suggested that mysterious human abilities such as motion, sensation, and thought could be caused by a physical process. On the other hand, connecting thought to electricity may not have been so disturbing at that time, since electricity was itself another mysterious process that could be readily equated with an unknowable "life force."

A later finding that brought neural function further into the physical, non-ethereal realm was the demonstration of neural delays by Hermann Helmholtz, the famous 19th century scientist. Helmholtz showed that stimulation of a nerve at different points led to contractions of the corresponding muscle at measurably different times. Not only was the speed measurable, but it was relatively slow (about 70 mph). Helmholtz's father wrote him a letter rejecting these findings as absurd. Helmholtz's father was sure that he could directly perceive the world, and that his thoughts were converted directly into actions. If he moved his hand and watched it, perception didn't fall behind. Research into the seeming seamlessness of experience is now actively pursued using careful timing of perception and brain electrical activity. An understanding of how the illusion of simultaneity arises despite varying delays of sensory signal processing would provide a partial solution to the mind–body problem.

Hodgkin and Huxley

By the time Hodgkin and Huxley got to the problem, much progress had been made. The resting potential was well described and the action potential had been described as a "negative variation" in this potential. (We now describe the action potential from the inside as a positive variation: positive inside, negative outside.) The squid axon had been picked out as an ideal experimental preparation due to its enormous size compared to other neurons. Squids, like other invertebrates, have unmyelinated axons — wires without insulation. Their axons are leaky and prone to signal loss, which has caused them to evolve extremely broad axons that conduct electricity better. From the experimentalist's point of view, this makes them big enough to see and to stick wires into.

Hodgkin and Huxley threaded a fine silver wire inside the axon. With this, they could measure the electrical potential inside and deliver enough current so as to maintain a particular voltage despite the efforts of the ion channels in the membrane to change it. This is called voltage clamp. They measured how much current was required to keep the voltage from changing. This told them how much current was being passed through the axon membrane and in which direction. They could then figure out which ions were responsible for which currents by doing the same experiments in sodium-free or potassium-free solutions. By running these experiments at many different voltages, they found out how the sodium and potassium currents grew and shrank with changes in membrane potential. They used these data to construct the parallel-conductance model (Fig. 12.2).

12.4 The parallel-conductance model

The parallel-conductance model is similar to the basic RC model of Fig. 11.4. Once again all points on the inside of the membrane are electrically connected via the cytoplasm (horizontal line at bottom). This is the point where we measure potential. The outside of the membrane is connected via the extracellular fluid (horizontal line at top) and is grounded, keeping it at 0 mV.

The circuit

The inside and outside of the membrane are connected via four parallel conducting pathways. On the left side are the membrane capacitor and the fixed membrane conductance. These two passive components are similar to those of the simple RC model. However, the resistor in this case is connected to a battery. Batteries are represented by two parallel lines of different lengths. The long line of the battery schematic indicates the positive pole. The passive conductance in the parallel-conductance model is known as the leak conductance. Because the channels carrying this current are not voltage-sensitive, the leak conductance remains the same at any voltage, providing a constant "leakiness" for current. The potential of

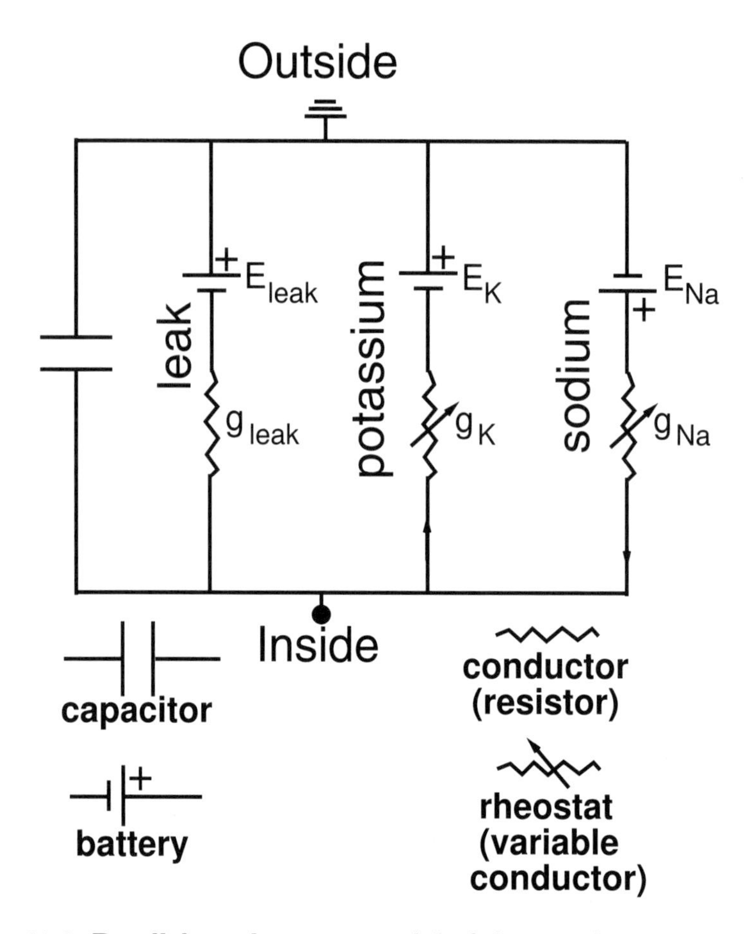

Fig. 12.2: Parallel-conductance model of the membrane. The various variable conductors (variable resistors, rheostats) connect the inside and outside. They are arrayed in parallel.

the battery associated with g_{leak} is E_{leak}. (E for electrical potential and V for voltage are synonymous but E is usually used for batteries.) The short line of the leak battery is connected to the inside of the membrane making the membrane inside-negative. E_{leak} is the major determinant of resting membrane potential.

On the right side of the circuit diagram in Fig. 12.2 are the two active branches: the sodium battery and conductance, and the potassium battery and conductance. Note that these two batteries are pointed in opposite directions. The potassium battery, like the leak battery, will make the membrane negative. The sodium battery will make the membrane positive. The conductance symbol under each battery has an arrow through it. This means that it is a variable (or controllable) conductor, also called a rheostat. The rheostat is the thing you turn to dim the room lights to set a romantic mood. In this case, the rheostats will be controlled not by the level of romance but by the level of the membrane voltage. Since the rheostat influences membrane voltage and the membrane voltage influences the rheostat, this will lead to positive or negative feedback loops, as will be described below.

At rest, the sodium and potassium conductances are turned off so that these two lines are not conducting. Under these circumstances, the two associated batteries have no effect on membrane voltage. If one of these conductors were to be turned all the way on (zero resistance), then the associated battery would dominate the membrane potential. If both of these conductors were turned on all the way at the same time, you would have the situation that you get when you connect the leads wrong while jumping a car battery — the battery will discharge massively, overheat, and blow up. Luckily, this doesn't happen in the brain.

Currents

Benjamin Franklin defined current as the flow of positive charge. In electronic equipment, current through wires is carried by negative electrons. Therefore, the direction of current in wires is opposite to the direction that charge flows. In biology, current is carried by ions from dissolved salts that move through water. Most of the ions involved are positive, like sodium (Na^+), potassium (K^+), and calcium (Ca^{++}). Chloride (Cl^-), a negative ion, is also important. Positive ions, with one or more superscript +, are called cations; negative ions, with superscript −, are called anions. Because calcium has twice the charge of sodium, movement of calcium ions will result in twice as much current. When dealing with cation flux, the direction of ion movement is the same as the direction of current.

Current is measured during voltage-clamp. To clamp a constant voltage onto the membrane, current is injected or withdrawn from the inside of the cell through an electrode. This current must exactly cancel out any currents that are passing through the membrane in order to prevent these membrane currents from changing the membrane potential. Current direction is defined with respect to the membrane, not the electrode. Inward current is positive charge going across the membrane from outside to inside. Outward current is positive charge going from the inside of the membrane to the outside.

Interpreting voltage-clamp results is a bit of an art. Sodium (positive ions) passing into a cell or chloride (negative ions) passing out of a cell would both be examples of inward current. Since sodium is at a higher concentration outside of the cell, an increase in sodium flux will result in an inward current. Potassium, being at higher concentration inside the cell, would be a typical outward current. When the membrane is stable (whether at RMP or at some other voltage imposed by voltage clamp), there is some balance of ongoing outward and inward currents. In addition to an increase in these baseline currents, it is possible for an experimental manipulation (a drug or a voltage step) to produce a decrease in the baseline currents. An increase in potassium current is an outward current, but a reduction in the baseline potassium current is measured as an inward current. Chloride reversal potential is near to RMP. Therefore, a chloride flux can be either inward or outward. Because chloride carries a charge of -1, a decrease in inward chloride flux is an inward current.

Calculations

The calculations for the parallel-conductance model are similar to those for the RC model except that we have to add in the batteries. As in the RC model (Fig. 11.4), Ohm's law gives a resistive current equal to conductance times membrane voltage. As before, this can be written as either $I_R = g \cdot V_{memb}$ or $V_{memb} = I_R \cdot R$. A battery in series with the conductor gives a voltage boost on that line. For the leak conductance: $V_{memb} = E_{leak} + I_{leak} \cdot R_{leak}$. E_{leak} is the value of the leak current battery, usually about -75 mV. We need to add up currents (Kirchhoff's law), so we turn the equation around to give $I_{leak} = g_{leak} \cdot (V_{memb} - E_{leak})$. Similarly, $I_{Na} = g_{Na} \cdot (V_{memb} - E_{Na})$ and $I_K = g_K \cdot (V_{memb} - E_K)$.

All of the currents add up to zero: $0 = I_C + I_{Na} + I_K + I_{leak}$. Therefore, $-I_C = I_{Na} + I_K + I_{leak}$. Substituting for the currents gives the parallel-conductance equation:

$$
\begin{aligned}
-C\frac{\Delta V}{\Delta t} &= g_{leak} \cdot (V_{memb} - E_{leak}) \\
&+ g_{Na} \cdot (V_{memb} - E_{Na}) \\
&+ g_K \cdot (V_{memb} - E_K)
\end{aligned}
$$

Because $I_C = C \cdot \frac{\Delta V}{\Delta t}$, positive capacitative current is a depolarizing current that makes the inside of the membrane more positive. In the parallel-conductance equation, as in the RC model, capacitative current is opposite in sign from the conductive currents. Therefore, a negative conductive current is a positive capacitative current and produces depolarization. Negative conductive currents are inward currents, involving the flow of current through ion channels from outside to inside. Notice that negative current has nothing to do with the sign of the ion that is carrying the current. Instead, it is an indication of the direction of current. Also notice that the sign change is confusing: a negative membrane current produces a positive voltage effect. The direction of negative current is an arithmetic consequence of measuring membrane voltage on the inside rather than the outside. In the literature, the phrase "membrane current" is used as a synonym for conductive current. Therefore, negative current flows in and depolarizes; positive current flows out and hyperpolarizes. However, it's worth remembering that there is zero total current flow: the inward conductive current is matched by outward capacitative current. It is actually the latter that is most closely associated with the depolarization.

At steady state, there will be no capacitative current since voltage is not changing: $0 = I_{Na} + I_K + I_{leak} = g_{leak} \cdot (V_{memb} - E_{leak}) + g_{Na} \cdot (V_{memb} - E_{Na}) + g_K \cdot (V_{memb} - E_K)$. Solving this equation for V_{memb} gives the resting membrane potential:

$$V_{memb} = \frac{g_{leak} \cdot E_{leak} + g_{Na} \cdot E_{Na} + g_K \cdot E_K}{g_{leak} + g_{Na} + g_K}$$

This is a version of the Goldman-Hodgkin-Katz (GHK) equation. It looks complicated but it just says that steady-state membrane voltage will be the weighted sum of the batteries, with the weighting provided by the conductance associated with that battery. Since g_{leak} is the dominant conductance at rest, it will have the greatest effect on determining RMP. If a conductance is turned off completely (e.g., $g_{Na} = 0$), the corresponding battery has no influence. If, on the other hand, a conductance is very high, then the other batteries will have very little influence, e.g., if $g_{Na} >> g_K$ and $g_{Na} >> g_{leak}$, then $V_{memb} \sim \frac{g_{Na} \cdot E_{Na}}{g_{Na}}$, hence $V_{memb} \sim E_{Na}$.

Where do the batteries come from?

The batteries are an indirect result of proteins that pump ions across the membrane. These ions then try to flow back "downhill," in the direction of their chemical gradient from high concentration to low concentration. Only a little current has to flow in order to set up an equal and opposite electrical gradient. The electrical gradient, *opposite* in direction to the chemical gradient, is the battery. This electrical potential is called the Nernst potential. It can be precisely calculated by knowing the concentrations of a particular ion inside and outside of the cell (see Glossary for definition). Each ion has its own Nernst potential. The value in millivolts of the Nernst potential is the strength of the battery that we use in the circuit diagram.

With its many ins and outs, the origin of the Nernst potential can be confusing. In Fig. 12.3, I show the origin of the sodium battery. Sodium is pumped from inside to outside (#1 in Fig. 12.3) by a protein that uses energy from ATP. The pumping leaves sodium concentration outside of the cell ($[\mathrm{Na}]_\mathrm{o} \sim 140$ millimoles) higher than it is in the cytoplasm ($[\mathrm{Na}]_\mathrm{i} \sim 10$ millimoles). The concentration difference across the membrane does not in itself lead to any *charge separation*, since sodium ions on both sides are appropriately matched with negatively charged proteins. Since there is more sodium outside, it "wants" to flow inside due to diffusion (#2 in Fig. 12.3). (Diffusion is what makes a drop of ink spread out in a glass of water; it wants to go where no ink has gone before.) As long as the selective channels for sodium remain closed, sodium cannot diffuse and the sodium concentration gradient has no effect on membrane potential.

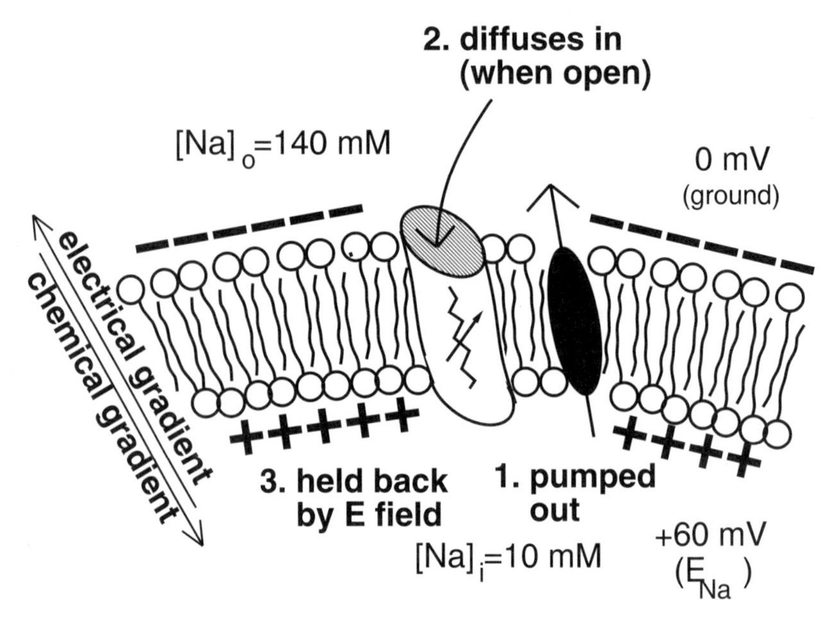

Fig. 12.3: The origin of the Nernst potential for example of Na^+. Na^+ is pumped out by an energy-consuming protein, flows back in through Na^+ channels, and is pushed back as it creates its own electric field. The sodium flux eventually reaches electrochemical equilibrium at a potential of about 60 mV. The superimposed rheostat sign is meant to indicate that the channel can be open or closed. This is a mixed metaphor: the symbol belongs to a different model.

When the sodium channel opens, sodium rushes down its concentration gradient. The negative proteins that are paired with the sodium ions cannot follow; they are not allowed through the sodium channel. This diffusion of sodium across the membrane leads to charge separation across the membrane, with unmatched sodium ions on the inside and unmatched negative protein molecules on the outside. The unmatched sodium ions inside the membrane will stay near the membrane, in order to be close to their lost negative brethren. This bunching of positives next to the inside of the membrane, with a corresponding bunching of negatives next to the outside, creates an electric field (#3 in Fig. 12.3) that opposes inward diffusion through the ion channels. This outward electric field is the sodium battery. The *inward* diffusive force and the *outward* electrical force reach a steady state (Nernst equilibrium) so that there is no net flow of ions and little need for continued pumping to maintain equilibrium.

The concentration difference between inside and outside can be directly translated into an electrical potential by using the Nernst equation. E_{Na} is approximately +60 mV. The positive plate of the sodium battery in Fig. 12.2 will be 60 mV relative to the negative plate. By contrast, potassium is at high concentration inside and low concentration outside. The potassium chemical gradient is outward so the electrical gradient is inward. The positive inward electrical gradient would be +90 mV if measured from the outside of the membrane, relative to a grounded inside. However, we always measure the potential on the inside, relative to ground outside, so the potassium potential (E_K) is about −90 mV.

All of the reversal potentials can vary slightly in different cells (different pumps) or under different conditions. For example, the sodium reversal potential can be slightly lower if you're sweating a lot: less sodium outside, less inward chemical gradient, less outward electrical gradient. The potassium reversal potential gets less negative under conditions where cells fire a lot. The constant firing allows potassium to build up outside of the cells. The accumulation of extracellular potassium reduces the outward chemical gradient, reducing the inward electrical gradient. Because resting membrane potential is largely determined by baseline potassium flux (the leak conductances conduct primarily potassium), this change in E_K changes E_{leak} and depolarizes cells. Even cells that have not themselves been firing are affected. In this way, extreme activity in one set of cells can depolarize neighboring cells and make them more excitable. This is one factor that can lead to the spread of uncontrolled activity in seizures.

12.5 Behavior of the active channels

To complete the Hodgkin-Huxley model, we have to describe the behavior of the sodium and potassium channels. We continue to look at the model from several different descriptive perspectives. First, we presented the electronics perspective and considered the channels as rheostats. Second, we talked about ion channels from a biophysical viewpoint, describing flux and electrochemical equilibrium. Third, we will describe the interaction of the channels in terms of positive and negative feedback systems. Fourth, we will return to ion channel behavior, describing their conductance properties in terms of somewhat mythical "particles." Fifth, we will study the equations that model the system. Sixth, we will run the simulation and demonstrate the model graphically.

This mess of descriptions includes different levels of organization (single channel vs. whole membrane), different approaches (electrical engineering vs. biology), and different methods of presentation (numerical vs. graphical). Doing modeling from so many directions would seem to make things harder rather than easier. It means that we have to not only understand concepts in biochemistry, biology, and electronics, but also be familiar with various tools provided by math and computer science. Additionally, we will see that not all of the descriptions are fully consistent with one another. For example, electronically we described the sodium channel as a rheostat. However, we also describe the sodium channel as having little on and off switches, making the channel sound more like a tiny subcircuit than like a rheostat.

The payoff for all of this hard work can be appreciated by considering the famous parable of the blind men who meet an elephant (one feels the trunk, one a leg, another a tusk, etc., and they give widely discrepant descriptions of the beast). None of them knows anything useful about elephants but, if they pool their knowledge, they may be able to create a passable picture. Similarly, coming at a model from many angles permits us to come closer and closer to understanding the thing itself. We have neither the concepts nor the mental capacity to allow us to wrap our brains around all of this complexity, and see nature as it really is. Instead we use these different models as tools to pick up different clues to this underlying reality. As we move back and forth between representations, we gain further insights.

There are other approaches to describing the Hodgkin and Huxley model, in addition to five or six that I use in this chapter. For example, one can use complex graphical representations that allow us to look at several dimensions of the dynamics at once. This is called a phase-plane representation. There are also other kinds of mathematical tricks that can help us understand the system better. An example of this would be descriptions of nullclines and of the space of solutions as a field. Going beyond the scope of the Hodgkin and Huxley model, we could look at levels of detail that are not directly considered in the model: we could describe the detailed molecular conformation changes that determine channel opening and closing or include the detailed physical chemistry of how salts and water interact.

Feedback systems

The Hodgkin and Huxley sodium and potassium channels are voltage-sensitive conductances. They go on and off with voltage change. As we describe the cycle of the action potential, we can speak in terms of positive feedback and negative feedback. Positive feedback occurs when the cycle of influence causes a change to produce more change in the same direction. If, for example, the faster you drive the more excited you get about driving fast (a positive feedback loop), you will tend to drive faster and faster. Positive feedback systems are not self-limiting; they are limited by something outside of the positive feedback system (i.e., a crash) or the opposition of some negative feedback system (e.g., prudence or the police). Negative feedback systems are self-limiting, since the cycle of influence produces a change that opposes the original change. Negative feedback is very common in biological systems: hunger makes you eat, food reduces hunger.

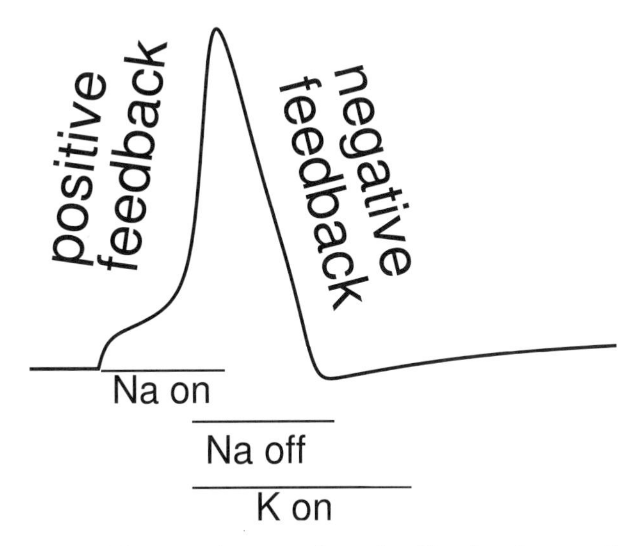

Fig. 12.4: Positive and negative feedback shape the action potential.

The Hodgkin and Huxley model uses two types of ion channels that are controlled by three types of switches. Turning on the sodium channel pushes voltage up (depolarizes) and turning on the potassium channel pulls it back down (hyperpolarizes). The channels are controlled by voltage. The sodium channel is controlled by both on and off switches. Increasing voltage turns on the sodium channel. This is a positive feedback loop. Increasing voltage also turns off the sodium channel, by means of a different switch. This is a negative feedback loop. The potassium channel only has one type of switch. Increasing voltage turns on the potassium channel, hyperpolarizing the membrane and providing a negative feedback loop.

The positive and negative feedback loops happen at different rates in the Hodgkin-Huxley model (Fig. 12.4). Initially, the action potential is triggered by a depolarization (movement of the potential in a positive direction, toward 0 mV). This can be a result of a current injection or a synaptic potential. This depolarization causes the sodium channel to switch on quickly. Current through the sodium channel will cause more depolarization, which will turn on the sodium channel more, which will cause more depolarization, which will turn on the sodium channel more ... This is the positive feedback that produces the upsweep of the spike as voltage rises rapidly.

The sodium channel's on-switch provides the positive feedback by activating the channel with depolarization. The off-switch provides the negative feedback by inactivating the channel during depolarization. While the sodium channel is being switched on, it is being more slowly switched off. The delay in negative feedback allows potential to rise about 100 mV over about a millisecond. Then negative feedback kicks in. The action potential reaches its peak as the sodium channel switches off and the potential heads back down toward resting potential. Further negative feedback is provided by the turning on of the potassium conductance. This connects the membrane to the negative potassium battery, which opposes the positive sodium battery.

Particle duality

Hodgkin and Huxley emphasized that their model was empirical, meaning that they matched the behavior of the action potential without needing to know what exactly was going on at the molecular level. They had no reason to expect that the details would correspond to what actually exists in the membrane. However, the model did provide a set of successful predictions at the membrane level. Most notably, the model predicted the existence of ion channels turning on and off independently. There was no way to demonstrate these ion channels until patch clamping was developed decades later.

Hodgkin and Huxley also had an implicit view of the functionality of switches controlling the ion channels. They called these switches *particles.* At the single-channel level, each particle was binary, taking on a value of 0 for blocking and 1 for nonblocking. The implicit concept portrayed the particles as physical objects that could block a single channel and prevent flow of current. Both the sodium and the potassium channel were described as having four particles. In the case of the potassium channel, each of the four particles behaves identically. Each of these potassium particles is called n. Mathematically, the presence of four particles per channel is denoted by a multiplication: $n \cdot n \cdot n \cdot n = n^4$. In the case of the sodium channel there are 3 m particles and one h particle, giving $m^3 \cdot h$.

When considering the single-channel level, each particle is binary-valued. However, at the population level, each particle takes on analog values between 0 and 1. This binary-analog particle duality is the kind of anomaly that can crop up as we move from one view of a model to another. Looked at from the bottom-up view, the particles appeared as little independent objects that could either block the channel and prevent conduction or get out of the way and allow conduction. If any particle is blocking a channel, that channel is closed and there is no flow through it. This view of the model gives the binary values: 1 for not blocking and 0 for blocking.

Looking at the model from a higher level (perhaps not high enough to be top-down, maybe middle-sideways), the particles in the Hodgkin-Huxley model represent a population and are represented by analog values. At the single-channel level, a channel is either opened or closed, 1 or 0. When we consider a large population of channels, some percent of this population will be open and the rest closed. We use an analog value from 0 to 1 to denote the percent of channels that are open. Similarly, particles take on a binary value, blocking or not blocking, when considered at the single-channel level but an analog value when considering the entire channel population. This analog value for the particle represents an unblocking probability or, equivalently, the percent of particles that are not blocking at a given time. Hence, to model a population of channels, we end up modeling a "population" of m particles, a "population" of h particles, and a "population" of n particles. This doesn't really make much sense because an m particle belongs to a channel. There is physically a population of channels, but not an isolable population of m particles.

This binary-analog particle duality is an example of a common problem in modeling. An aspect of a model that makes literal sense at one level is manipulated for computational or organizational convenience at another level. Or, as in this case, an aspect of a model that was arrived at empirically reveals interesting predictions when its detailed implications are looked at. Either way, the back and forth between levels alternately obscures and illuminates. In the present case, binary particles are conceptually helpful at the single-channel level, making predictions about what may be going on in individual channel molecules. Although the prediction of a physical blocker has not been borne out for the particular channels that Hodgkin and Huxley studied, other potassium channels have been described in which a piece of a protein physically occludes the channel.

At the membrane level, individually modeling a vast population of individual channels with individual particles was hopeless in Hodgkin and Huxley's day. It is computationally feasible nowadays but would not be practical for large simulations. The m population makes little sense at the channel level but describes the action potential extremely well. Thus, this compromise is both useful and used.

Particle dynamics

The population of m particle moves gradually from near 0 toward 1 during depolarization. This turns the sodium channel on. At the same time as the sodium channel is being turned on through the movement of the m particle from 0 to 1, it is also starting to turn off through the movement of the h particle from 1 to 0. h changes more slowly than m. Eventually h approaches 0 and sodium channel conductance drops to a small value. The n particle goes from 0 to 1 at about the same rate at which h goes from 1 to 0. This increases the potassium conductance, giving the potassium battery more influence over the circuit. The potassium current opposes and eventually reverses the spike depolarization.

m moves quickly to unblock the sodium channel, while h moves slowly to block it and n moves slowly to unblock the potassium channel. Since m moves so much faster than anyone else, sodium channel unblocking dominates at the beginning. m increases, unblocking the channel and providing the positive feedback. At the same time, the slower n movement is leading to some unblocking of the potassium channels, increasing potassium current and reducing this depolarization through negative feedback.

During the action potential downswing, everything reverses. m heads toward 0. n heads toward 0 more slowly. h goes back toward 1.

12.6 The particle equations

Now let's look at the equations that describe m, h, and n. I show the underlying functions graphically here; the full set of functions is given under "Hodgkin-Huxley equations" in the Glossary. Conveniently, the same basic form is used for all of the particles. The value of the sodium and potassium conductances are products of a maximal conductance value and the values of their associated particles. Conductance is represented by g and maximum conductance by $\overline{g}$, which is called either "g-bar" or "g-max." For sodium, conductance $g_{Na} = \overline{g}_{Na} \cdot m^3 \cdot h$. For potassium, $g_K = \overline{g}_K \cdot n^4$. If all the particles were set to 0, then $g_{Na} = g_K = 0$ and we would be left with only the passive membrane. If all the n particles were set to 1, then potassium current would be maximal: $g_K = \overline{g}_K$. If all the m particles were set to 1, but the h particle was 0, then g_{Na} would still equal 0. As we will now see, the parameterizations for these particles involve asymptotes that do not allow them to ever reach 0 or 1, but they can get close.

Each particle is indirectly parameterized by voltage. At the negative resting membrane potential, the m particle of the sodium channel is near 0 (off) and the h particle is near 1 (on). During the action potential m will go from 0 toward 1 and h will go from 1 toward 0. m will change faster than h will. We are dealing with rates of change, so once again we use calculus and differential equations to describe the process.

The discrete form of the differential equation for m is $\tau_m \cdot \frac{\Delta m}{\Delta t} = m_\infty - m$, a standard form for a first-order differential equation. The passive membrane equation, introduced in Chap. 11, can also be written using this standard form: $\tau_{memb} \cdot \frac{\Delta V}{\Delta t} = V_{max} - V$. In the charging curve of Fig. 11.5, we noted that V approaches V_{max} as the steady state. Similarly, m will approach m_∞. (The "∞" denotes steady-state — the value that would be reached after an infinite amount of time.) In the case of the charging curve, we noted that τ_{memb} determines how fast the membrane will approach V_{max}. Similarly, τ_m determines how fast m will approach m_∞.

As with the charging curve, the solution to the differential equation for m will be an exponential. However, in contrast with the fixed parameters of the passive membrane equation, the parameters of the particle equations are not constant. They are functions of voltage (Fig. 12.5) and the voltage keeps changing. Let's take m as an example. Since voltage is changing, m is chasing a moving target, m_∞. If the membrane is being depolarized, m_∞ is getting bigger and m is moving toward a bigger number. τ_m is also changing

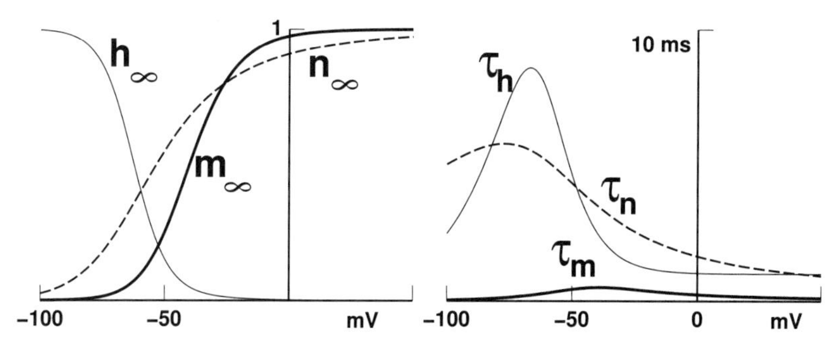

Fig. 12.5: Control parameters $m_\infty, h_\infty, n_\infty, \tau_m, \tau_h, \tau_n$ are themselves parameterized by voltage. (τ values are given at a squidish 6.3° C.)

with voltage, speeding up or slowing down the rate at which m is moving toward m_∞. τ_m triples with the first 30 mV of depolarization from rest, slowing down the rise of m and thereby providing another, relatively minor, negative feedback. The equations for h and n are described by equations of the same form. m, n, and h are all simultaneously following their own moving targets at their own changing rates.

The time constant curves show that τ_m is much smaller than τ_n or τ_h at all voltage levels. This means that m will always move faster than h or n. Specifically, τ_m is less than a half millisecond at all voltages. This allows m to follow m_∞ with a time lag of about a millisecond (two to three time constants). τ_n and τ_h are in the range of 2 to 10 ms. Grossly, these particles will lag behind their infinity values by about 10 to 20 ms. The rates for all of these processes, like other chemical reactions, are dependent on temperature. Action potentials at ocean temperature (e.g., in a cold-blooded squid) take place much slower than they do in a warm squid or a warm person. The τ's all have to be adjusted to take account of temperature. The adjustment factor for channels and for other active proteins is called Q_{10} (pronounced "q-ten").

State variables define a state

Having introduced all the components, it is apparent that this is a highly complex system of interacting parts. It is worth stepping back and summarizing the interactions. There are four differential equations, one each for V, m, h, and n. We have represented these differential equations in their discrete form in order to handle them numerically. V, m, h, and n are the four state variables of the system.

We previously discussed the concept of *state* in the context of neural activity and neural representation. In the case of linked differential equations, one defines the state of a dynamical system by noting the values of all of the state variables. The Hodgkin-Huxley system is a four-dimensional system. The state at any time can be given as a point in four-dimensional space. You could draw that point on a hypersheet of four-dimensional graph paper. Another well-studied system, the earth and the moon, is also a four-dimensional system. The earth–moon system's dimensions are position of the moon, velocity of the moon, position of the earth, and velocity of the earth. The fact that the earth and moon exist in three-dimensional space is unrelated to its four dynamical dimensions. By the way, if we add in the sun, we have a six-dimensional dynamical system with the positions and velocity of each of the three objects. This is the famous three-body problem.

The dynamics of astronomy and axons becomes complicated and interesting due to the interactions of the state variables. The differential equations are not independent. They interact. Something that happens to one state variable will eventually affect all of them. If a meteor hits the moon, it will affect the earth. Depending on the equations, the time constants, the feedback loops, interactions with other bodies, etc., this effect may become apparent in several weeks or only after many millennia.

Interlocking or linked differential equations are hard to grasp. All of these feedback systems can make it impossible to figure out who is doing what to whom when. The behavior of each state variable is influenced by its history and by the history of the other state variables. When state variable A changes, it influences state variables B, C,... These changes then feed back and change element A in turn. Chasing these state variable interactions around multiple feedback cycles produces proverbial chicken-and-egg predicaments. On the bright side, as linked ODEs go (ODE = ordinary differential equation), the Hodgkin and Huxley ODEs aren't so bad. Everything is linked through only one state variable, voltage. This means that we can sometimes view events as if voltage were "controlling" everything, even though voltage is itself controlled by the current through the sodium and potassium channels.

Taken in these terms, the genesis of the action potential can be viewed as the story of four state variables, each chasing its steady-state values. As an example, we can start with a current injection, I_{in}. V will start to chase $V_{max} = I_{in}/g_{memb}$. As V rises, m_∞ rises and m will chase m_∞ with a slight delay. m rising causes rising g_{Na} causes rising I_{Na}. The additional current and conductance will push $V_{max} = I_{total}/g_{total}$. Notice that the increase in sodium conductance is actually opposing the increase in sodium current, another minor negative feedback. Meanwhile h is chasing h_∞ toward smaller-and-smaller numbers. This is slowly turning off the sodium channel even as the m particle is turning it on. Additionally, n is chasing n_∞ and turning on the potassium channel, pulling the potential down toward the negative value of the potassium battery.

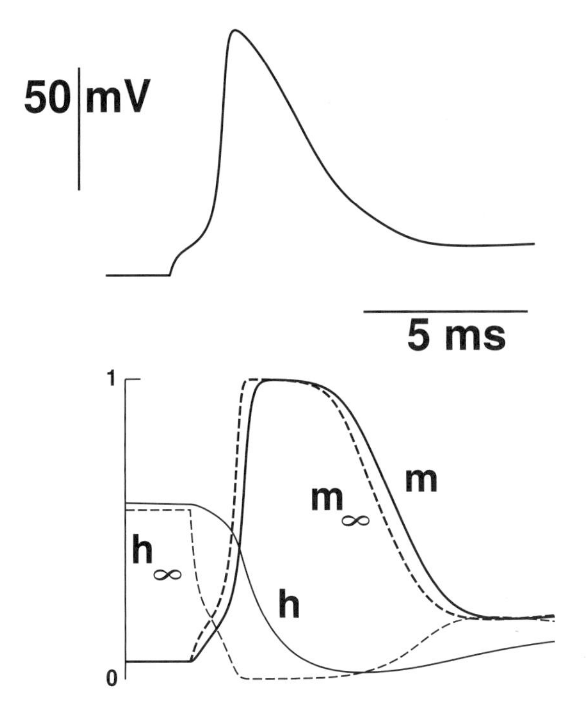

Fig. 12.6: Simplification of Hodgkin-Huxley model, without potassium channel. m **follows** m_∞ **and** h **follows** h_∞**. Voltage rises during sodium activation and falls during inactivation.**

12.7 Simulation

The proof of the pudding is in the simulation. To simplify the description, I start with a partial model, initially leaving out the potassium channel. Without n, we are dealing with a three-dimensional system that would allow us to map the state in our regular three-dimensional space if we wanted to. Instead, I use two-dimensional plots to map each state variable against time. Fig. 12.6 also shows the steady state values m_∞ and h_∞ as they change with time. Note that m follows m_∞ with a lag of about 1 ms, while h follows h_∞ with a longer lag. At the end of the simulation m has caught up with m_∞. h still lags h_∞ since τ_h is considerably greater at this V value. The peak of the τ_h curve, about 8.6 ms at 6.3°C, is at -67 mV, which places it near resting membrane potential. Therefore, the relaxation of h to h_∞ is relatively slow at this voltage.

The vocabulary for describing changes in active channels employs a set of confusingly similar words. Using V as a control variable that turns the other state variables on or off, we can construct the following table of descriptors for sodium channel control:

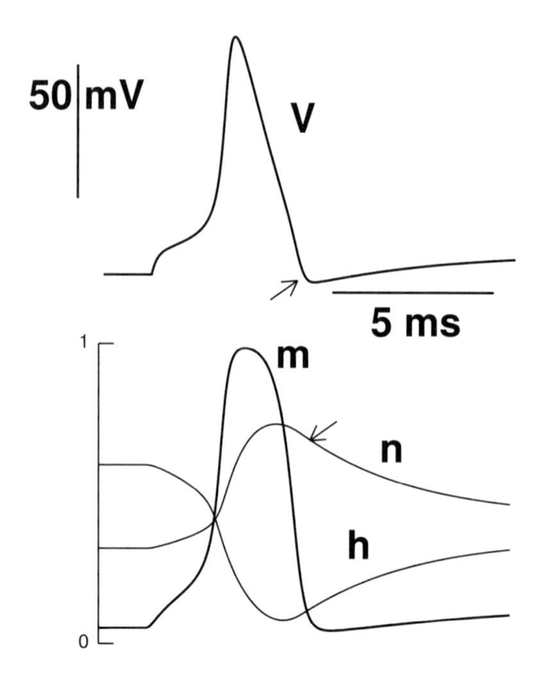

Fig. 12.7: The full Hodgkin and Huxley simulation showing all state variables. Arrows indicate afterhyperpolarization (AHP) and its cause.

	$\Uparrow V$	$\Downarrow V$
m	activation ($\uparrow m$)	**de**activation ($\downarrow m$)
h	**in**activation ($\downarrow h$)	**dein**activation ($\uparrow h$)

Activation (of m) and *dein*activation (of h, note the double negative) are both needed in order to open the channel. Activation occurs with depolarization, and deinactivation occurs with repolarization. Without deinactivation, a second spike cannot occur. Similarly, either deactivation (of m) or inactivation (of h) can close the channel. Inactivation terminates the sodium influx near the peak of the action potential, while deactivation is simply the resetting of m during repolarization. Using this vocabulary we can provide yet another description of the cycle of Fig. 12.6. During the upswing m activates and h inactivates. During the downswing, m deactivates and h deinactivates.

Putting the potassium channel back into the model, the behavior changes slightly (Fig. 12.7). Activation of the potassium channel turns on an outward current that helps pull the voltage back down toward rest after the peak of the spike (repolarization). The lag of n behind n_∞ causes the potassium current to remain on a little longer so that the voltage overshoots

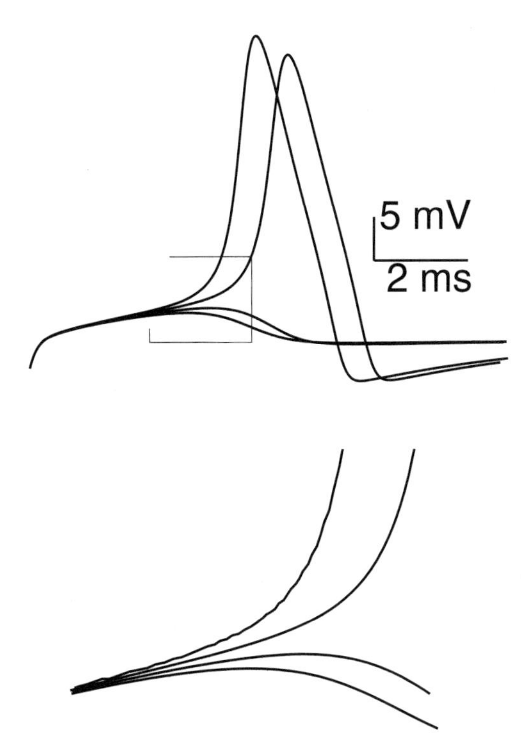

Fig. 12.8: Identifying the spiking threshold for a Hodgkin-Huxley model. The area around threshold is expanded below (rectangle).

the resting potential, producing an afterhyperpolarization (AHP, arrow in Fig. 12.7) as the potassium channel slowly deactivates (arrow on n). (Note that the potassium channel shows deactivation but no inactivation — there is an activation particle n but no inactivation particle.)

12.8 Implications for signaling

The genesis of the action potential gives it several important properties that have implications for data transmission. These can be illustrated using the Hodgkin and Huxley model. These properties of neuron signal generation can provide bottom-up clues for constructing network models.

The threshold and channel memory

The action potential has a threshold (Fig. 12.8). A current injection that does not reach the threshold does not generate a spike. At the threshold, inward (sodium) current exceeds outward (potassium) current and positive feedback kicks in. In Fig. 12.8, the threshold for firing is about -51 mV.

From the perspective of neural network theory, this threshold could be taken to be the sharp threshold of a binary activation function. This would allow the neuron to add up its inputs and then provide a rapid signal indicating whether or not sufficient excitation had been received. However, in contrast to standard neural network theory, the Hodgkin and Huxley threshold is not a fixed value.

The three channel particles, m, h, and n, all respond with a lag. This lag provides a simple form of memory. Something that happened in the past can be "remembered," while the m, h, or n state variables catch up with their steady-state values. The afterhyperpolarization (Fig. 12.7) is an example of this. The AHP reflects firing history — it's only present after the neuron has fired.

This history is not always immediately reflected in the membrane potential but can be held hidden in the state variables, inaccessible to experimental detection. For example, a hyperpolarizing input provides immediate inhibition. The hyperpolarization opposes any depolarization that would push the potential up to threshold. However, after the hyperpolarization ends, h is left at a relatively high and n at a relatively low value for a brief period of time. This pushes the effective threshold down closer to rest, making it easier to fire the cell. A subsequent depolarization will open the sodium channel more, and the potassium channel less, than it otherwise would (Fig. 12.9). Similarly, a preceding depolarization, which is immediately excitatory, will have a late effect that is inhibitory.

From a neural network perspective, this membrane memory could be tuned to allow the neuron to respond preferentially to certain sequences of inputs. In this simple case, an optimal stimulation would involve an IPSP followed by an EPSP after an interval of two to three times τ_n at RMP. A neuron has dozens of channel types, allowing the construction of more complex responses that can build up over relatively long periods of time. A novel firing pattern could be the result of some combination of inputs occurring over several seconds. This would allow the use of very complex, hard-to-interpret coding schemes.

Rate coding redux

Having speculated about complex history-dependent coding schemes, I now wish to return to the comforting simplicity of rate coding. In Fig. 11.13 we showed that slow potential theory explains the transduction from a presynaptic rate code to a postsynaptic depolarization plateau; increasing input rate gave an increased depolarization, due to increasing current flow. Using the Hodgkin-Huxley model, we can complete the sequence of signal transductions by showing that a depolarizing current injection converts to increasing firing rate within a certain range (Fig. 12.10). The greater the

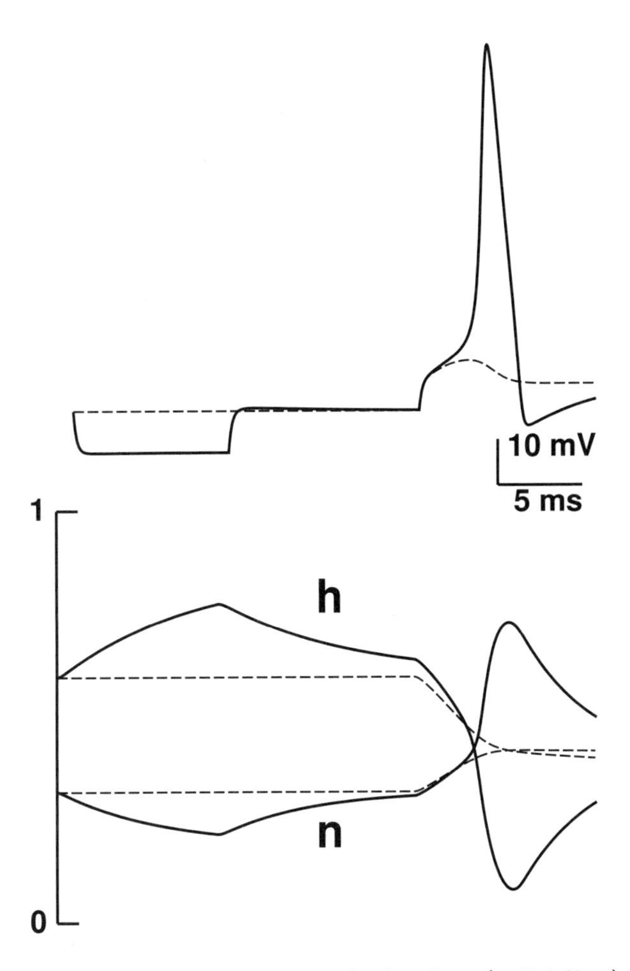

Fig. 12.9: With preceding hyperpolarization (solid line), h is elevated and n is depressed allowing a small depolarization to fire the cell 10 ms later. In the absence of the hyperpolarization, the same small stimulus is subthreshold (dashed line).

current, the greater the firing rate. Outside this range, there is no spiking. Below, there is a threshold for repetitive spiking. Above, spiking is blocked.

The trace at the bottom of Fig. 12.10 (0.88 nA) illustrates activity just below the threshold for continuous repetitive spiking. Below 0.84 nA, this Hodgkin-Huxley model produces only one spike. (By *this* Hodgkin-Huxley model, I mean a parallel-conductance model using sodium and potassium channels parameterized using the Hodgkin-Huxley equations, but with different specific parameters describing the *infinity* and *tau* curves.) This model produces one spike all the way down to the spiking threshold, about 0.2 nA. On the other end, high current injections produce higher rates.

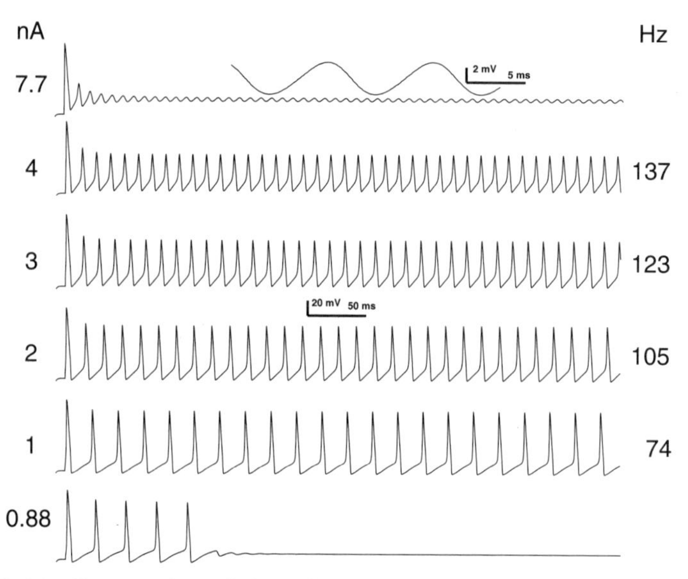

Fig. 12.10: Increasing firing frequency with increasing current injection. At and below 0.88 nA, there is no continuous repetitive spiking. At and above 7.7 nA depolarization blockade is seen.

The spikes become smaller and smaller. This contradicts what I said earlier about spikes being stereotyped and of constant amplitude. In fact, spike size does carry information about spike rate. It does not appear that this amplitude information is used however. In Fig. 12.10, a 4-nA current injection gives a measurable 137-Hz spike frequency. The spikes at this rate are only about half the size of the spikes produced by a 1-nA current injection.

As we go to higher and higher injections, the spikes get less and less spike-like as we gradually pass over to the low-amplitude oscillation that is characteristic of depolarization blockade. Depolarization blockade occurs when the voltage gets so high that the h particle remains near 0. This means that the sodium channel does not deinactivate. Since the sodium channel is continuously inactivated, it is not possible to generate spikes. For example, in the top trace, with 7.7 nA of injected current, the tiny oscillation has an amplitude of about 4 mV and frequency of about 165 Hz. Examination of state variables demonstrates that this oscillation is based on an interaction between V and m without substantial contribution from n and h. This is the dynamics of depolarization blockade, not the dynamics of neural spiking.

Using the Hodgkin and Huxley model of neuron spiking, we can compare this realistic input/output (I/O) curve with the sigmoid (squashing) curve, the idealized input/output curve used in artificial neural network modeling (Fig. 12.11). Both curves are monotonically increasing, meaning they only

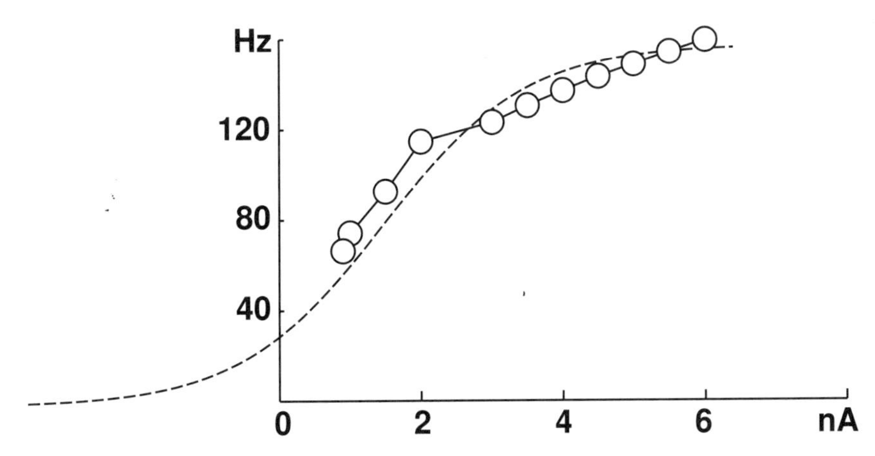

Fig. 12.11: An I-f curve from a Hodgkin-Huxley model compared to the standard sigmoid I/O curve of artificial neural networks.

go up. Although it does not asymptote, the realistic I-f (current-frequency) curve in Fig. 12.11, like the sigmoid curve, does show some reduction in slope with higher input values. However, the sigmoid curve covers all input values, while the realistic I-f curve (current-frequency curve) only has outputs for a certain range of inputs. In math-speak, this means it has a limited domain (x-axis) and limited range (y-axis). Below this domain, there is a floor effect; above, there is a ceiling effect. By altering the Hodgkin and Huxley parameters we can move the ceiling, the floor, and the precise relationship between current and frequency. However, these measures are not independent, so that if you try to move the floor down, the ceiling and slope of the I-f relation (the gain) will change as well. This means that it is not possible to precisely tune a Hodgkin-Huxley model to produce exactly the response one might want for a particular network model.

12.9 Summary and thoughts

The Hodgkin-Huxley model of the action potential is the most influential computer model in neuroscience and as such remains a touchstone of computational neuroscience. It's a dynamical model that arises from the interaction of four time-dependent state variables — V, m, h, and n. Of these only V, voltage, is directly measurable. The others are putative populations of switches that turn sodium and potassium channels on and off.

Electrically, the Hodgkin-Huxley model is the basic membrane RC circuit with two conductances added in parallel. Hence the circuit is called the parallel-conductance model. The two added conductances are the active sodium and potassium conductances. These conductances are active because they change with change in voltage. A controllable resistance (conductance) is called a rheostat. Each of the conductances, including the passive "leak" conductance, is attached to a battery. The battery potential (voltage) depends on the distribution of the particular ion that flows through its own selective conductance. This Nernst potential is the electrical field that holds back the chemical flow of the ion across the membrane down its concentration gradient.

The spike is the result of a set of interacting feedback loops. Depolarization activates sodium channels ($\uparrow m$) producing positive feedback with further depolarization. This is the upswing of the spike. Following this, two negative feedback influences kick in. The sodium channel starts to inactivate ($\downarrow h$). Additionally, activation of the potassium channel actively pulls the potential back toward and past the resting membrane potential.

The Hodgkin-Huxley model can be used to see how action potential behavior will influence neural signal processing and signal transduction. For example, the neuron has a threshold for action potential generation that can be altered by preceding inputs in a paradoxical way. An earlier excitatory input will raise the threshold, producing a late inhibitory influence. A preceding inhibitory input will lower the threshold, resulting in a relatively excitable state.

Repetitive action potential firing is possible over only a limited range of inputs. Too little input produces no spikes or only a few spikes. Too much input produces depolarization blockade with a low amplitude oscillation. This limited range makes it difficult to use standard Hodgkin-Huxley model dynamics for rate coding in neural network models. Adding in the dynamics of other channels that are present in neurons makes it possible to get a wider range of firing frequency.

13
Compartment Modeling

13.1 Why learn this?

In the previous chapters, I took two steps away from artificial neural network simplifications toward more complex neural realities. First, I demonstrated how membrane properties allow a neuron to do temporal integration. Then I showed how active channels produce the action potential, a complex signal. In this chapter, I build on these concepts in order to produce a full modern model neuron.

Most neurons have sodium and potassium channels similar to those that Hodgkin and Huxley found in squid axon. Neurons also have many other ion channel types. In the case of both potassium and calcium channels, there are large families of channels with different thresholds and time courses for activation and inactivation. Some of these channels are activated by the presence of ligands. These ligands may be other ions (e.g., a calcium-sensitive potassium channel) or chemicals of various types. The ligands may bind the channel from within the cytoplasm or, in the case of synapses, from outside of the cell. The ligand sensitivity may be accompanied by voltage sensitivity as well. Sophisticated multistate Markov models have been developed to describe these channels in great detail. For simplicity, these various channels types are often modeled using modifications of the Hodgkin-Huxley formalism.

Complex branching dendritic trees introduce additional complexity compared to the simple parallel-conductance model. Dendritic trees are modeled using compartment modeling. Each compartment is an RC or parallel conductance circuit. Individual compartments are connected by resistors. Different inputs (synapses) can be located in different compartments, placing them at different locations in the dendritic tree. In general, inputs that are further away from the spike-generating zone will have less influence on output than those that are closer. However, active ion channels can also be located at different points in the dendritic tree. Inward currents, such as those produced by sodium or calcium channels, would give excitatory synaptic inputs a boost. Outward or shunting conductances, produced by potassium or chloride channels, would tend to decrement excitatory inputs.

There are various types of synapses that provide inputs into neurons. Not surprisingly, they are far more complicated that the simple *weight times state* multiplication connection that we used in artificial neural network models. Most synapses are chemical and introduce complicated issues of chemical kinetics, diffusion, and breakdown. We will not attempt to address this level of modeling here but will utilize a simple variation on the Hodgkin-Huxley model to describe synapses.

By the end of this chapter, we will be dealing with a complex single-neuron model. From a numerical processing point of view, these single neuron models are actually bigger than many artificial neural network models. As I've previously suggested, it is entirely possible that the real single neuron might also be bigger from an information processing perspective — single neuron as CPU rather than transistor. However, with all of this complexity, such models of single neuron electrical activity still only scratch the surface of the information processing potential of the cell. Specifically, there remains a big gap between simulation and reality due to the lack of data and modeling techniques for handling chemical interactions within the cell.

The neuron is a giant chemical plant (well, a tiny chemical plant with lots of different chemical reactions). Much of this chemical activity is standard cell metabolism that looks the same in the liver or kidney. However, other chemical messenger networks are involved in neural information processing. Neurotransmitters are chemicals that transmit information across chemical synapses between neurons. Some of these transmitters cause the release of other chemicals that pass the message along inside of the cell. These secondary chemicals are called second messengers. Second messengers may then cause the release of third and fourth messengers. The various higher-level messengers may include positive and negative feedback loops, producing chemical networks of great complexity. Such chemical networks can be modeled. However, the complexity of these chemical models makes it difficult to incorporate them into standard electrical models of single cells, much less into networks of these cells.

13.2 Dividing into compartments

Compartment modeling is used widely in biology, typically to follow concentrations of chemicals. In general, one divides some biological entity (e.g., a person) into compartments. In the model, the chemical we're measuring is at the same level everywhere within a single compartment. The different compartments exchange the chemical among themselves depending on their connectivity. A typical example comes from pharmacology. We can describe the distribution of a drug in terms of the compartments of the human body: blood, bone, fat, muscle, urine, etc. If we are treating a bone infection, we place penicillin intravenously in the blood compartment. Since "blood" is a single compartment, we assume, somewhat unrealistically, that the drug instantly mixes with all of the blood so that the drug concentration is the same everywhere in that compartment. We then look at how quickly this compartment exchanges penicillin with bone, urine, and muscle, the first being the target location, the second being the primary route of excretion (ground in the electrical analogy), and the third being an alternative location where drug may be shunted away. We can use the compartment model to find out how much penicillin will get to the bone, how fast it will get there, and how fast it will leave. This enables us to calculate a good dosage and schedule in order to achieve adequate penicillin concentration to kill the germs.

In neural modeling, we are dealing with voltages instead of concentrations and with current instead of chemical flux. The single compartment is equipotential (syn. isopotential) — there is one voltage everywhere in that compartment. This is equivalent to the instantaneous mixing referred to in the previous example. If the compartments are too large, equipotentiality cannot be assumed and the numerical simulation will produce a poor approximation. Each compartment corresponds to a section of membrane. The single compartment model must include at least capacitance and leak conductance.

In addition to the obligatory capacitance and leak conductance, active components will also be added to some compartments. This requires additional rheostat/battery branches, similar to the sodium and potassium branches of the parallel-conductance model (Fig. 12.2). Some of these rheostats will be synapses as explained below. Others will be similar to Hodgkin-Huxley sodium and potassium channels, usually readjusted to more closely match the kinetics of the ion channels found in a particular cell type of a particular non-squid animal. Other types of sodium and potassium channels, as well as calcium channels, may be included. Each of these will have the ion-appropriate battery. Each will have a rheostat with some kind of dynamics. Rheostat dynamics may be parameterized by equations similar to the Hodgkin and Huxley equations or by some other kinetic scheme. Rheostat sensitivity to second messengers may be included in addition to, or instead of, voltage sensitivity. Potassium channels that are sensitive to calcium are commonly found in neurons. Although complex chemical modeling is not usually done in compartment models, internal calcium concentrations are sometimes included in order to model the activation of these calcium-sensitive potassium channels.

An individual compartment is equipotential but will typically differ in voltage from neighboring compartments. In this way, different parts of the neuron can be doing different things at the same time. For example, one part of the neuron may be spiking while another part is hyperpolarized.

Building the model

To build a model of a real neuron, we measure the lengths and diameters of dendrites and measure the soma. A made-up picture of a neuron is shown in Fig. 13.1 (left). Neurons are filled with dye to make their dendrites easily visible; the Golgi staining method colors neurons black. Using a microscope linked to a computer, one can trace a neuron. This is hard work since neurons are not typically flat and therefore must be traced through multiple sections that don't always match up well, due to differential shrinkage of the tissue during processing.

Using these measurements, we can assign cylindrical compartments (Fig. 13.1 center). The soma or cell body is in reality a roughly spherical or pyramidal structure. However, it's just going to end up as a bunch of conductances and a capacitance in the circuit model. Therefore, we can represent it as a cylinder in the intermediate model as long as we preserve its surface area. (More surface area will translate into more capacitance and more conductance.)

In the context of numerical integration, the spatial division of dendrites into small cylinders of length Δl is analogous to the temporal division of duration into subdivisions of Δt. The quality of any numerical integration depends on making the time steps Δt short enough so that the change in state variables during a single time step is insignificant. Similarly, the quality of a compartment model depends on making the spatial steps (cylinder length) small enough that the change in voltage over the length of the cylinder is insignificant. This is the assumption of equipotentiality. Choosing smaller cylinders or smaller time steps will give a better approximation but will increase simulation time.

Within the dendrite, current flows through the cytoplasm inside. Cytoplasm is glop with mobile ions that carry current. Cytoplasmic resistance is a significant impediment to flow of current along the dendrite. Wider cylinders allow more ions to pass, giving less resistance, less signal drop-off, and faster signal transmission. This is the reason that squid axons are so wide. If a dendritic cylinder is narrow, it will have relatively low conductance (high resistance) and a greater voltage drop-off along its length. The voltage drop-off can be calculated using Ohm's law. The difference in voltage from end to end equals longitudinal current times longitudinal re-

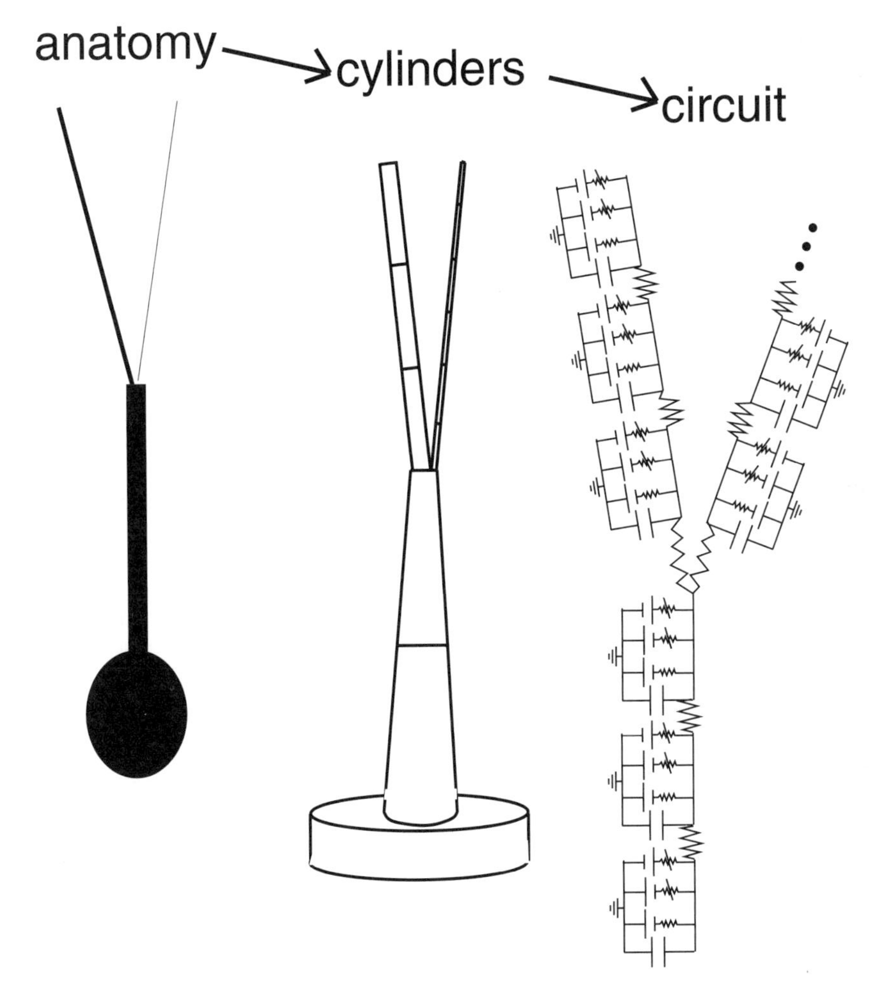

Fig. 13.1: Three representations of a neuron model: faux Golgi anatomy (kind of what a cell looks like under the microscope); cylindrical compartmentalization; equivalent circuit model.

sistance: $V_+ - V_- = I_l \cdot R_l$. (Notice the use of subscripts for spatial intervals compared to the use of superscripts for temporal intervals.) Narrow cylinders have high longitudinal resistance and therefore must be divided into a larger number of compartments. This is illustrated in the two branches in the cylindrical representation in Fig. 13.1.

Having chosen the cylindrical compartmentalization of a dendrite, we can put together an equivalent circuit model for each compartment and connect them with resistors. Each compartment will have capacitance and leak conductance. Some may have additional conductances representing voltage-

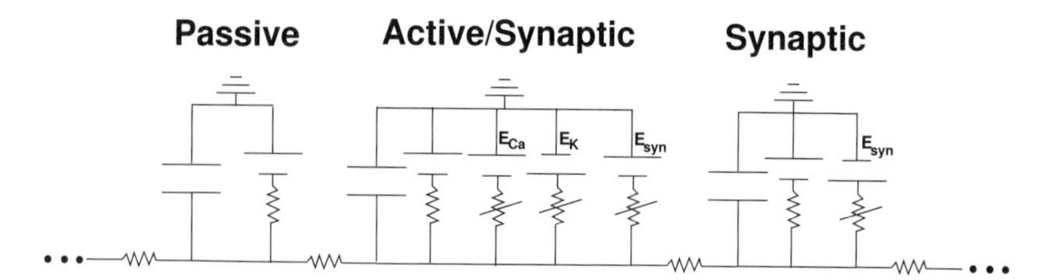

Fig. 13.2: Three example compartments from a dendrite model. On the circuit diagram, different rheostat branches are distinguished only by the directions of their batteries; I have added labels to identify them further.

sensitive channels and synapses. In the circuit of Fig. 13.1, the ground symbol indicates the extracellular side of each compartment. For simplicity, I have illustrated each compartment as a standard parallel-conductance circuit.

When putting together a model of a particular neuron, neurophysiological data would likely suggest the inclusion of additional conductances in some compartments. Other parallel rheostats would be included to represent synapses. Active conductances might be omitted from some compartments in the interest of computational speed and efficiency. In Fig. 13.2, I show three compartments from a neuron model. The passive compartment to the left has capacitance and leak resistance but no rheostats. The second compartment is labeled to indicate that it has active potassium and calcium channels as well as an inhibitory synapse. The synapse can be identified as inhibitory by noting that the negative pole of the battery connects to the inside of the membrane. The third compartment has an excitatory synapse. Without the labels, there is nothing at the circuit diagram level to distinguish a synaptic (ligand-gated) channel from a voltage-gated channel. All of that detail is in the parameterizations, the details of the equations describing the rheostats.

Individual compartments in a multicompartment model are connected to neighboring compartments by a resistor. This resistor represents the cytoplasmic axial resistance, also called longitudinal resistance, between the center of the two connected compartments. As mentioned above, wider dendrites will have less resistance. However, wider dendrites will typically be represented with longer cylinders so that total distance, and total resistance, between centers will be similar. Where two compartments represent cylinders of different widths, resistances sum from cylinder center to the connecting point on either side of the junction. A branch point, where two dendrites come off a single parent dendrite, is represented with two resistors as shown in Fig. 13.1.

The whole massive circuit in Fig. 13.1 or Fig. 13.2 can be simulated using the same methods we previously detailed: 1) Kirchhoff's law, to sum up all currents at intersections; 2) Ohm's and capacitance laws, to determine currents across passive components; 3) Hodgkin-Huxley equations and parameters, to determine behavior of particles for active components; 4) numerical integration, to solve the differential equations. This is a big project in itself, likely to delay any simulation project for months to years. For this reason, it is generally best to use a software package that is designed to handle solutions of differential equations. Since these packages exist, I will not further detail the mathematical and programming complexities of compartment-model calculations. These details aside, a basic knowledge of compartment modeling remains useful for understanding how to set parameters and handle simulations.

13.3 Chemical synapse modeling

We previously utilized square wave (Fig. 11.8) and alpha function (Fig. 11.12) current injections as inputs to our models. The former was a realistic model of current injected through an electrode by an experimentalist. The latter was an intermediate model that had the approximate form of a synaptic current. Chemical synaptic responses are actually conductance changes rather than current injections. A synaptic conductance change will permit current flow. As I show, sometimes it is the synaptic conductance change, rather than the synaptic current, that affects postsynaptic activity.

The Hodgkin-Huxley model describes membrane rheostats controlled by voltage. Chemical synapses are membrane rheostats controlled by neurotransmitters. When modeling a network, these neurotransmitters are not explicitly included. Instead, a spike in a presynaptic neuron is used to activate a rheostat in the postsynaptic neuron. The postsynaptic rheostat represents an ion channel triggered by a postsynaptic receptor. Depending on the polarity of the battery associated with that rheostat (inward positive or inward negative), that synapse could be made excitatory or inhibitory (Fig. 13.2).

As with most things in nature, chemical synaptic transmission is terribly complex. Synaptic simulation can be made terribly complex as well. A full synaptic model would have to consider chemical diffusion, reuptake, breakdown, and interaction with proteins before one even gets to a receptor. Then, at the receptor level, one would consider the fact that receptors are proteins that typically have multiple states (each of which could be included in a more complex model, e.g., a Markov model). This gives them properties such as activation, deactivation, inactivation, and deinactivation as well as desensitization (channel turns off if the chemical ligand has been sitting there too long), allosteric interactions (other chemicals binding at different sites can alter the response to the main neurotransmitter), and multiple binding sites (more than one molecule must bind for the channel to open). Then, there's the fact that the synapse can grow and change. This depends not only on classical learning effects but also on more low-level adaptation to the chemical environment. For example, receptors will typically upregulate (more receptors will be inserted into the membrane) if they do not see much neurotransmitter for a while and downregulate in the converse condition. They can also change shape and shift the location of receptors.

Of course, we want to ignore all of that and focus on the basics: a chemical is released and binds, the associated channel opens (activates) and then closes (deactivates). For simplicity, we will have the channel open instantly. This means that the conductance will go instantly from 0 to g_{max} and the current will also shoot up instantly. However, the voltage will not shoot up since the membrane capacitance provides a brake. We will then use an exponential decay for conductance inactivation. This corresponds to the simple differential equation (discrete version): $\tau_g \cdot \frac{\Delta g}{\Delta t} = -g$.

In Fig. 13.3, excitatory and inhibitory synaptic responses are compared. The same conductance is used in both cases. However, the synaptic reversal potentials (E_{syn}) differ. For this reason the excitatory postsynaptic current (EPSC) is negative (inward and depolarizing), while the inhibitory postsynaptic current (IPSC) is positive (outward and hyperpolarizing). This is the result of the expression for current: $I_{syn} = g_{syn} \cdot (V - E_{syn})$. In the case of the excitatory synapse, the reversal potential (E_{syn}) is more positive than resting potential. For the inhibitory synapse, the reversal potential is more negative than the resting membrane potential (RMP). Although the conductances in Fig. 13.3 are the same in both the excitatory and inhibitory cases, the magnitudes of both postsynaptic currents and postsynaptic potentials differ. The differing size of EPSC and IPSC reflects the difference between membrane voltage and E_{syn} in each case: $V - E_{syn}$. E_{excite} is far from rest, while E_{inhib} is close to rest. This potential difference is called the driving force for the current. In Fig. 13.3, the initial driving force ($RMP - E_{syn}$) is $-70 - -20 = -50$ mV for the excitatory case, and $-70 - -90 = 20$ mV for the inhibitory case. As membrane voltage moves toward E_{syn}, driving force gradually decreases. A very large synaptic conductance can drag membrane voltage all the way to E_{syn}.

Shunting inhibition

Classically, inhibitory synapses produce a hyperpolarizing IPSP. However, some inhibitory synapses do not change the membrane potential. This is known as shunting inhibition. Shunting inhibition occurs when $E_{syn} = RMP$. Activation of a shunting synapse alone will not cause any current to flow: $I_{syn} = g_{syn} \cdot (\text{RMP} - E_{syn}) = 0$. Therefore, there will be no change in membrane potential. However, in the presence of an excitatory input, the shunting influence reduces depolarization. Inward current that would otherwise pass outward through the capacitor and depolarize the membrane

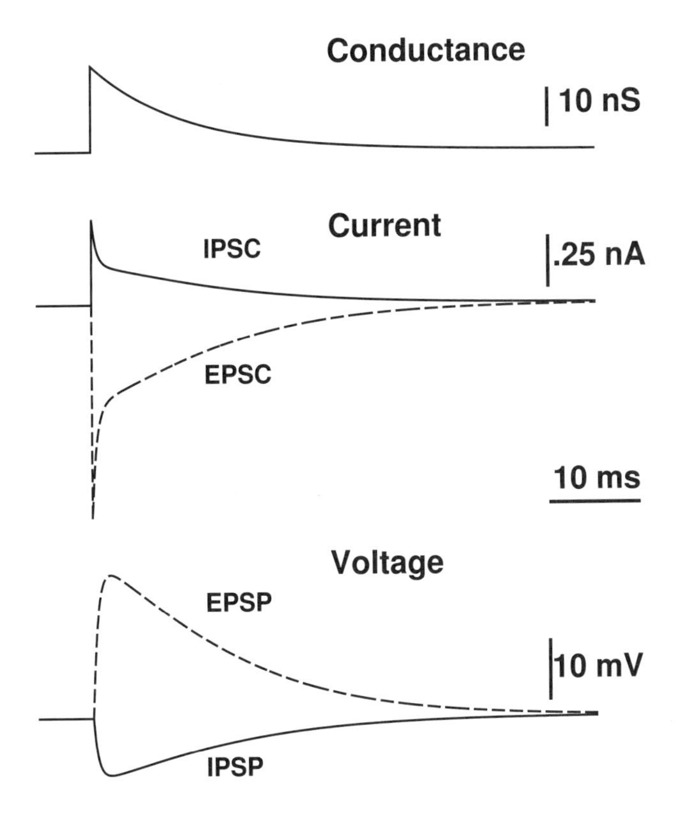

Fig. 13.3: Simple model of a chemical synapse: Conductance activates instantly, then inactivates with time constant τ_g. Current flows inward (negative) for excitation (dashed line), outward (positive) for inhibition (solid line). Inward current produces a depolarizing EPSP (dashed line); outward current produces a hyperpolarizing IPSP (solid line).

is instead passed outward (shunted) through this synaptic short circuit and made ineffectual.

Consider the two examples in Fig. 13.4. A hyperpolarizing IPSP and a depolarizing EPSP roughly add up. In this case they produce a small compound PSP (compound since it is the summation of postsynaptic potentials), consistent with the fact that a large EPSP was being added to a smaller IPSP. This synaptic interaction may be regarded as being approximately additive, comparable with the linear summation of inputs used in artificial neural network models. However, a check of the numbers reveals that these PSPs are not strictly additive. In the top traces of Fig. 13.4, the peak values are 23.4 mV depolarization for the EPSP and −9.3 mV hyperpolarization for the IPSP. The numbers sum to +14.1 mV, but the

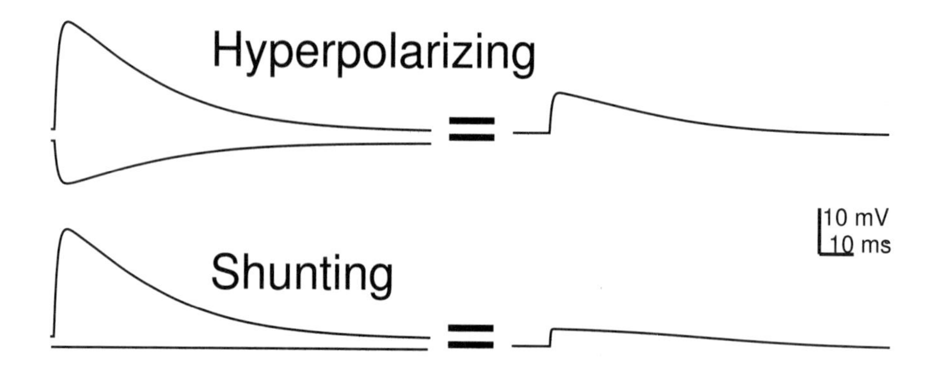

Fig. 13.4: A large EPSP and small IPSP add together to produce a smaller EPSP. Each demonstration is the result of three separate simulations: excitation alone (left top), inhibition alone (left bottom), compound PSP (right).

compound postsynaptic potential is only +9.7 mV. The lost millivolts are due to the shunting that accompanies all synaptic activation.

Shunting is present in association with any conductance change, whether synaptic or due to voltage-sensitive channels. Shunting is most dramatic in cases where the voltage shift is minimal. In the lower trace of Fig. 13.4, I used a shunting conductance 10 times greater than the excitatory conductance. The +23.4 mV EPSP is reduced to a 4.2-mV compound PSP. By increasing the shunting conductance, one can make the resulting compound PSP smaller and smaller. Shunting inhibitory synapses are sometimes called multiplicative synapses to illustrate the notion that a zero IPSP "times" an EPSP gives a zero result. Actually, the compound PSP will never go to zero, no matter how large one makes the shunting inhibition. Without the depolarization from rest, there is nothing to shunt. The concepts of additive and multiplicative inhibition can be used to map the neurobiology onto the simple arithmetic of artificial neural networks. However, they are only crude approximations.

GABA and glutamate

There are many neurotransmitters. Each neurotransmitter is associated with many different receptors. Each receptor may mediate many different actions. A description of what is know about synaptic responses fills books. I will be brief.

Neurotransmitters are somewhat arbitrarily classified as either a classical neurotransmitter or a neuromodulator, based on their time scale of action. The implicit hypothesis is that neuromodulators act on relatively long time scales, while classical neurotransmitters are quick and are involved in moment-to-moment information processing. However, given the myriad actions of most neurotransmitters, it is likely that many have some modulatory and some classical functionality. Furthermore, it may be that the time scales of neurotransmitter actions are a spectrum, and cannot be readily classified into two distinct groups. In Chap. 14, I discuss acetylcholine, a neuromodulator that is believed to influence learning. For now, I focus on the two major classical neurotransmitters, GABA and glutamate.

The major inhibitory neurotransmitter in the central nervous system is GABA. The major excitatory neurotransmitter is glutamate. There are two types of GABA receptors, conveniently called $GABA_A$ and $GABA_B$. $GABA_A$ is directly connected to an ion channel selective for chloride. Chloride is distributed so that its Nernst potential is near to the resting potential. Therefore, the $GABA_A$ effect is primarily shunting. It is also a relatively fast PSP, going on and off over about 5 to 10 ms. The $GABA_B$ receptor, by contrast, is complexly coupled via second messengers that have a variety of effects. A hyperpolarizing effect on the membrane is produced via a link to a potassium-selective channel. Because of the time needed for the second-messenger linkage and the properties of the associated channel, $GABA_B$ is relatively slow, lasting about 100 ms.

On the excitatory side, with glutamate as the neurotransmitter, the receptors have somewhat more obscure names: AMPA and NMDA (defined in the Glossary). As with many receptor names, these are based on sensitivity to alternative agonists — chemical ligands that are particularly good at activating the receptor. As in most cases, these are artificial drugs and are not present in the body, where glutamate is the natural ligand. The AMPA receptor responds rapidly. The NMDA receptor is slower. These two receptors are well studied since together they are responsible for the biological analogue of Hebbian learning, long-term potentiation (LTP). The properties of Hebbian learning through LTP are largely based on the response profile of the NMDA receptor.

Both AMPA and NMDA receptors are linked directly to ion channels, rather than to second messengers. Both have a reversal potential (E_{syn}) of approximately 0 millivolts (about 70 mV depolarized from rest). The Hebbian rule states that a synapse is strengthened when it simultaneously registers pre- and postsynaptic activity. The NMDA receptor can make this simultaneous measurement because it has both ligand and voltage sensitivity. The sensitivity to an external chemical ligand allows it to register presynaptic activity. Presence of glutamate indicates presynaptic activity since the glutamate is released with the arrival of a presynaptic action potential. The NMDA channel's voltage sensitivity gives it a way of measuring postsynaptic excitation. However, postsynaptic depolarization does not necessarily lead to postsynaptic spiking, so LTP-mediated synaptic enhancement can take place without increased postsynaptic activity. This suggests that LTP may not be a classical Hebbian mechanism.

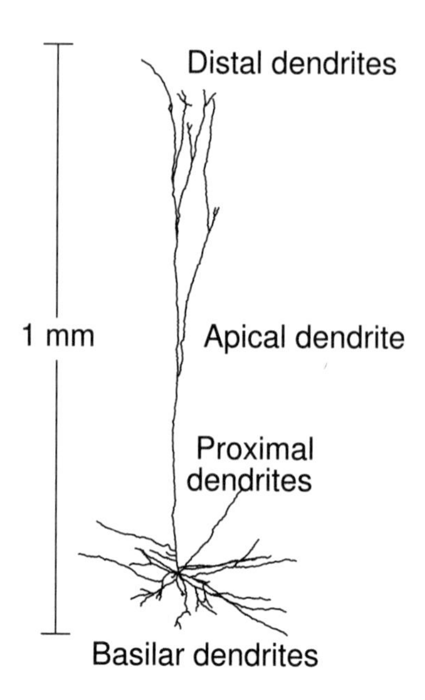

Fig. 13.5: A neocortical pyramidal neuron.

The voltage sensitivity of the NMDA channel is a little different than that of the Hodgkin-Huxley channels. In the Hodgkin-Huxley channels, the voltage-sensitive mechanism is an intrinsic part of the channel protein that shifts with voltage change. By contrast, NMDA voltage sensitivity is due to blockade of the pore by a magnesium ion (Mg^{++}) coming from outside of the cell. This ion will sit in the pore unless it is kicked out due to depolarization. With the magnesium out of the way, the pore will conduct ions if it has also been activated synaptically by glutamate, the neurotransmitter.

LTP is much studied. It has been found that the expression of LTP is due to a change in AMPA conductance rather than NMDA conductance. Thus the NMDA channel acts as the Hebbian trigger, while Hebbian expression is handled by AMPA. In addition to depolarization, local calcium entry is important in producing the Hebbian change. There has been much debate as to whether the Hebbian strengthening takes place at the postsynaptic receptor level or at presynaptic neurotransmitter release sites or both.

13.4 Passive neuron model

Fig. 13.5 shows a tracing of a cortical pyramidal neuron. Such neurons are called "pyramidal" because the soma is somewhat pyramid-shaped (not shown in figure). The long dendrite at the top, emerging from the apex of the pyramid, is called the apical dendrite. The hairy dendrites at the base are called basilar dendrites. Locations are referenced with respect to the soma: a distal dendrite is far from the soma and a proximal dendrite is close to the soma.

Passive compartment models have long been the standard model for assessment of response properties. In a passive model, each individual compartment has only capacitance and leak conductance, with no active voltage-sensitive conductances. Synaptic conductances may be included as inputs to passive models. Synapses are active elements, in that they involve variable responses to a ligand. Despite this, the model is still considered passive. The response of a passive model does not depend on prior activity history.

Passive neuron properties have been a subject of debate for decades. The leakiness of neuronal membranes can't be measured directly. It can only be inferred from responses to current injection. Unfortunately, connecting an electrode to a cell damages the cell, making it hard to know whether subsequent measurements are accurate. Some electrodes are stuck right through the membrane (cell impalement). This makes a hole. Current leaks through the hole. Other electrodes solve this problem by sealing tightly on the membrane (whole-cell patch). However, this actually produces a bigger hole that goes from the cell into the electrode itself. Current leakage isn't a problem, but chemical leakage is. This type of recording changes cell membrane function by altering the second messengers inside the cell. Both approaches introduce artifact. This is the classic experimental dilemma — probes that mess up the thing that you're probing.

The debate about passive properties has been remarkably passionate. Perhaps this is because a conclusion would be a clue in a larger debate about the processing role of the neuron: simple or complex, transistor or CPU? If neuronal membranes are very tight (high resistance), then the neuron is electrotonically compact. This is also expressed by saying that the cell has a long length-constant. (This can be confusing: a cell with a *long* length-constant is *compact*. The length-constant is long so the cell is electrotonically small by comparison.) A long length-constant means that signals coming in from anywhere will easily reach the soma and be able to trigger action potentials. In the case of a compact cell, even large neurons might act like the simple processing units of the artificial neural network models. Conversely, if neurons are leaky (high conductance, short length-constant), then signals will be handled differently depending on where they arrive in the cell. In a situation where single inputs are ineffective, complicated spatiotemporal input combinations might be required to produce cell activity. This would suggest complex coding and decoding strategems.

Recent studies have suggested that many, perhaps all, neurons have active membranes with voltage-sensitive channels. Some active dendrites that have been studied appear to generate inward currents that increase depolarizing potentials, boosting the EPSP down to the soma and help-

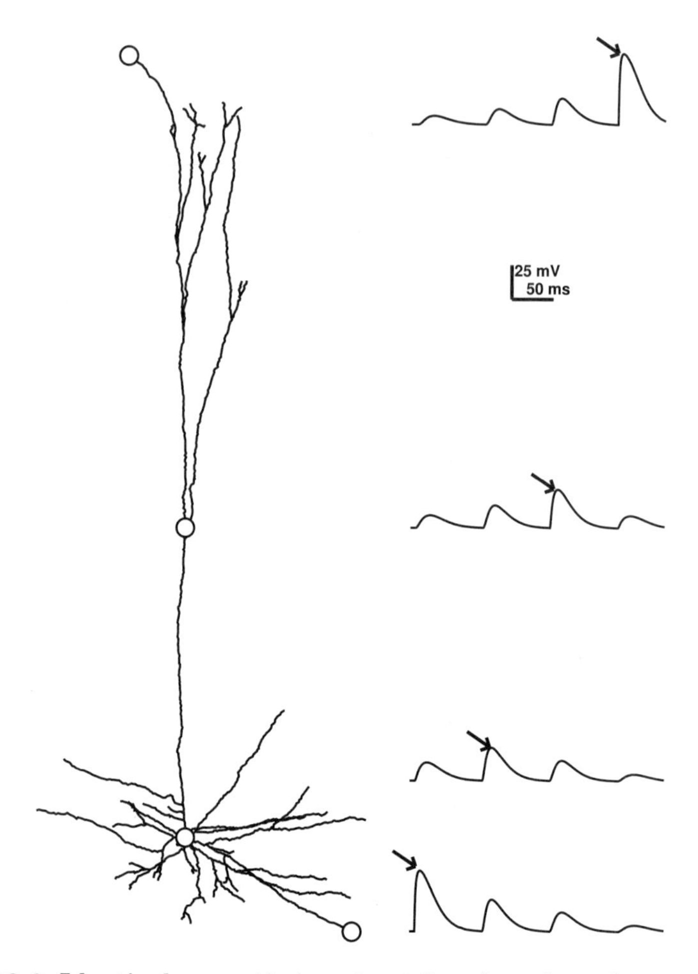

Fig. 13.6: Identical synaptic inputs at four locations demonstrate variation in signal spread. Position of trace shows recording location. EPSPs are activated from bottom to top. Arrow shows which EPSP in the sequence was initiated at that recording location.

ing the action potentials head back up. Conversely, some dendrites may have depolarization-activated channels that mediate outward current. Such channels would reduce excitatory potentials and thereby decrease the excitability of the cell. It is also possible that some neurons may have varying mixtures of boosters and reducers located near particular synapses as an adjunct synaptic-weighting scheme.

Synaptic responses

Distal synaptic inputs will generally have less effect on the soma than will proximal inputs. If we initiate an EPSP at different points in this neuron, we can see how much signal spreads to other parts of the neuron (Fig. 13.6). In the passive neuron model, membrane conductance is the major parameter that will determine signal spread. If membrane conductance is high (membrane resistance low), then the membrane is leaky, hence the length-constant is short, hence the cell is electrically large. Signals will not get very far. Conversely, if the conductance is low (membrane resistance high), then the membrane is tight, the length-constant is long, and the cell is compact. In Fig. 13.6, I used a low membrane conductance so that signals could propagate from one end of the neuron to the other.

As expected, there is a fall-off in the signal along the dendritic tree. Notice, however, that the EPSPs initiated at the ends are larger than those triggered at the soma or halfway up the basal dendrite. Remember that the same synaptic conductance was used at each location. The difference in response is due to the fact that current has only one way to go at a sealed end and therefore tends to "pile up" instead of leaking away in both directions down the dendrite. In electrical terms, these terminal locations have relatively high *input impedance.* Impedance refers to any obstacle to current flow. In a dendrite, resistance provides a constant impedance, while capacitance provides a frequency-dependent impedance.

In addition to increasing the size of an EPSP at a distal dendrite, high terminal input impedance also reduces the drop-off of the EPSP going away from the soma, compared to the drop-off of an EPSP going toward the soma. These two effects, the increased size of a distal EPSP and the increased drop-off toward the soma, tend to cancel out. Therefore, a distal input arriving at the soma is about the same size as a somatic input arriving at the distal dendrite. High distal input impedance produces a large distal EPSP (top trace, 4th EPSP). This large EPSP is substantially decremented when seen at the soma (3rd trace, 4th EPSP). Going in the other direction, the somatic EPSP is relatively small due to lower input impedance (3rd trace, 2nd EPSP) but signal fall-off from the soma to the distal dendrite is relatively small (top trace, 2nd EPSP). In this way, input-impedance differences are balanced by transfer-impedance asymmetry in passive models. David Jaffe and Ted Carnevale have described this as "passive normalization," a possible mechanism to equalize synaptic inputs arriving at different locations in the dendritic tree.

As we saw in our discussion of shunting synapses, leakiness of the membrane can increase due to synaptic activation. It may be that neurons that have very tight membranes when measured *in vitro* are much leakier *in vivo* due to the constant barrage of synaptic signals. A pyramidal cell can have tens of thousands of synapses. Large numbers of excitatory and inhibitory synaptic signals could balance each other so as to produce no net change in membrane potential, yet make the neuron leaky. This would substantially decrease signal spread in the dendrites.

13.5 Back-propagating spikes and the Hebb synapse

According to the classical Cajal neuron doctrine, the action potential, generated in the soma, spreads forward to the axon but not back into the dendrites. Lately, there has been considerable interest in the spread of spikes back up the dendrite. This is referred to as a back-propagating action potential. Although it is named so partly in reference to the back-propagation artificial neural network algorithm (Chap. 9), the back-propagating spike and back-prop algorithm have nothing in common other than their backwardness.

It is generally assumed that back-propagating action potentials require active dendrites. We can test this assumption in a model by comparing spike back-propagation with passive dendrites to back-propagation with active dendrites. In Fig. 13.7, the soma is active (Hodgkin-Huxley channels) but the dendrites are passive. A synapse is activated at the soma (rightward arrow). An action potential, generated in the soma, propagates up the dendrite, broadening and decrementing as it goes. At the distal dendrite the action potential is barely apparent (oblique arrow). Even if one sets R_{memb} to a very high value, increasing the length-constant so as to make the neuron as compact as possible, the spike only dents the membrane potential in the distal dendrite.

Repeating the simulation of Fig. 13.7 with active dendrites (Hodgkin-Huxley channels) produces a far more robust back-propagating action potential (Fig. 13.8). In this case I placed the synapse in the distal dendrites (downward arrow) rather than in the soma. While inhibitory synapses are often on or near the cell soma, excitatory synapses tend to be located more distally. With the passive dendrites of Fig. 13.7, a distal synapse would not be able to fire the cell. In Fig. 13.8, the EPSP is augmented by the active channels and easily reaches the soma. After the neuron fires, the action potential heads back up the dendritic tree. The earlier passage of the EPSP activated and began to inactivate the sodium channels in the dendrites. This partial inactivation slightly distorts and delays the back-propagating spike (oblique arrow).

Recall that the back-prop algorithm propagates error signals backward across synapses. Spike back-propagation propagates an outgoing signal from the soma back into the synaptic arrival zone. Although back-prop, the spike, is not *back-prop*, the algorithm, it does have implications for learning. Hebb's rule states that synaptic strength increases when there is a coincidence of presynaptic activity with postsynaptic activity. Postsynaptic activity can be interpreted in different ways, but Hebb's original

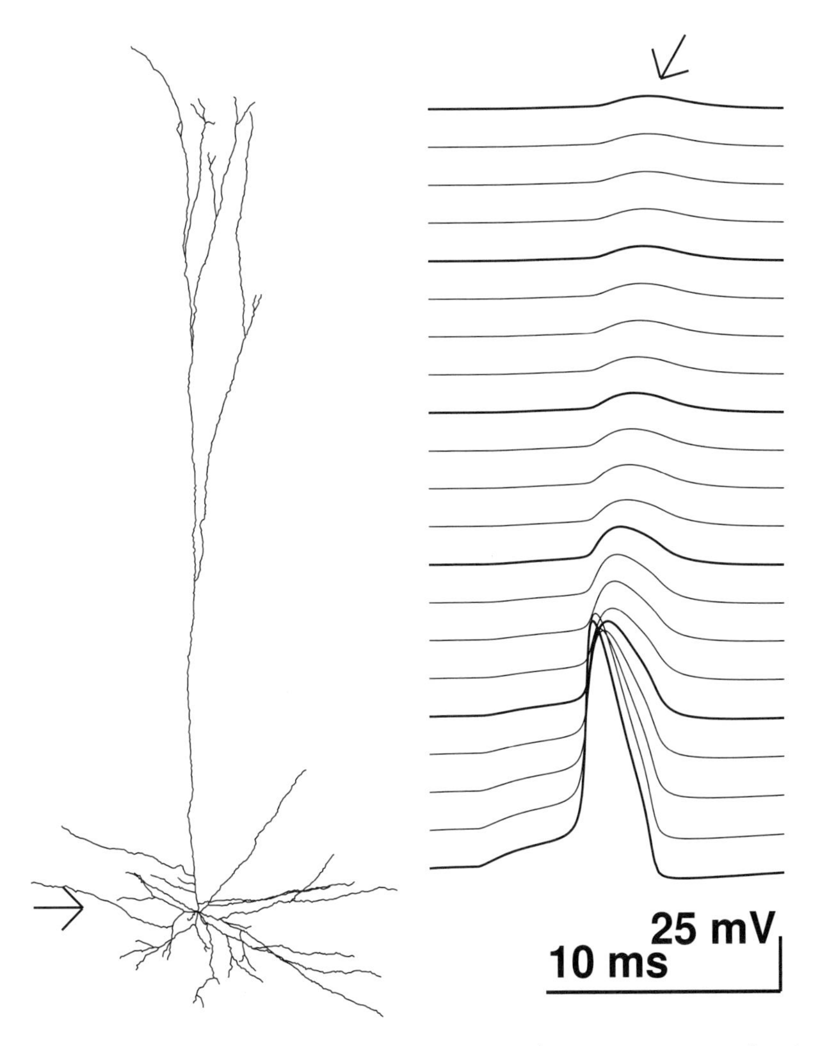

Fig. 13.7: Synaptic activation at the soma (rightward arrow) triggers an action potential that then passes up the dendritic tree. Each trace is positioned next to its location on the dendritic tree. Only the soma compartment is active; all dendrite compartments are passive.

idea was that postsynaptic activity is the activity coming out of the postsynaptic neuron. Back-propagating action potentials provide a way for this output signal to get back to the synaptic zone so that pre- and postsynaptic activity can be simultaneously monitored.

As mentioned above, the phenomenon of LTP and the NMDA receptor have been intensively studied as a biological example of the Hebb synapse.

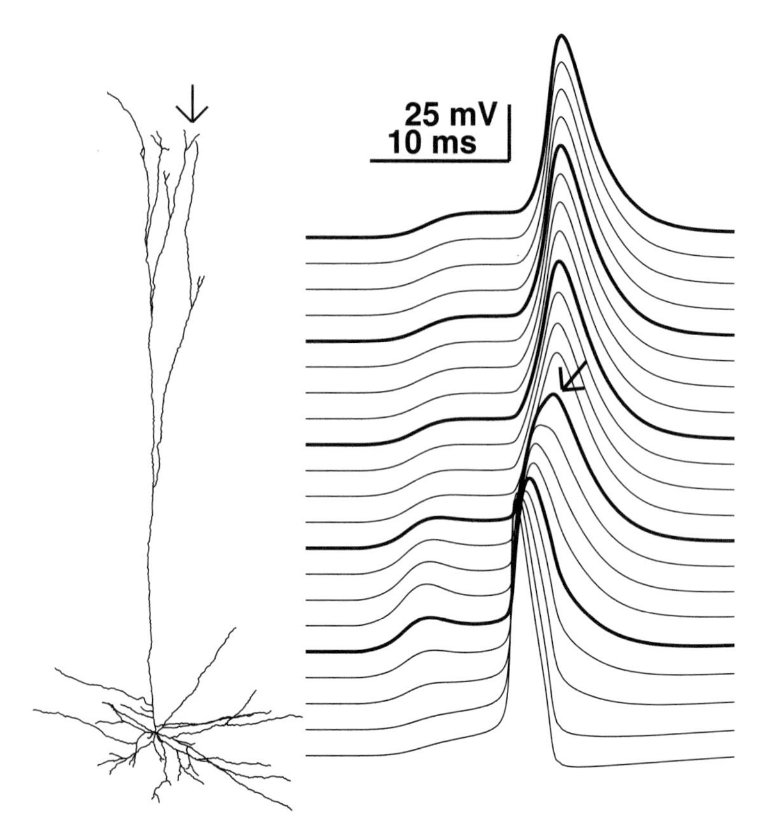

Fig. 13.8: Back-propagating action potential. Down arrow: EPSP location. Oblique arrow: delay in back-propagating spike.

We can now put together models that connect input and output in the way Hebb originally envisioned. As described above, NMDA serves as a coincidence detector by being sensitive to both intracellular voltage and extracellular ligand. The ligand is glutamate and the voltage sensitivity comes from channel blockade by magnesium. Active dendritic channels permit the action potential to propagate back up the apical dendrite. The depolarization from the back-propagating action potential relieves the magnesium block. In the presence of glutamate, NMDA will signal for synaptic strength augmentation via the AMPA receptor.

If we assume that augmentation will be proportional to total NMDA current, we can look at how the concentration of dendritic active channels will affect this NMDA signal. We would expect that increased sodium-channel density will lead to a bigger back-propagating spike, which will produce more NMDA current. Modeling shows this to be true for a limited range of values. However, when the density starts to get too high, the NMDA current starts to decline. This is due to two factors. The high-density case gives a larger but shorter back-propagated action potential. The shorter duration of the action potential means less time for the incoming current. Remember that the incoming current is dependent not only on the conductance but also on a driving force dependent on the distance between the voltage and the reversal potential for the channel. Therefore, a higher amplitude spike reduces driving force and also reduces incoming current.

13.6 Summary and thoughts

Compartment modeling combined with Hodgkin-Huxley-like parameterizations of ion channels and synapses is the state of the art for detailed neural simulation. Each compartment is a parallel-conductance model. Different rheostats can be used in different compartments to allow representation of excitatory and inhibitory synapses, as well as a wide variety of ion sensitive channels with different dynamics. In general, chemical processing in neurons is omitted from these models, although some will include internal calcium concentrations in order to accurately model calcium-sensitive potassium channels. Synapses are modeled by triggering a rheostat based on activity in a presynaptic cell. There are a variety of common neurotransmitter receptors that are modeled, included $GABA_A$, $GABA_B$, AMPA, and NMDA.

There remains considerable uncertainty regarding the passive properties of neurons. This may have implications for the processing power of the single neuron. If neurons are tight (high resistance, long length-constant, electrotonically compact), then the neuron will tend to treat all signals similarly and sum them in the manner of an artificial neural network sum-and-squash unit. If neurons are leaky (high conductance, short length-constant, electrotonically large), then different signals will have very different effects on the neuron, suggesting more processing complexity. It is likely that some neurons are compact and that others are not. Of course, the physical size of the neuron will also play a role: little neurons are likely to be electrotonically compact. The amount of synaptic input will also be an important factor because synaptic activation increases conductance. Neurons that receive constant synaptic bombardment will be leakier and electrotonically larger. There are also neuromodulators that will alter the electrotonic properties of a neuron.

This active debate about passive properties may be irrelevant for the many large neurons that have active dendritic channels that boost depolarizing signals. There is still relatively little known about these dendritic channels, some of which may actually mute, rather than boost, signals. Depolarizing currents have been shown not only to boost EPSPs on their way down to the soma but also to permit action potentials to back-propagate up the dendrites from the soma. This has implications for the biological instantiation of the Hebb rule, since back-propagating spikes relay the neuron's output signal back to the synapse, allowing input and output to be simultaneously assessed at the synapse.

There is a remarkable conservation of pyramidal cell dendrite length across different mammalian species — about 1 mm. Pyramidal cells are the major projection cells of cortex. Some researchers have hypothesized that pyramidal cells have kept their particular size due to some particular signal integration advantage involving temporal and spatial summation. Knowledge of the detailed topography of active channels and synaptics in the apical dendrite might finally reveal what neurons really do.

14

From Artificial Neural Network to Realistic Neural Network

14.1 Why learn this?

As noted toward the beginning of this book, the field of computational neuroscience, still bound by its hybrid conception, has yet to really integrate computation and neuroscience. The computer science side, represented by artificial neural networks, continues to come up with the big concepts. The neuroscience side continues to generate often confusing, if not confounding, data. Reconciliation would require that the concepts of artificial neural networks be directly applied to realistic neural network modeling and from there to direct measurement in brain circuits.

In this chapter, I give an example of artificial neural network learning theory being brought to the biological level. Similar examples could be taken from work in the visual system. I chose this particular example because it also helps illustrate some issues and problems in computational neuroscience. The focus of this chapter is as much on the theory of theory as it is on the theory of learning and the particular model discussed.

14.2 Hopfield revisited

Hebb's rule for learning through activity-dependent changes in synaptic strength is one of the few concepts commonly accepted all the way from artificial modeling through realistic modeling to experimental neuroscience. Hebb's rule is used to form cell assemblies. The cell assembly hypothesis is also widely accepted. To directly demonstrate cell assemblies experimentally, it would be necessary to record from very many cells simultaneously in an awake, behaving animal. This is not technologically feasible. Development of a biologically realistic Hebb assembly model would make it possible to predict and then search for experimentally testable biological corollaries of Hebb's theory. Because the Hopfield model is the best demonstration of how cell assemblies would form, Hasselmo and other researchers have been trying to port the Hopfield artificial neural network into a realistic neural network implementation.

The standard implementation of Hopfield's algorithm for learning in a content-addressable memory is given in Chap. 10. We start with N vectors, each representing a different memory. Each vector is multiplied by itself using the outer-product rule. This produces a separate matrix for each memory. Each matrix can map an incomplete memory, a part of the original vector used to form that matrix, onto the whole memory — making it content-addressable. Next, the N outer-product matrices are added together to form a final summed matrix. This summed matrix is the connectivity matrix for the Hopfield network. The product of this connectivity matrix and a partial or degraded copy of any of the N memories will produce the full, corrected memory.

Implemented in this way, the Hopfield network is nonbiological. Outer-product and matrix multiplications require global processing algorithms to pool and coordinate information over arrays of values. Biologically, this would require that populations of neurons share and exchange information instantaneously. Real neurons are local processing elements that operate sequentially in real time.

Focusing on learning, the problem with this standard artificial neural network implementation is that learning in the Hopfield network does not take place over time, but instead just happens instantaneously as the vectors are multiplied and matrices summed. Actually, Hopfield's original description of his network did not emphasize this implementation. Instead he talked about asynchronous updating: individual synapses would be updated at different times in response to activity in the network. This interpretation of local interactions is more biological than the standard global implementation. However, when we try to implement the network in this way, we run into a problem.

That problem is cross-talk: presentation of new memories activates old memories. The network learns the *combination of old and new* instead of learning the new pattern independently. In Fig. 14.1, two overlapping patterns are presented sequentially: AB and AC. Ideally, Hebbian learning should potentiate the $A \leftrightarrow B$ and $A \leftrightarrow C$ synapses so that activation of unit B will recall AB and activation of unit C will recall AC. Instead, we end up with a mix. First, AB is learned. Then AC is presented. This activates B via the potentiated AB synapse. Because both B and C are now active at the same time, the $B \leftrightarrow C$ synapses are also potentiated. How-

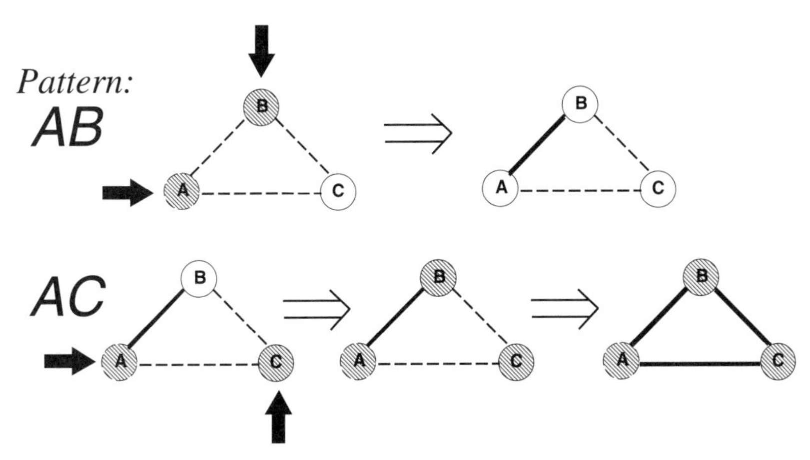

Fig. 14.1: Cross-talk spreads activity during learning. Top: activation of AB (arrows) potentiates $A - B$ synapse (bold line). Bottom: subsequent activation of AC potentiates $A - B$ synapse but B triggers C and $B - C$ synapse is potentiated. (Dashed lines: naive synapses; cross-hatched circles: active cells).

ever, B and C are not found together in either of the memories. They were mistakenly coactivated because of cross-talk during learning. Now, when B is activated during recall, it will activate the entire network instead of just the pattern of which it was a part.

Instead of having constituent assemblies, the whole network has become one big cell assembly. The different patterns have not remained separate, but instead blend into each other. This makes sense when we consider that a cell assembly is a content-addressable memory that is designed to complete a partial input. If there is any overlap between a new memory and an old memory, the new memory will trigger the old memory and the new memory will not be learned independently. Instead, the system will learn a combined memory that contains all the features (the union) of both memories together. As another example, let's say the system was being trained on faces. Instead of learning a new face distinct from other faces, the system would instead reinforce its knowledge of anatomy — the nose lies below the eyes, the mouth lies below the nose. Not only that, but if it were presented with a beard it would overgeneralize, remembering that all faces have beards. This would not be a very useful memory system.

14.3 Suppression model for reducing interference

The solution to the interference problem is to make learning and recall distinct phases so that they don't get in each other's way. Turning off

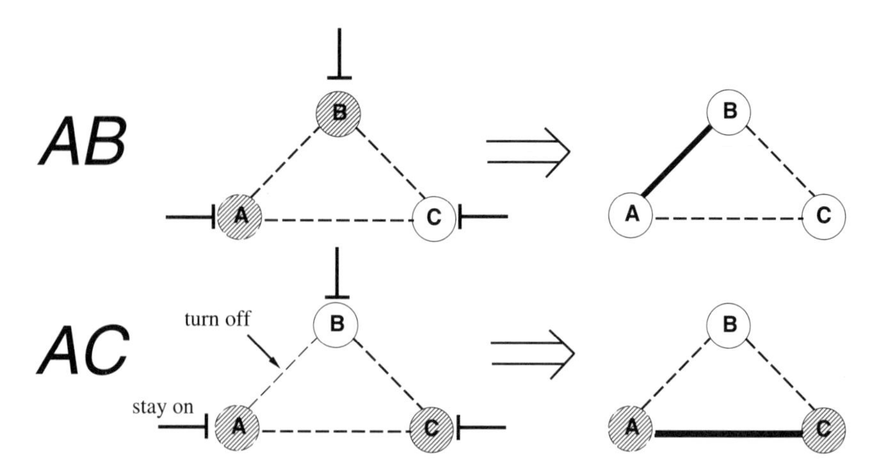

Fig. 14.2: Suppression model: turning off associative synapses during learning prevents cross-talk. Symbols and patterns as in Fig. 14.1.

synaptic connections during learning would prevent the new memory from bleeding over into the old one. However, turning off all synapses would also prevent the initial activations of units by the external presentation of the memory. Therefore, it is necessary to turn off or reduce the intrinsic, associative synaptic connections while preserving the incoming afferent connections (Fig. 14.2). Although the turned-off associative synapse will not activate the postsynaptic unit, it can still be potentiated. The evidence of this synaptic potentiation cannot be seen until the synapse is turned on again.

Hasselmo and his colleagues recognized this problem and tracked down mechanisms that could turn off these associative synapses during learning. They identified acetylcholine (adjective: cholinergic) as a modulatory neurotransmitter that was known to reduce the amount of glutamate released at some synapses. The finding of damaged cholinergic cells in Alzheimer dementia had previously suggested that acetylcholine might have an important role in learning and memory. The researchers suggested that acetylcholine was necessary to avoid cross-talk during learning. The critical experimental observation, now confirmed in more than one brain area, was that cholinergic activation primarily suppresses associative synapses, leaving afferent synapses intact.

14.4 A digression into philosophy

Modern philosophy of science hypothesizes that science progresses by paradigm shifts, major changes in theory such as occurred when the earth-centered system of Ptolemy was replaced by the sun-centered system of Copernicus. A corollary of this theory of theories is that theory should lead experiment: all experiments should be driven by specific hypotheses. I'm certainly in favor of developing explicit hypotheses. However, there is an interplay between theory and data that is not entirely captured by the notion of hypothesis-driven research. We can see in the present example how theory (cross-talk in Hebb assemblies) suggested a search for data (neuromodulatory suppression of associative synapses). This was an hypothesis-driven search for facts.

Classical theory of science is derived from physics. In physics, a new observation can cleanly disprove a governing hypothesis and perhaps suggest a new hypothesis. Biology ain't physics. In biology, if a new fact doesn't fit into the hypothesis, the hypothesis will be bent and shaped in order to make it fit. If the fact still doesn't fit, that fact will be ignored. This may be awfully sloppy intellectually, but unfortunately experiments are rarely conclusive in biology. The results *in vitro* may differ from the results *in vivo*; the results in cat may be different from the results in rat; the results under urethane anesthesia may not look like the results under halothane anesthesia; the results from electron microscopy may appear to contradict the results from light microscopy. This partly reflects the many artifacts that can pop up in the complex technology of data acquisition. It is also related to the notion, introduced early in this book, that biology is a hack. Things are cobbled together from bits and pieces of old stuff in order to get the job done. This means that a particular experiment may uncover vestigial mechanisms that are not currently in use. It also means that two different species may be doing the same thing in different ways through homologous evolution (e.g., the independent development of wings in the birds and the bees, and the bats).

Because of this intrinsic sloppiness, biological theorizing must constantly battle the odd couplet of excessive sloppiness and excessive neatness. You do not want to be so attached to your theory that your dead fingers cannot be pried loose no matter how much evidence is adduced against you. However, you need to be confident enough so that you aren't immediately dislodged from defense of your poster by hecklers at the trade fair.

In neuroscience, the search for facts does not always require that someone do an experiment. In many cases, needed data will turn up after a search through the literature. There are plenty of orphan facts out there waiting for their theory to come in. Similarly, once the fact is fit into the hypothesis, more orphan facts raise their hands and demand to be let in as well. When they were originally hatched, these orphan facts were probably associated with some kind of hypothesis. In many cases, however, this old hypothesis has been long forgotten. The orphan facts now live out their lives in journal limbo, hypothesis-free, waiting for a brave theorist to give them meaning and purpose.

In this case, the Hopfield network provided a hypothetical need for synaptic suppression. A search turned up acetylcholine as a candidate mechanism. Following this hypothesis-driven fact-finding, we can now do fact-driven hypothesis-building. Specifically, we can look at the many additional orphan facts about acetylcholine and use these orphan facts to extend the Hopfield hypothesis.

14.5 Acetylcholine has multiple effects

In the jargon of pharmacology, drugs can be clean or dirty. Clean drugs have a single effect. They bind exclusively at one spot and leave everything else alone. Dirty drugs are all over the place. They bind at multiple places and have multiple effects. From the perspective of pharmacology, clean is good and dirty is bad. You want specificity: your penicillin should kill the bacteria and leave your own cells alone.

From the perspective of biology, dirtiness appears to be the rule rather than the exception, a part of the aforementioned biological sloppiness. Biologically, endogenous "drugs" are ligands that float from one location in the body to another. Such ligands include not only neurotransmitters but also hormones and a variety of immune agents. Such endogenous compounds invariably turn out to have multiple receptors. In some cases this multiplicity of receptors just reflects the reuse of the same ligands in a variety of unrelated roles. As mentioned previously, biological chemicals are reused endlessly — various metabolic and genetic agents are recycled as neurotransmitters. However, it seems probable that multiple neurotransmitter actions will often turn out to be functionally significant, that the many cellular changes effected by release of one neurotransmitter will all contribute to a single effect at the network level. A particular neurotransmitter in a particular part of the brain would then have a particular functional role. This would not preclude its having a different role in a different part of the brain or in different states, such as sleep and wakefulness.

In the present example, acetylcholine is hypothesized to place the network into learning mode. One way that it does this is by depressing intrinsic, while preserving afferent, synapses. We then assume that other cholinergic effects also contribute to learning. These effects include 1) suppression of inhibitory synapses, 2) suppression of spike adaptation, 3) augmentation of the NMDA component of all synapses, and 4) increased propensity to long-term potentiation (LTP).

Fig. 14.3 shows some of acetylcholine's effects graphically. Impinging on one cell, these various effects will be expected to interact in complex ways. There appears to be a direct effect on Hebbian learning: acetylcholine lowers the threshold for producing LTP and can even produce LTP by itself when given in larger doses. This could be a direct action of acetylcholine on

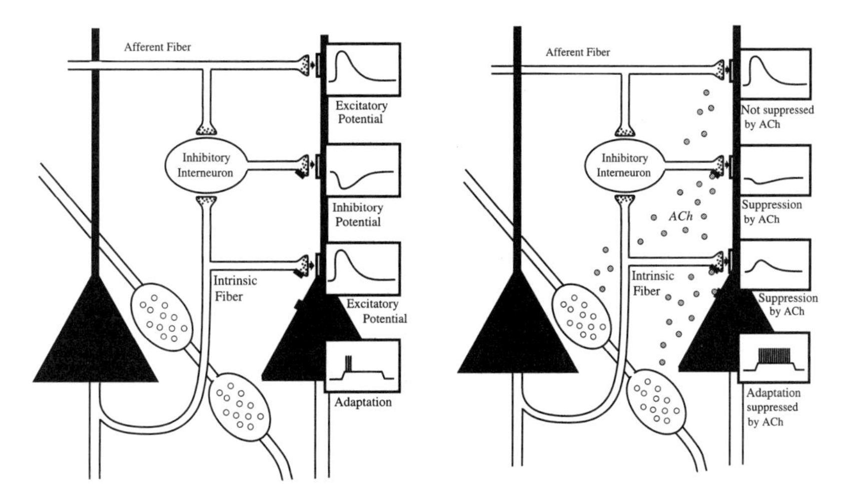

Fig. 14.3: Cellular effects of acetylcholine.

NMDA synapses or might be a secondary effect. For example, acetylcholine-mediated increased postsynaptic firing might secondarily enhance LTP.

Ideally, we would want to examine all of the known acetylcholine effects in concert using both the experiment and the model at the cellular level. This would tell us how the cells' overall input/output functions will be altered. We could then place the altered cells into the network and test learning. Typically, modeling projects have a more modest scope. In this case, it is easier to study just one of the cholinergic effects in isolation. This single-effect modeling may fail if the learning effect is an emergent property due to several individual acetylcholine effects acting together. Nonetheless, the pursuit of single effects is simpler and may provide incremental improvements in network learning and in our learning about the network.

The dual-matrix hypothesis

One of the cholinergic effect mentioned above, augmentation of the NMDA component of synapses, is one that my colleagues and I have looked at. I retell the story in order to contrast the relative clarity of a completed story with the preliminary confusion of modeling research.

The biological version of the Hebbian synapse utilizes two receptors. The NMDA receptor acts as monitor of pre- and postsynaptic activation through its dual ligand and voltage sensitivity. The AMPA receptor serves as the output device. Activity across AMPA is augmented following Hebbian activation. If NMDA were solely a monitor and a triggering device, it would communicate only with the AMPA receptor or with presynaptic mechanisms, leading to greater transmitter release. However, NMDA is primarily connected to an ion channel that depolarizes the postsynaptic cell. Thus it is not only a monitor and trigger, but also a synaptic communicator.

The dual role of NMDA has dual implications. First, it sets up a conflict of interest. NMDA both depolarizes the postsynaptic cell and is responsible for monitoring depolarization. This creates the potential for positive feedback at the synapse: NMDA could depolarize and then call for further depolarization through AMPA augmentation. The further depolarization would permit greater NMDA current. However, the time constants for synaptic augmentation are relatively long so that this positive feedback is not explosive. Additionally, other mechanisms reduce synaptic strength and control runaway depolarization.

The second implication has to do with the dual activation through AMPA and NMDA triggered by glutamate release. The network may be viewed as having two connectivity matrices. AMPA strength is augmented and decremented though Hebbian mechanisms. NMDA strength remains unchanged. Even if they start out the same, the effective connectivity of the two networks would end up being different. Additionally, AMPA strength may be zero at some synapses. These are called silent synapses since, absent postsynaptic depolarization, the synapse will not do anything. Such connections would be represented in the NMDA network alone unless synaptic potentiation took place.

In addition to the dual synaptic receptors, there is another major discrepancy between the Hopfield network and experimental reality. The Hopfield network is fully connected (Fig. 14.4, top). The biological network is not; it is sparsely connected (Fig. 14.4, bottom). In the Hopfield network, any two neurons that fire together will become connected. In biology, a single neuron connects with relatively few of the many other neurons in a particular brain area. Neurons that are not connected anatomically will not become newly connected as a result of their activity. (Axonal sprouting can occur, but it is not believed to be Hebbian.) The sparse cell-to-cell connectivity in the brain limits the formation of cell assemblies.

Now we can connect the dots. The matrix is sparse, making it harder to store information in the network. Each connection is precious since there are so few of them. However, each connection is effectively two connections. In our study, we hypothesized that the second, unaugmented, NMDA connection would preserve this precious connectivity by allowing activity during learning when the AMPA connectivity is suppressed. We did modeling to test this hypothesis. In line with our hypothesis, a network with dual connectivity worked better than the network with mono connectivity.

True confessions

I've told a lovely story about the dual-matrix hypothesis. Alas, it is not the true story of the modeling. As with most science, we did not really set out to discover what we did discover. I've told the story in reverse, with the

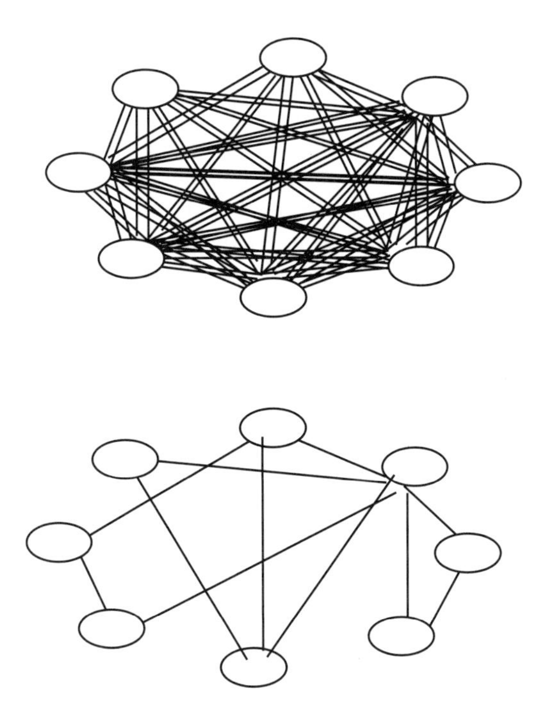

Fig. 14.4: Top: Connectivity in the Hopfield network is full and reciprocal (thin lines). Bottom: In the brain the cell-assembly layer is sparsely connected.

explanation first and the model last. Actually, the simulations were done as explorations. Various facts, opinions, and phenomena congealed along the way.

The real story is as follows. We took a Hopfield network. We put in some biological features. The network worked better with them than it did without them. We then went back and figured out why it worked better. Finally, the story made some sort of sense to us. After suitable writing and rewriting, it made some sense to non-us individuals as well. Above, I pointed out that biology is not physics. To that, I will add that biological modeling (and science in general) is not the same as engineering. As you set out to do one project, you may end up with a different one.

14.6 Summary and thoughts

In this chapter, I tried to show how we can now begin to close the circle of theory and experiment. Artificial neural network theory suggests how a content-addressable memory can be built. Efforts to implement this artificial neural network in a realistic network reveal problems with the algorithm. A work-around is developed that fixes the algorithm but requires an additional mechanism. The additional mechanism is found experimentally. It is the neurotransmitter acetylcholine. Other effects of acetylcholine are then explored for their possible contributions to learning. One such effect (NMDA augmentation) helps solve another problem with the biological version of the original artificial neural network algorithm (sparseness).

One advantage of realistic neural networks compared to artificial neural networks is that they include detailed results that can allow pharmacological predictions. For example, a model of acetylcholine in learning could be used to determine the relative importance of the various acetylcholine effects. This information could suggest which acetylcholine agonists would be more likely to preserve learning in patients with early Alzheimer disease.

I favor using artificial neural network ideas to develop realistic neural networks that make experimental predictions. An alternative approach is to skip the realistic neural network step and go straight from artificial neural network to brain measurements. This requires that we accept literally the assumptions of artificial neural network modeling, most notably the assumption of rate coding. fMRI is a physiological method that can measure correlates of metabolic activity in the awake behaving person or animal. Metabolic activity will increase with increased firing rate. As fMRI resolution improves toward the single neuron level, it may be possible to look directly into the brain for evidence of artificial neural network algorithms.

Having now stretched neural modeling about as far as I am able to take it, I would like to step back and again consider the brain more directly. All of our models are barely rubbing the edges of this enormous *terra incognita*. Columbus and crew set foot on an unknown island at the middle of the end of the earth. Maybe it looks like India. Maybe it smells like India. Is this really India?

15
Neural Circuits

15.1 Why learn this?

In prior chapters, I've mostly generalized about brain areas and brain cells — talking about typical neurons and generic network architectures. Most of the brain doesn't look anything like these generic versions. Each brain area has its own peculiar cells that have idiosyncratic dendritic morphology and differing complements of voltage-sensitive ion channels. Various brain areas also have very different circuitry. Some brain areas have odd types of connections (e.g., dendrodendritic or axoaxonic) that aren't covered in standard brain theory. In this chapter, I describe a few brain circuits. For the most part, I only give gross connectivity without giving any specifics about the forms or spiking patterns of the cells involved.

15.2 The basic layout

The hip bone is connected to the knee bone and the knee bone is connected to the hip bone. This basic principle of biological reciprocity is often respected in the brain as well. Neocortical areas typically project back to areas from which they receive projections. Principal cells of a brain area are generally taken to be those cells that are largest and that project to other brain areas. The principal cells of most areas are excitatory. There are two major brain areas, cerebellum and basal ganglia, where the principal cells are inhibitory.

Microscopically, the brain is made up of neurons and glia. In anatomy and pathology, the opposite of "microscopic" is "gross," meaning able to be seen with the unaided eye. Grossly, the brain is made up of white matter and gray matter. The white matter is made up of myelinated axons, the wires. The gray matter contains the neuronal cell bodies and the neuropil, the tangle of dendrites and axons that makes up local circuitry.

Grossly, mammalian brains have an outer gray matter cortex that forms a rind around the brain. Below that is white matter made up of projections to and from cortex. Below that are the deep nuclei or ganglia of the brain, irregular balls of gray matter. Cortex and these various nuclei and ganglia are the major circuits for modeling.

15.3 Hippocampus

The hippocampus is the seat of episodic memory. It is one of the most heavily investigated areas of the brain. There is a famous hippocampus-less man who is known only by his initials — H.M. Hippocampus was removed from both sides of his brain to cure his epilepsy. This left him with some odd memory problems that have been intensively studied over the past several decades. He can remember things that happened before his hippocampectomy (removal of his hippocampus) but cannot form new complex memories. However, this does not mean that he cannot learn. He can learn complex games and unusual tasks like mirror writing (writing while watching your hand in a mirror). Afterward, he cannot remember having learned these things. When he is later asked to mirror write, he is surprised to find that he can do it easily. He cannot learn to find his way around the place where he lives or to recognize people whom he sees every day. However, if a person is consistently nasty to him, he will avoid that person. However, he will be unable to identify the person in a lineup or say why he doesn't like him.

It appears that the hippocampus is responsible for things that are newly learned, but that information is then gradually passed to neocortical areas. LTP, the biological form of the Hebb synapse, is prominent in the hippocampus and has been particularly well studied there. The hippocampus shows distinct states of activity, ranging from slow theta oscillations to high amplitude sharp waves. It may be that this propensity to produce sharp waves explains why the hippocampus is a common site of seizure initiation.

Activity in the hippocampus and nearby areas appears to proceed around a loop: entorhinal cortex → dentate gyrus → CA3 → CA1 → subiculum → entorhinal cortex (Fig. 15.1). (CA stands for *cornu ammonis* but it is rarely written or said in full.) All of these projections are excitatory. Entorhinal cortex receives information from much of the neocortex and projects back to the neocortex in turn. The most intensively studied section of this loop is

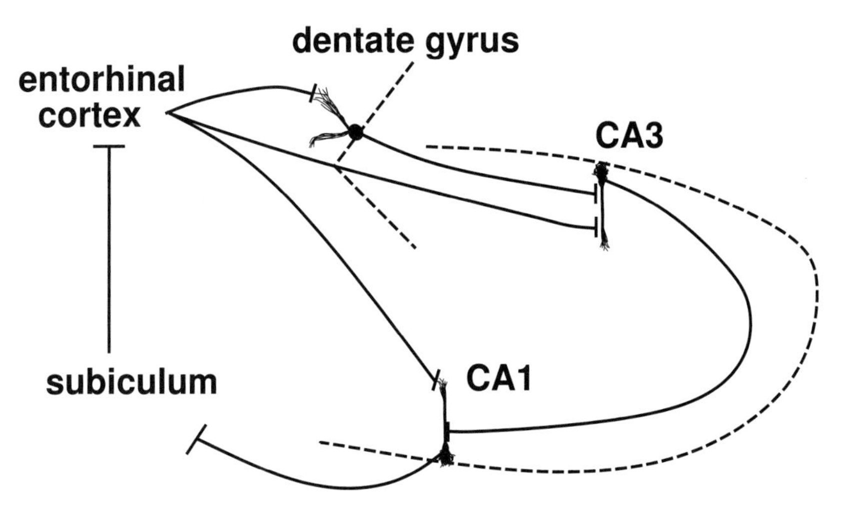

Fig. 15.1: Hippocampus. Dashed lines represent location of cell bodies (cell body layers); solid lines are connections. The aggregation of cells making up the sharply angled "blades" of dentate gyrus and the loop of cornus ammonis (CA) are easily seen with a low-power microscope.

called the trisynaptic pathway. It runs from the *perforant path* output axons of entorhinal cortex via the *mossy fibers* to CA3 and thence via *Schaeffer collaterals* to CA1.

The perforant path is the major input pathway from entorhinal cortex to hippocampus proper. Perforant path projects most strongly to dentate gyrus but also projects directly downstream to CA3 and CA1. Dentate gyrus granule cells are infrequently firing neurons that receive perforant path inputs and then project via mossy fibers. The mossy fibers form powerful inputs onto the proximal apical dendrites of pyramidal cells of area CA3. CA3 neurons are heavily interconnected. The relatively dense connectivity matrix of this area has led some to speculate that CA3 might form an attractor memory network similar to the fully interconnected Hopfield network.

CA3 projects in turn via Schaeffer collaterals to CA1. These are collaterals (branches) of axons that connect the CA3 neurons to each other. (There is also a CA2 between CA3 and CA1 but it appears to just be a transition zone.) CA1 has sparse interconnections. The Schaeffer collateral synapses are the site where NMDA-dependent LTP has been most fully described. CA1 projects to the subiculum, which projects to the entorhinal cortex, completing the loop.

15.4 Thalamus

The thalamus lies at the center of the brain and likely has a central function as well. It is made up of a variety of small nuclei. Most of these have reciprocal (back-and-forth) connections with a particular area of overlying cortex. The ones that are connected with a sensory area are considered relay nuclei because they connect a peripheral sensory area to primary sensory cortex. For example, the retina projects to a thalamic nucleus called the *lateral geniculate nucleus*, which projects in turn to the primary visual area of occipital cortex. Other thalamic nuclei are reciprocally connected with a motor area or are connected to hippocampus and other memory areas. The moniker "relay nucleus" is probably an oversimplification. Presumably, the thalamus is doing something to incoming messages, not just receiving them and sending them on to the cortex. One theory, made popular by Francis Crick, is that the thalamus might play a role in directing attention by activating a particular area of cortex when immediate processing is required.

When a person or animal is awake, thalamic cells produce fast spikes. In certain stages of sleep, the thalamus produces infrequent bursts that add up to slow oscillations. In addition to being seen in sleep, similar slow oscillations also turn up in absence epilepsy, a peculiar epileptic syndrome of children that causes brief periods of unconsciousness that look like simple daydreaming. This slow oscillatory pattern of thalamic activity has been much more extensively modeled than has the fast firing of the awake state.

An oscillation can be produced through the interactions of thalamocortical cells and reticularis cells. Thalamocortical cells are the principal excitatory cells of thalamus. They have reciprocal projections with cortex. Thalamocortical cells are the relay cells in sensory nuclei. Relay cells receive information from the sensory periphery and send it up to cortex. Reticularis cells are inhibitory cells that lie next to the main nuclei in a thin layer called the thalamic reticular nucleus. Reticularis cells have reciprocal connections with thalamocortical cells. There is also a population of local inhibitory interneurons in most thalamic nuclei.

Fig. 15.2 is an approximate circuit diagram of a generic sensory thalamic nucleus. Information comes in from the outside to the interneurons and the relay cells. It goes up from the relay cells to the cortex. The relay cells are also influenced by descending projections from cortex and by inhibitory connections from reticularis cells and interneurons. There are no direct connections between relay cells. It is unclear whether the back-projection from cortex involves cell to cell or area to area reciprocal connectivity.

One of the most peculiar aspects of thalamic circuitry is the existence of synaptic triads, shown in the inset in Fig. 15.2. The triads are found where the sensory input arrives in the thalamus and forms an excitatory connection with a dendrite of a relay cell and with a dendrite of an interneuron. The interneuron then makes an inhibitory connection with the relay cell

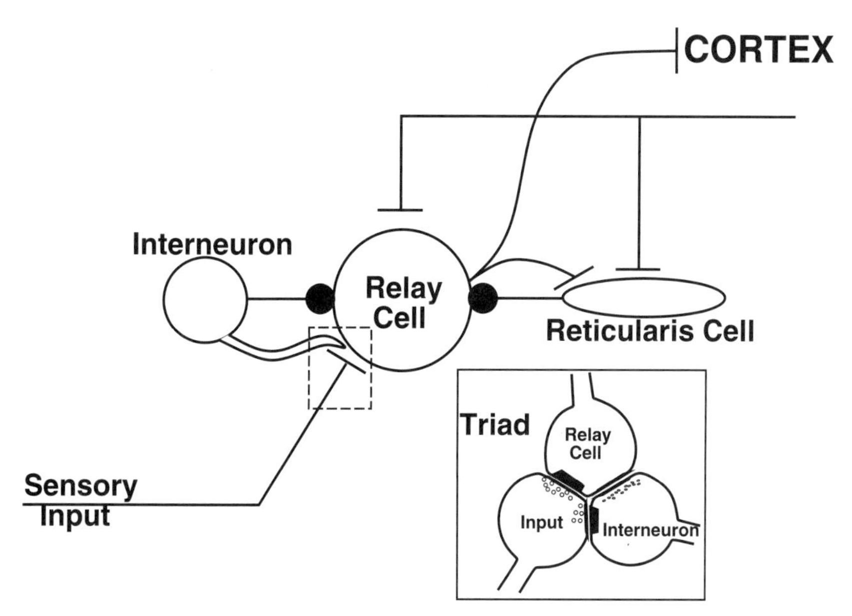

Fig. 15.2: Thalamus. Synaptic triad (rectangle and inset) involves interaction of thalamocortical and interneuron dendrites with the sensory afferent. It is unknown whether an axon from that interneuron would project to the same thalamocortical neuron, as shown here.

at the same location. This is a dendrodendritic synapse from interneuron to relay cell, which would be expected to provide a very rapid, highly localized, feedforward inhibition. It is unclear whether this immediate triad inhibition is the sole effect of interneuron activation at this dendrite. It is possible that the interneuron excitation could propagate from the dendrite down to the interneuron soma. In this case, this same input might also activate an action potential that would travel to the axon and inhibit relay cells (perhaps even the same relay cell) via axodendritic synapses.

15.5 Cerebellum

The cerebellum is the little brain that is piggy-backed on the back of the big brain. It has its own cortex and its own deep nuclei just like the big brain does. However, the major projecting cells of cerebellum, called Purkinje cells, are inhibitory rather than excitatory. The cerebellum is involved in coordinating movement. People with damage to the cerebellum cannot reach and grab things — they are likely to either miss the thing that they are reaching for or knock it over. They may also have problems walking, tending to weave as if drunk. In fact, reversible cerebellar damage is

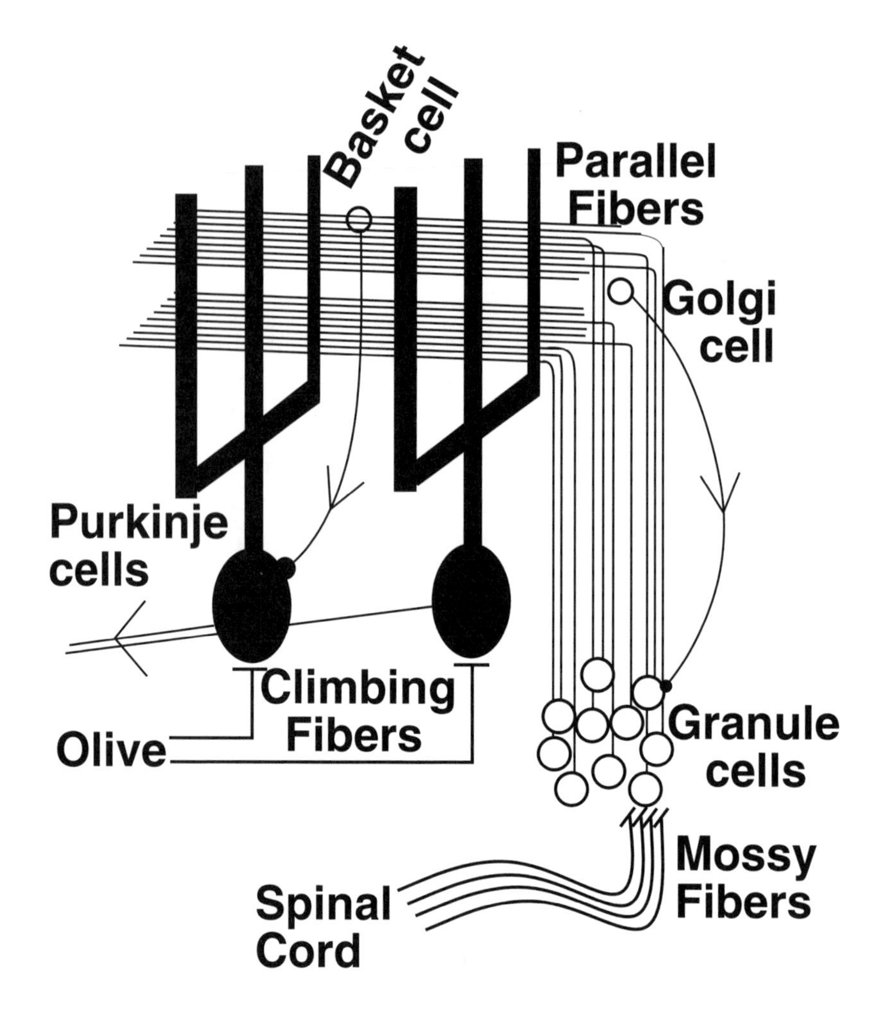

Fig. 15.3: Cerebellum. Inputs and parallel fibers are excitatory. Other intrinsic connections are inhibitory.

the reason for the gait, speech, and nose-touching abnormalities of alcohol intoxication.

In the human, the cerebellum has about 10 million Purkinje cells that collectively receive more than a trillion synapses. That's a lot of wiring. The most numerous inputs to the Purkinje cells come from granule cells, whose excitatory projections look like telephone lines as they run across connecting to about 100 Purkinje cells in sequence (Fig. 15.3). These telephone lines are called parallel fibers. Major sensory inputs to the cerebellum arrive from the limbs via the spinal cord and thence the mossy fibers to produce excitatory synapses on granule cells. Additionally, there is a powerful excitatory input from a brainstem structure called the inferior olive. The excitatory inputs from the olive are called climbing fibers because they climb up and wrap around the Purkinje cell main axon like ivy around a tree trunk.

There are also a variety of inhibitory cells involved in this circuit. The most prominent of these is the Purkinje cell itself, which provides an inhibitory input to the deep cerebellar nuclei. There are also basket cells, which are activated by parallel fibers and produce lateral inhibition on Purkinje cells. Another set of inhibitory cells are the Golgi cells, which get inputs from parallel, mossy, and climbing fibers, and feedback to inhibit granule cells.

15.6 Basal ganglia

The basal ganglia are a group of deep nuclei anterior to the thalamus. This nuclear group is involved in motor execution and planning. Like the cerebellum, the principal cells of basal ganglia are inhibitory. The basal ganglia accept input from much of the cortex and funnel this input down into relatively few cells. In the rat, for example, about one billion cortical inputs converge onto only about one million striatal cells (the striatum is another name for the putamen and caudate). These spiny stellate cells of striatum project in turn onto only about 100,000 cells of globus pallidus pars externa. The convergence from order $1 \cdot 10^9$ cortical cells to order $1 \cdot 10^5$ globus pallidus cells represents a convergence ratio of about 10,000 over these two synaptic steps.

The basal ganglia circuitry is generally described as having two major pathways, called the direct pathway and the indirect pathway. In Fig. 15.4, I have shown the direct pathway as solid lines and the indirect as dashed lines. Cortex projects to putamen and caudate in both cases. The direct pathway leads from there to substantia nigra pars interna and thence to thalamus. There are projections from thalamus to cortex, closing the loop. The direct pathway involves excitation (cortex to striatum) $\rightarrow$ inhibition (striatum to globus pallidus) $\rightarrow$ inhibition (globus pallidus to thalamus) $\rightarrow$ to excitation (thalamus to cortex), making it overall excitatory around the loop.

Again starting in caudate/putamen, the indirect pathway gets to the thalamus via a somewhat more circuitous route through three way-stations: globus pallidus pars externa, subthalamic nucleus, and globus pallidus pars interna. The indirect pathway from cortex to cortex involves excitation $\rightarrow$ inhibition $\rightarrow$ inhibition (globus pallidus pars externa to subthalamic nucleus) $\rightarrow$ excitation (subthalamic nucleus to globus pallidus pars interna) $\rightarrow$ inhibition $\rightarrow$ excitation. The three inhibitory projections would make it overall inhibitory around the loop.

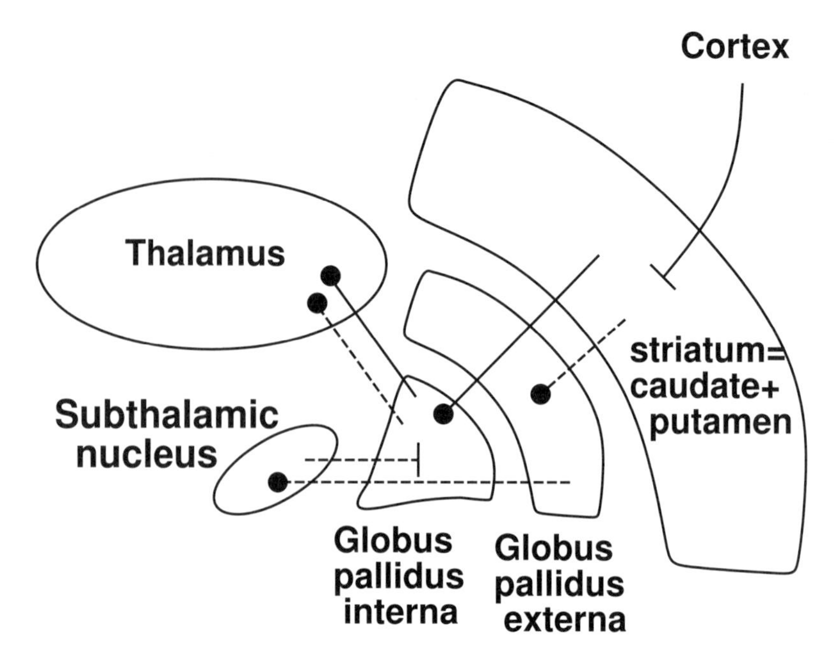

Fig. 15.4: Basal ganglia. Solid lines: direct pathway; dashed lines: indirect pathway.

Feedback loops can only work through inhibitory links if the cells being projected to are active. In the rate-coding paradigm, the cells would be tonically active and the inhibitory projections would reduce this rate of firing. In fact, both globus pallidus and thalamic cells, the two targets of the inhibitory projections, do have relatively high firing rates. Spiny stellate cells of caudate and putamen, on the other hand, fire rarely.

As with all such simplifications, the division of basal ganglia circuitry into two pathways leaves out several other pathways and subpathways. There is intrinsic connectivity in the striatum, where both GABA and acetylcholine serve as neurotransmitters. There is also a well-described and important interaction with the substantia nigra. The substantia nigra has two parts. Substantia nigra pars reticulata is just an extension of globus pallidus and has the same connectivity. Substantia nigra pars compacta, on the other hand, has dopaminergic cells that project up to the striatum. These are the cells that are lost in Parkinson disease, a disease that can be reversed, at least initially, by replacing the lost dopamine with drugs.

15.7 Neocortex

When people think about the brain, they usually first consider cortex, that magnificent edifice that separates us from the lowly beasts of field and forest, or at least from the nonmammalian beasts of field and forest. Another result of this corticocentrism is that neocortex is perhaps the most modeled areas of brain, although many such models are just generic artificial neural networks padded with a brief mention of cortex in the introduction to a paper.

Despite all this attention, a single basic cortical circuit has not been delineated. This failure is largely due to the complexity of cortex. It may also be due to the fact that different areas of cortex appear to have somewhat different wiring suited to their function. Brodmann, a famous late-19th century anatomist, was able to separate out 52 areas of cortex based on morphological criteria alone. There are likely many more areas if one considers functional specialization, although it is not clear to what extent functional differences require different wiring.

Neocortex has six layers. Layers 3 and 5 house large projecting pyramidal cells. Layer 4 is generally an input layer. Other layers have intersecting dendrites and axons. Such cell-sparse areas are called neuropil. There are other, smaller, excitatory cells in addition to the large pyramidal cells. There are also a variety of inhibitory interneurons of varying sizes and shapes. Some of these inhibitory cells have large dendrites that reach up to higher layers just as the pyramidal cells do. There is an unusual interneuron, the chandelier cell, that synapses on pyramidal cell axons. It has generally been assumed that chandelier cells can shut down pyramidal cell output but it has not been possible to demonstrate this.

Visual cortex is the most thoroughly studied neocortical area. Primary visual cortex is called V1 in cat and Brodmann area 17 in monkey and human. It is also called striate cortex because of the prominent white stripe (stria of Gennari) in layer 4. This stripe is white due to the myelinated fibers that arrive there from lateral geniculate nucleus, the visual area of thalamus. Many models of visual cortex have been developed. Most of these models seek to explain the responses of cortical neurons to particular forms of visual stimulation. They generally do not take account of the detailed circuitry of cortex or the particular firing patterns of different cell types.

Fig. 15.5 is my take on a basic neocortical circuit diagram. It is loosely based on a model put forth by Kevan Martin and Rodney Douglas. Input from the thalamus arrives in layer 4 and projects onto basal dendrites of layer 3 cells and apical dendrites of layer 5 cells. Excitatory cells in each layer interact with each other and with the cells of the other layer. Inhibitory neurons form networks of mutually connected cells within each layer and also interact with the pyramidal cells of that layer. Outputs project from layer 5 to the thalamus and beyond. There are also outputs from low-lying pyramidal cells in layer 6.

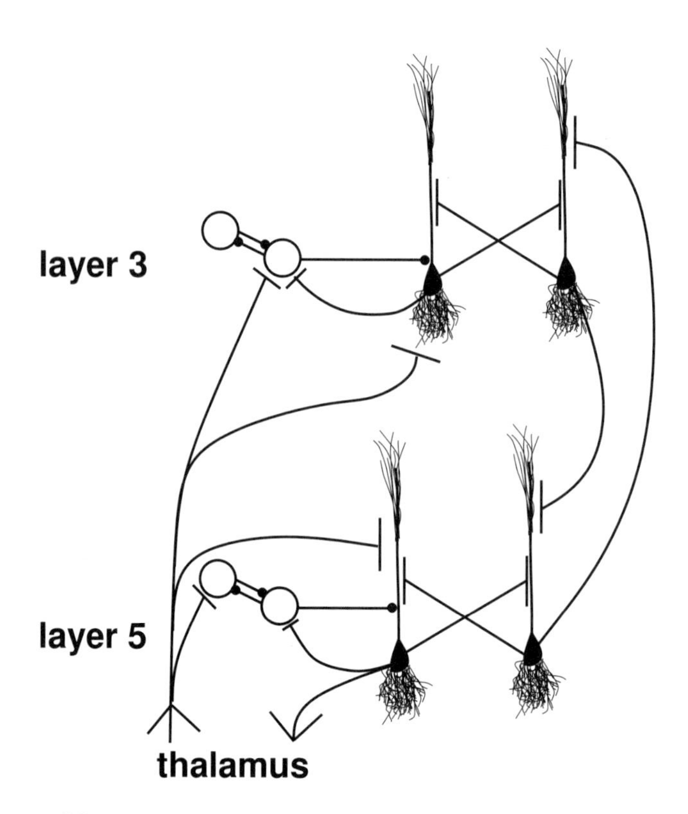

Fig. 15.5: Neocortex.

15.8 Summary and thoughts

Although I've tried to be fairly specific about connectivity in this chapter, I've only scratched the surface of the enormous variability of brain areas. For example, I presented a generic thalamic circuit. Actually, the thalamus is made up of different nuclei, and these different nuclei have somewhat different circuitry. In some nuclei of some animal species there are no thalamic interneurons, so the circuitry necessarily differs from the generic norm that I've shown.

I'm concluding on a note of mystery. The variety of circuits and cells in the brain is remarkable. From an engineering perspective, as a brain designer, one would prefer to build this complex system in a hierarchical manner, starting with a small set of circuit elements and organizing these in an only slightly larger set of fundamental subcircuits. This way things can be kept neatly organized and understandable. Conflicts between different circuit elements are kept to a minimum because the elements understand each other, and they share common assumptions about input, output, and processing methods.

The brain, like the eukaryotic cell, was built more in the spirit of a junkyard scavenger than of a space shuttle designer. The different pieces were originally developed for different purposes and gradually evolved so as to work together in a productive manner. Our knowledge of the brain has progressed to the point where different brain areas can be assigned general functions. A few things can be said about connections and their strength and importance. So how does it all work?

16 The Basics

16.1 Why learn this?

It's hard to be comfortable learning a new field whose routine chores are intimidating or mysterious. Although some aspects of computational neuroscience are based on fairly high-level math, most of what is needed can be handled with algebra. Most of algebra can be explained in plain words. William Clifford, an algebra reformer of a century and a half ago, said that "algebra, which cannot be translated into good English and sound common sense, is bad algebra." He was actually complaining about the linear algebra that is presented below. Despite its shortcomings, I've tried to present even linear algebra in good English.

As mentioned previously, the other saving grace of computational neuroscience is the computer itself. It's always been the case that a picture or demonstration is worth approximately $1 \cdot 10^3$ words. The computer can easily generate $1 \cdot 10^3$ demonstrations, allowing it to quickly outpace the paltry $1 \cdot 10^4$ words of this chapter. Most of the sections below have companion programs. The presentation here can also be readily supplemented by standard programs such as Octave, Matlab, Maple, or Mathematica.

In the following, I cover a variety of useful applied math skills. I start with the handling of units since these express the distinction between math and applied math — applied math is about specific things. I then discuss the binary number system, used to work with computers at the machine level. I cover some of the basic rules of linear algebra, frequently used in artificial neural network implementations. From there, I switch from the discrete to the continuous and talk about calculus, reviewing methods of numerical integration and briefly talking about analytic calculus as well. Finally, I consider some essential electricity concepts, particularly the effects of resistance and capacitance.

This chapter is made to be skimmed rather than read. Selective sections can then be read in more detail as needed to supplement information contained in the preceding chapters.

16.2 Units

Units and unit conversion is a nuisance. Different neural simulators use different units. Different physiology papers give results in different units. A great program for playing with units is the Unix "units" program, which I used to check all of the calculations here.

Scientific notation

In science, we encounter lots of numbers that are too big or too small to represent in the usual way. For example, 752400000000 and 0.00000000007524 are not easy to read. Instead, we separate significant digits from magnitude by using scientific notation. The significant digits can be written as a number between 1 and 10. The magnitude is written as a power of 10. Then, 752400000000 is written $7.524 \cdot 10^{11}$ and 0.00000000007524 is written $7.524 \cdot 10^{-11}$. 7.524 is the *mantissa* in both cases, but the exponents differ. For numbers greater than 1, the exponent can be readily determined by counting the digits to the right of the most significant digit (7 here). For numbers less than 1, count the zeros from the decimal point and add one.

Starting with Fortran, computer languages, and now computer lingo, has used the letter "e" (or "E") as a shorthand to mean exponent of 10. Instead of $7.524 \cdot 10^{11}$, we can write 7.524e11; for $7.524 \cdot 10^{-11}$, we write 7.524e−11. The "e" here, pronounced "e," has nothing to do with $e \sim 2.71828$, the base of the natural logarithm.

Numerical prefixes

An alternative to scientific notation, sometimes called engineering notation, is the use of numerical prefixes in the metric system. These numerical prefixes express the power of 10 that would be the second part of the scientific notation. They are used just before some type of unit, such as meter or amp. Some of these are widely used, for example, kilo and milli in kilometer and millimeter. With the exception of centi (10^{-2}), the commonly used prefixes express exponents that are multiples of 3: e.g., 10^3, 10^6, 10^9.

Therefore, the significant digits will be given as numbers between 1 and 1000. Above 1000, you go to the next prefix up. For example, 32432.5 meters would be written as 32.4325 km (kilometers), while .0324325 meters would be written as 32.4325 mm (millimeters). These conversions are easy to do if you just think about shifting the decimal point left or right by three places.

prefix	name	magnitude	original meaning
T	tera	10^{12}	monster
G	giga	10^{9}	giant
M	mega	10^{6}	large
k	kilo	10^{3}	thousand
c	centi	10^{-2}	hundred
m	milli	10^{-3}	thousand
μ	micro	10^{-6}	small
n	nano	10^{-9}	dwarf
p	pico	10^{-12}	a little bit

Here's a table of prefixes used in computer science and neuroscience. There are more of them, all the way up to 10^{24} (yotta; I guess 10^{72} is "yotta, yotta, yotta") and down to 10^{-24} (yocto). Except for μ for micro, they are generally easy to remember as the first letter of the full prefix. Small numbers are small letters and big ones (beyond kilo) are capitals: note the difference between "m" (milli) and "M" (mega).

The meanings of these prefixes is usually changed when dealing with the bits and bytes of computers. For the prefixes from kilo and above, each is taken to be a power of 1024 (2^{10}) instead of a power of 1000. Therefore, a megabyte (MB) is actually 1,048,576 (1024^2) bytes. More awful yet, this special computer rule is used only for bits and bytes measured directly, but is not used for information transfer rates. Specifically, bits per second (bps) does not use this convention. 12 kbps is 12,000 bps, which would actually transfer only $\frac{12000}{1024}$kb $\sim$ 11.7 kb in 1 second. 12 kbps is not equal to 12 kb per second! This is all such a mess that people screw it up all the time. When 3.5-inch floppy disks doubled in density so as to hold 1440 kB, this was marketed as a "1.44 MB disk." However, it actually has $1440 \cdot 1024 = 1,474,560$ bytes, which is correctly written as $\frac{1,474,560}{1,048,576} \sim 1.4$ MB.

Units and abbreviations

Because computational neuroscience makes use of many fields of study, we end up using lots of different units that can be hard to keep track of. The abbreviation for a phenomenon will differ from the abbreviation for that phenomenon's units, but is sometimes the same as the abbreviation of something else's units. For example, farads (F) are the units of capacitance (C). However, C as a unit is short for coulombs, the unit of charge, whose symbol is Q. Current, another C word, uses I as a symbol and is measured in amperes (A). Case matters: s is seconds, but S is siemens, the measure of conductance. Synonyms add to the confusion. Voltage, V, is also called electrical potential, E.

Below is a table of abbreviations. I've ordered them to put together similar or identical abbreviations that could be mixed up. I've focused on units commonly used in computational neuroscience, so I don't bother mentioning that "c" is also the speed of light. I give units as they are commonly used in computational neuroscience rather than giving the basic SI unit. For example, cell diameters are generally measured in micrometers (microns), not in meters.

In realistic neural modeling, subscripts are frequently needed to qualify symbols. For example, membrane resistance (R_m or R_{memb}) must be distinguished from longitudinal cytoplasmic resistance, which is represented by R_a or R_l. When referring to membrane resistance and capacitance, a lower-case letter is used to represent the resistance or capacitance per unit length: r_m in ohm-cm or c_m in farads per cm. An upper-case letter is used to refer to specific resistance or capacitance, which is given per a unit area of membrane: R_m in ohms times square centimeter and C_m in farads per square centimeter.

phenomenon	abbrev.	unit	unit abbrev.
computer memory	RAM	byte	B
data		bit	b
charge	Q	coulomb	C
capacitance	C	farad	F
current	I or i	ampere, amp	A
voltage, potential	V or E	volt	V
energy	E	joule	J
volume	V	milliliter, ...	ml, cc
area	A	squared length	μm^2, cm^2
velocity	v	meter per second	m/s
resistance	R	ohm	Ω
conductance	g	siemens, mho	S
time	t	second	s
frequency	f	hertz	Hz, /s
period	T	second	s
temperature	T	degree	°C, °K
diameter, length	d, L	micron	μm
chemical amount	n	mole	mol
concentration	[x]	molarity	M

Another set of confusing abbreviations comes from common numbers and constants. For example, i is used in mathematics to represent the $\sqrt{-1}$, the basic imaginary number. However, i is needed in electronics for current, so electrical engineering uses j for $\sqrt{-1}$. Here are some number symbols and constants.

name	abbrev.	value	comments & examples
imaginary number	i or j	$\sqrt{-1}$	j in electronics, i in math
circle ratio	π	3.1415926...	circumference/diameter
base of natural log	e	2.718281828...	$\ln(e) = 1$
base 10 exponent	e or E	none	1e3=1E3= $1 \cdot 10^3 = 1000$
Faraday	F	96485.341 A s	a mole of charge
gas constant	R	8.314 J/(mol K)	PV=nRT, R=k·Av
Boltzmann's constant	k	$1.38 \cdot 10^{-23}$ J/K	similar to gas constant
Avogadro number	Av	$6.022 \cdot 10^{23}$	molecules in a mole

Unit conversions

Unit conversions can be hard to do in your head, but they are easy to do on paper if you remember one rule: always multiply by 1. Given an equivalent measure, you can form a ratio that equals 1. For example, 5280 feet = 1 mile. Therefore, $\frac{1 \text{ mile}}{5280 \text{ feet}} = \frac{5280 \text{ feet}}{1 \text{ mile}} = 1$. Each time we multiply by 1, a unit that is in the numerator at one step is in the denominator in the next or vice versa. These units cancel each other and can be crossed out.

Knowing that 1 inch is approximately 2.54 cm, we can convert 50 million millimeters into miles as follows:

$$\frac{5 \cdot 10^7 \text{mm}}{1} \cdot \frac{1 \text{ cm}}{10 \text{ mm}} \cdot \frac{1 \text{ in}}{2.54 \text{ cm}} \cdot \frac{1 \text{ ft}}{12 \text{ in}} \cdot \frac{1 \text{ mile}}{5280 \text{ ft}} = \frac{5 \cdot 10^7 \cdot 1 \cdot 1 \cdot 1 \cdot 1 \text{ mile}}{10 \cdot 2.54 \cdot 12 \cdot 5280} \sim 31 \text{ miles}$$

Here the mm in the numerator in the first fraction cancels with the mm in the denominator of the second fraction and so on. We are left with only miles.

We frequently have some measure that's given as a ratio (e.g., miles per hour) and have to determine one of the measures given a value for the other one. For example, how long does it take to drive 1 inch at 60 mph?

$$\frac{1 \text{ inch}}{1} \cdot \frac{1 \text{ hr}}{60 \text{ miles}} \cdot \frac{1 \text{ mile}}{5280 \text{ ft}} \cdot \frac{1 \text{ ft}}{12 \text{ in}} \cdot \frac{60 \text{ min}}{1 \text{ hr}} \cdot \frac{60 \text{ sec}}{1 \text{ min}} \cdot \frac{10^6 \mu s}{1 \text{sec}} = \frac{1 \cdot 1 \cdot 1 \cdot 1 \cdot 1 \cdot 60 \cdot 60 \cdot 10^6}{1 \cdot 60 \cdot 5280 \cdot 12 \cdot 1 \cdot 1 \cdot 1} \sim 947 \mu s$$

Notice that I turned the fraction for 60 mph upside down to provide the second fraction. I also put all the given values down first and then converted each of them, rather than insisting that unit matching should be done at each neighboring fraction. The answer is just short of 1 ms, about the duration of an action potential.

Another example, common in neuroscience, is to calculate the total membrane capacitance (c_m, in farads) of a cylindrical compartment given specific capacitance (C_m, in farads per unit area) and the size of the compartment. The commonly accepted value for specific capacitance is 1.0 μF/cm^2. We will use a compartment 30 microns long and 20 microns wide. The surface area of a cylinder is $\pi\cdot$ length $\cdot$ diameter ($\pi \cdot L \cdot d$), so in this case area $= \pi \cdot 30\ \mu\text{m} \cdot 20\ \mu\text{m} \sim 1885\ \mu\text{m}^2$. A ratio for a squared or cubed volume can be obtained by simply raising the ratio for the unit conversion to the appropriate power ($1^2 = 1^3 = \ldots = 1$). In this example, since $\frac{1 \text{ cm}}{10 \text{ mm}}$ provides a ratio for cm to mm distance conversion, we can square this to get the ratio $\frac{1^2 \text{ cm}^2}{10^2 \text{ mm}^2}$ as a conversion for area. Similarly, we would cube it to get a conversion factor for volume.

$$\frac{1885\ \mu\text{m}^2}{1} \cdot \frac{1 \text{ mm}^2}{(10^3)^2\ \mu\text{m}^2} \cdot \frac{1^2 \text{ cm}^2}{10^2 \text{ mm}^2} \cdot \frac{1\ \mu F}{1 \text{ cm}^2} \cdot \frac{1\ F}{10^6\ \mu F} = 1.885 \cdot 10^{-11} F = 18.85 \cdot 10^{-12} F = 18.85\ pF$$

Here, as in many cases, we didn't know ahead of time the overall magnitude of the answer, so we originally expressed it in farads. We then change the exponent from -11 to -12, dividing by 10, and multiply the mantissa by 10, in order to get an exponent that corresponds to a standard prefix of engineering notation. If, for some reason, we wanted to work with mF instead, we could easily convert to get $18.8 \cdot 10^{-9} mF$.

Dimensional analysis

Dimensional or unit analysis is an assessment as to whether units that are being used together really belong together. If you try to convert miles to feet, that's fine; but if you try to convert miles to seconds, you can immediately see that you have an error in units. Such conversion error are less obvious when you deal with complicated conglomerations of units. For example, it turns out that a calorie per second can be converted into volt-amps (it's about 4.2 of them) since they are both units of power. (Volt-amps means volts times amps, not volts minus amps.) In Chap. 11 we add $\frac{C}{\Delta t} + g$. It's only possible to add numbers if the units for both addends are the same. In fact, $\frac{1F}{1s} = 1S$.

To make working with units a little easier, SI units are neatly matched up so that the major electrical and mechanical laws work out when you use standard units. Ohm's law states that V = IR; 1 volt (V) equals 1 amp-ohm (1 A-Ω). These can then be used for unit conversions by producing a ratio of 1: $1 = \frac{V}{A \cdot \Omega}$. $R = \frac{1}{g}$: 1 ohm equals $\frac{1}{\text{siemens}}$; $1 = S \cdot \Omega$. $Q = CV$ so 1 coulomb (C) is 1 farad-volt (F-V); $1 = \frac{C}{F \cdot V}$. $I = C \cdot \frac{\Delta V}{\Delta t}$ so 1 amp equals 1 farad-volt/second; $1 = \frac{A \cdot s}{F \cdot V}$. Of course, all these units divide out equally well if you make them all into milliunits or microunits or kilounits.

16.3 Binary

The details of care and handling for binary numbers seem trivial and may remind some readers of bad days in third grade. However, comfort with binary is useful for understanding the computer as a machine. In addition to their use in regular algebra (addition and subtraction mostly), binary numbers are also used in Boolean algebra (combinations of true and false). To handle binary numbers, it is also helpful to know octal (base 8) and hexadecimal (base 16).

Translating back and forth

Numbers can be represented in different forms depending on the value used as the basis for the number system. The noncomputer world generally uses base 10, decimal. A base specifies how many different symbols are available. Base 10 has 10 symbols: 0,1,2,3,4,5,6,7,8,9. Base 8 (octal) has eight symbols: 0,1,2,3,4,5,6,7; base 16 (hexadecimal) has 16: 0,1,2,3,4,5,6,7,8,9,A,B,C,D,E,F; base 2 (binary) has two: 0,1. Binary is hard to work with when doing arithmetic by hand because the numbers get so long. Translation from binary into octal or hexadecimal (also called "hex") is easy, so these are commonly used by programmers.

Modern number systems all use place notation. This means that the location of a numeral specifies its value. With any of the bases, the position of a numeral relative to the radix point determines a power of the base to be used as a multiplicative factor. In the decimal system, the radix point is called the decimal point. Every position to the left of the point is a power of the base counting up from zero. Every position to the right of the point is a power of the base counting down from negative one.

The base is sometimes indicated with a subscript: e.g., $10_8 = 8_{10}$. To identify the value of a number, you simply raise the base to the power appropriate to the place and then multiply by the numeral that's in that place. The number $257_{10} = 2 \cdot 10^2 + 5 \cdot 10^1 + 7 \cdot 10^0$. $257_8 = 2 \cdot 8^2 + 5 \cdot 8^1 + 7 \cdot 8^0 = 128_{10} + 40_{10} + 7_{10} = 175_{10}$. $257_{16} = 2 \cdot 16^2 + 5 \cdot 16^1 + 7 \cdot 16^0 = 512_{10} + 80_{10} + 7_{10} = 599_{10}$. Similarly, $1001.101_2 = 1 \cdot 2^4 + 1 \cdot 2^0 + 1 \cdot 2^{-1} + 1 \cdot 2^{-3} = 9.625_{10}$.

Translation of numbers between bases is relatively straightforward. For example, let's translate 532 into base 2. It's easiest to do if one refers to a table of powers of 2 in base 10. Translation into base 2 is easier than other bases since we only have to multiply each place by 1 or 0.

Powers of 2

power	value		power	value
0	1		6	64
1	2		7	128
2	4		8	256
3	8		9	512
4	16		10	1024
5	32		11	2048

The first step is to determine how many places will be needed. Looking at the table of powers, 532 is between 512 and 1024, so the base 2 representation will have a 1 in the 9th place (counting from a 0th place) and must have 10 places. 512 goes into 532 one time, so we put a 1 in the 9th place of the base 2 number. With 512 out of the way there is only 20 left $(532 - 512)$. The biggest power of 2 that can go into this is 16, so there is a 1 in the 4th place. This leaves 4, which is taken care of with a 1 in the 2nd place. All of the rest of the places get zeros. The number is 1000010100_2.

Another algorithm for converting to base 2 uses repeated division by 2 to determine place values in reverse order, from right to left. If a number is even, then the right-most (0th) place will have a 0 in it, and if it is odd, there will be a 1 there. This generalizes so that if we keep dividing by 2 we can use the remainder as the value for that place. Using 532 as an example again and working right to left:

$$
\begin{array}{r|cccccccccc}
532 & & & & & & & & & & 0 \\
266 & & & & & & & & & 0 & 0 \\
133 & & & & & & & & 1 & 0 & 0 \\
66 & & & & & & & 0 & 1 & 0 & 0 \\
33 & & & & & & 1 & 0 & 1 & 0 & 0 \\
16 & & & & & 0 & 1 & 0 & 1 & 0 & 0 \\
8 & & & & 0 & 0 & 1 & 0 & 1 & 0 & 0 \\
4 & & & 0 & 0 & 0 & 1 & 0 & 1 & 0 & 0 \\
2 & & 0 & 0 & 0 & 0 & 1 & 0 & 1 & 0 & 0 \\
1 & 1 & 0 & 0 & 0 & 0 & 1 & 0 & 1 & 0 & 0
\end{array}
$$

After completing this process the number can be read from left to right at the bottom: 1000010100_2. This same process works for translation to other bases as well — repeatedly divide and list remainders from right to left.

Addition and subtraction

The trick in adding or subtracting in a base other than base 10 is to notice when you'll have to carry (in addition) or borrow (in subtraction). In binary, there's no room to grow within a place so you're always carrying and borrowing.

$$\begin{array}{r}1\\+1\\\hline 10\end{array}\qquad\begin{array}{r}10\\-1\\\hline 1\end{array}\qquad\begin{array}{r}1111\\+1011\\\hline 11010\end{array}\qquad\begin{array}{r}10101\\-1010\\\hline 1011\end{array}$$

Octal and hex

When working with computers it's very convenient to work in octal (base 8) or hexadecimal (base 16) instead of decimal, since it is easy to translate from these into and out of binary. Hexadecimal utilizes 16 unique digits (radix 16): the 10 standard arabic numerals followed by the letters A to F. Hex is easily translated into binary since every hex numeral (called a hexit maybe?) corresponds to a sequence of 4 bits. Similarly every octal digit (0–7) corresponds to a sequence of 3 bits. Therefore, you can easily translate into and out of octal or hex on a digit-by-digit basis by using this table.

Octal to binary and hex to binary conversion

octal digit	binary value	hex digit	binary value	hex digit	binary value
0	000	0	0000	8	1000
1	001	1	0001	9	1001
2	010	2	0010	A	1010
3	011	3	0011	B	1011
4	100	4	0100	C	1100
5	101	5	0101	D	1101
6	110	6	0110	E	1110
7	111	7	0111	F	1111

To translate AF_{16} to binary just copy the digits off of the table: 10101111_2. Translating this binary number into octal requires rearranging the bits in groups of three starting at the right: **0**10 101 111. The left-most bits have to be left padded with zeros to complete the triplet. Then the octal number can be read off of the table: 257_8. Translating from any of these into decimal cannot be done with a simple table readout but must be multiplied out by hand: $AF_{16} = 10 \cdot 16^1 + 15 \cdot 16^0 = 175_{10}$ or $257_8 = 2 \cdot 8^2 + 5 \cdot 8^1 + 7 \cdot 8^0 = 175_{10}$. Nowadays, octal is less commonly used than hex because the modern byte (the elemental unit of computer memory) is typically 8 bits long, hence 2 hex numerals.

Boolean algebra

Boolean algebra uses bits as truth values. T, true, is 1, and F, false, is 0. The three major operators are AND ($\wedge$), OR ($\vee$), and NOT ($\sim$). NOT is also represented by placing a bar over a value: $\overline{\mathrm{T}} = \mathrm{F}$. In C, and some other computer languages, AND is '&,' OR is '|,' and NOT is '!'.

These are logical operators and the results can generally be understood by using logic. For the result of an AND to be *true*, both of the arguments must be true. For the result of an OR to be *false*, both of the arguments must be false. The following is the *truth table* for AND and OR.

A	B	$A \wedge B$	$A \vee B$
T	T	T	T
T	F	F	T
F	T	F	T
F	F	F	F

This same table can be generated by using standard arithmetic by defining any positive number as *true* and only zero as *false*. Then AND can be calculated using multiplication, and OR can be calculated using addition.

A	B	A	B	$A \cdot B$		$A + B$	
T	T	1	1	1	$\Rightarrow$ T	2	$\Rightarrow$ T
T	F	1	0	0	$\Rightarrow$ F	1	$\Rightarrow$ T
F	T	0	1	0	$\Rightarrow$ F	1	$\Rightarrow$ T
F	F	0	0	0	$\Rightarrow$ F	0	$\Rightarrow$ F

Boolean algebra shares many of the familiar rules of arithmetic. The commutative and associative laws are both valid for AND and OR. Boolean algebra has other properties that differ from those of arithmetic. One of the most useful of these is De Morgan's theorem, which states that $\overline{A \wedge B} = \overline{A} \vee \overline{B}$ and that $\overline{A \vee B} = \overline{A} \wedge \overline{B}$.

A popular Boolean operation in the world of neural networks is XOR, exclusive or, symbolized by $\veebar$. Here's the truth table for XOR:

A	B	$A \veebar B$
T	T	F
T	F	T
F	T	T
F	F	F

16.4 Linear algebra

Linear algebra is a basic tool for studying large systems. In this section, I present just the tools directly applicable to simulating neural networks and thereby omit most major concepts. For the newcomer to matrices, even the basics will seem confusing at first. However, the notions we need consist of a few rules that will become familiar after playing with some examples.

What is algebra? Why linear?

Algebra is the study of equations with one or more unknowns. In linear algebra we restrict ourselves to equations for straight lines (e.g., $y = 2 \cdot x + 1$). This means that none of the variables will be raised to a power or multiplied with one another, and that we won't use any of the transcendental functions such as sin, cos, exponential, or logarithm.

From this perspective, linear algebra would seem to be very simple. Complexity arises because we don't only want to handle a single linear equation at a time. Instead, we want to handle many linear equations simultaneously. To do this, we use matrices (plural of matrix) and vectors. In a linear algebra course, one first learns how to use linear algebra to solve simultaneous equations. This does not concern us here, since we are interested only in using vectors and matrices as a tool for artificial neural networks.

As explained in Chap. 6, a vector is an ordered set of numbers. A vector is defined by its length N: the number of numbers in the vector. A standard vector is a column vector, drawn from top to bottom. A vector drawn from left to right is a row vector. A column or row vector is like a column or row of a spreadsheet; a matrix is like the whole spreadsheet. A matrix is a two-dimensional array of numbers with every row having the same length as every other row and every column the same length as every other column. A matrix is defined by its dimensions, $M \times N$, where M is the number of rows and N is the number of columns (important: rows then columns). A matrix can be formed by putting together M row vectors of length N or equivalently by putting together N column vectors of length M. If $M{=}N$, the matrix is called a square matrix. M and N can take on any integer values from 1 on up.

A sparse matrix refers to a matrix with a lot of zeros in it. Sparse weight matrices are often encountered in neural networks. The weight matrix represents all possible connections. If most neurons are only connected to a small subset of the entire population, then the weight matrix will be sparse.

The *transpose* of a matrix is the result of flipping it over the diagonal that runs from the top left to the bottom right corner. The symbol for transpose is a superscripted 'T.' For example:

$$\begin{pmatrix} -5 & -7 \\ -25 & -8 \\ 10 & 14 \end{pmatrix}^{\mathrm{T}} = \begin{pmatrix} -5 & -25 & 10 \\ -7 & -8 & 14 \end{pmatrix}$$

A vector is really just a special case of a matrix for which one of the two dimensions is 1. A column vector is an $N \times 1$ matrix and a row vector is a $1 \times N$ matrix. Therefore, the transpose of a column vector is a row vector and *vice versa*. By convention, $\vec{x}$ is assumed to represent a column vector. $\vec{x}^T$ would be the same values as a row vector. A scalar can be considered a 1×1 matrix. Since vector and scalars can be considered members of the matrix family, the rules of arithmetic for matrices can also be used for handling vectors and scalars.

Elements of a matrix are typically referenced by an addressing scheme based on their row, column position. For example, in a matrix A, the number in the first row and fourth column of A is known as a_{14}. Note that the order for the numbering is *row, then column*, corresponding to $M \times N$ for the definition of matrix size. Since we often use a computer language for manipulating matrices, it is worth noting that different computer languages use different conventions for array offsets. In some languages arrays enumerate locations from zero. In C, for example, the first location in a matrix would be A[0][0]. In some other languages arrays start counting locations from 1. The first location in a Fortran matrix would be denoted A(1,1). An additional difference between these languages is that Fortran stores arrays by columns, while C and most other languages store by rows.

Addition and subtraction

Vector and matrix addition is very simple. To add two matrices, both matrices must be of identical dimensions. If not, the operation is undefined. Each element of the matrix is added to the corresponding element in the other matrix. The element in the first row, first column of matrix A is added to the element in the first row, first column of matrix B, and the result is placed in the first row, first column of the solution matrix. Consider the two matrices below. Both are 3×4 matrices, so we can add them.

$$\begin{pmatrix} 5 & 1 & 1 & 10 \\ 2 & 2 & 4 & 18 \\ 1 & 1 & 4 & 15 \end{pmatrix} + \begin{pmatrix} 2 & 7 & 0 & 12 \\ 3 & 6 & 1 & 1 \\ 2 & 3 & 4 & 5 \end{pmatrix} = \begin{pmatrix} 7 & 8 & 1 & 22 \\ 5 & 8 & 5 & 19 \\ 3 & 4 & 8 & 20 \end{pmatrix}$$

Subtraction is carried out in an identical manner: each element in the second matrix is subtracted from the corresponding element in the first matrix.

Dot product

The dot or inner product of two vectors is a scalar (a number). The two vectors must be of identical length. The vector on the left must be a row vector and the one on the right a column vector. When placed the other way around, column on the left and row on the right, the result is the outer product, described below. To find the inner product, multiply each entry of the first (row) vector by the corresponding entry of the second (column) vector and then add up all of these products. For example:

$$\begin{pmatrix} 5 & 3 & 1 & 10 & -4 & 15 \end{pmatrix} \cdot \begin{pmatrix} 2 \\ -6 \\ 4 \\ 8 \\ 2 \\ 1 \end{pmatrix} =$$

$$5 \cdot 2 \; + \; 3 \cdot -6 \; + \; 1 \cdot 4 \; + \; 10 \cdot 8 \; + \; -4 \cdot 2 \; + \; 15 \cdot 1 \quad = 83$$

Orthogonality

Orthogonal vectors are vectors that are at right angles to one another when graphed. On a two-dimensional graph, you graph vectors by drawing a line from the origin (0,0) to the x, y coordinates given by the two numbers in the vector. In general, the easiest way to determine whether two vectors are orthogonal is to see if their dot product is zero. This works for higher dimensional vectors where drawing a picture is either impractical (i.e., three dimensions) or impossible (i.e., four or more). Here are two examples of orthogonal vectors. The first example is two-dimensional and can be readily seen to be orthogonal from a graph (Fig. 16.1); the second example is six-dimensional and cannot.

$$\begin{pmatrix} 5 & 2 \end{pmatrix} \cdot \begin{pmatrix} -3 \\ 7.5 \end{pmatrix} = 0$$

$$\begin{pmatrix} 5 & 1 & -1 & 10 & -40 & 15 \end{pmatrix} \cdot \begin{pmatrix} -2 \\ -1 \\ 4 \\ 8 \\ 2 \\ 1 \end{pmatrix} = 0$$

Outer product

For the inner (dot) product, we take a row vector times a column vector and get a scalar as the result. For the outer product, we take a column vector times a row vector and get a matrix as a result. For the inner product the two vectors must be the same size. For the outer product they do not need to be. Using a column vector of size M and a row vector of size N, we can form the outer product by putting M row vectors on top of one another and multiplying each one by the corresponding value in the column vector, thus producing an $M \times N$ sized matrix. We can equally well put N of the

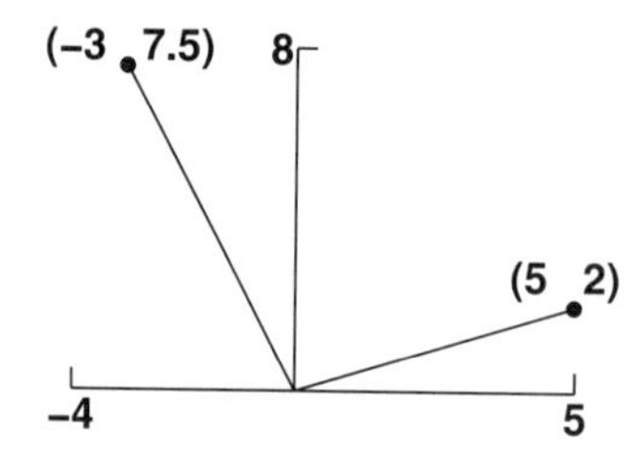

Fig. 16.1: Orthogonal vectors (5,2) and (−3,7.5) are at right angles to each other in a two-dimensional graph. Each vector is graphed by drawing a line segment from point (0,0) — the origin — to the point represented by the vector used as (x,y) coordinates.

column vectors next to each other and multiply each by the corresponding value in the row vector.

$$\begin{pmatrix} -2 \\ -1 \\ 4 \end{pmatrix} \begin{pmatrix} 5 & 1 & -1 & 10 \end{pmatrix} = \begin{pmatrix} -10 & -2 & 2 & -20 \\ -5 & -1 & 1 & -10 \\ 20 & 4 & -4 & 40 \end{pmatrix}$$

This can be done by writing out a pile of rows:

$$\begin{array}{rcc} -2 & \cdot & \begin{pmatrix} 5 & 1 & -1 & 10 \end{pmatrix} \\ -1 & \cdot & \begin{pmatrix} 5 & 1 & -1 & 10 \end{pmatrix} \\ 4 & \cdot & \begin{pmatrix} 5 & 1 & -1 & 10 \end{pmatrix} \end{array}$$

or a series of columns:

$$\begin{pmatrix} -2 \\ -1 \\ 4 \end{pmatrix} \cdot 5, \quad \begin{pmatrix} -2 \\ -1 \\ 4 \end{pmatrix} \cdot 1, \quad \begin{pmatrix} -2 \\ -1 \\ 4 \end{pmatrix} \cdot -1, \quad \begin{pmatrix} -2 \\ -1 \\ 4 \end{pmatrix} \cdot 10$$

Matrix multiplication

To add matrices, the matrices have to be of identical dimension. The rule for multiplication is more complicated. To multiply matrices, the number of columns for the matrix on the left must be the same as the number of rows for the matrix on the right. This means that matrix multiplication is not directly commutative — often the commuted operation is undefined because the dimensions don't match up.

Remember that a row vector of length N is a $1 \times N$ matrix. A column matrix of length N is an $N \times 1$ matrix. To multiply a vector times a matrix (vector on left side), the vector must be a row vector. To multiply a matrix times a vector (vector on right side), the vector must be a column vector. In neural networks, we generally just use right-side column vectors. Right-side multiplication is done by taking the dot product of each row of the matrix with the entire column vector. Each dot product result becomes a single element in the resultant column vector:

$$\begin{pmatrix} -1 & -2 & 3 & -2 \\ -5 & -1 & 1 & -1 \\ 2 & 4 & -7 & 4 \end{pmatrix} \cdot \begin{pmatrix} 5 \\ -1 \\ 2 \\ 3 \end{pmatrix} = \begin{pmatrix} \begin{pmatrix} -1 & -2 & 3 & -2 \end{pmatrix} \cdot \begin{pmatrix} 5 \\ -1 \\ 2 \\ 3 \end{pmatrix} \\ \begin{pmatrix} -5 & -1 & 1 & -1 \end{pmatrix} \cdot \begin{pmatrix} 5 \\ -1 \\ 2 \\ 3 \end{pmatrix} \\ \begin{pmatrix} 2 & 4 & -7 & 4 \end{pmatrix} \cdot \begin{pmatrix} 5 \\ -1 \\ 2 \\ 3 \end{pmatrix} \end{pmatrix} = \begin{pmatrix} -3 \\ -25 \\ 4 \end{pmatrix}$$

For any two matrices, the element in row a and column b of a resultant matrix is the dot product of row a of the left-side matrix with column b of the right-side matrix. This is why the left-side matrix row-length (number of columns) must be the same as the right-side matrix column-length (number of rows). It works out easily if you look at first at matrix sizes using the "row $\times$ column" convention. The sizes for an arbitrary matrix multiplication are $M \times N \cdot N \times P = M \times P$. Any element X_{ab} of the new $M \times P$-sized matrix is the dot product of row a from the left matrix and column b from the right matrix. Similarly, a dot product is multiplication of matrices size $1 \times N \cdot N \times 1 = 1 \times 1$ — the 1×1 matrix is a scalar.

Here's a full matrix multiplication for its entertainment value:

$$\begin{pmatrix} -1 & -2 & 2 & -2 \\ -5 & -1 & 1 & -1 \\ 2 & 4 & -4 & 4 \end{pmatrix} \cdot \begin{pmatrix} 5 & 1 \\ -1 & 1 \\ 2 & 3 \\ 3 & 5 \end{pmatrix} = \begin{pmatrix} -5 & -7 \\ -25 & -8 \\ 10 & 14 \end{pmatrix}$$

16.5 Numerical calculus

"A calculus" is a small stone that forms in the gallbladder or urinary tract; "the calculus" is a technique for solving problems involving change. A calculus is often painful; the calculus may be as well. A calculus can be treated medically or surgically; the calculus can be treated analytically or numerically. In this chapter, I treat it numerically and painlessly.

Biological relevance aside, it would probably be fair to say that calculus is the foundation of modern mathematics. The funny thing about modernity in mathematics is that it's been going on for about 300 years.

Infinitesimals

Since we are interested in computer modeling, we don't need all of the fancy calculus symbols. This makes things a lot easier, enabling us to use basic algebra to explain the concepts and solve the problems. Nonetheless, I introduce some of the symbols in order to make it easier to read other books where calculus is used.

Calculus is used to describe change. The Greek letter delta means change. Δ is the capital letter; δ is the small letter. Our letter "d" is also used. Usually, we will be talking about change over time. Time is represented as t, so change in time is $\Delta t = t_{\text{now}} - t_{\text{before}}$. The original application of calculus was mechanics, the physics of how objects move in time: falling apples, cannonballs, the moon, and so on. We use the variable x to represent the position of an object. So the change in position from location a to location b is $\Delta x = x_b - x_a$. Once we have measured both change in position (Δx), and change in time (Δt) we just divide to get average velocity: $v = \frac{\Delta x}{\Delta t}$. (Notice that v is used to represent velocity, while V is used to represent volts.) For example, to check the accuracy of your speedometer, you measure the time it takes you to drive 1 mile. Let's say it takes 1 minute. Then $\Delta x = 1$ mile; $\Delta t = 60$ seconds; $\frac{\Delta x}{\Delta t} = \frac{1 \text{ mile}}{1 \text{ minute}}$. We now just have to change units to show that a mile per minute is 60 miles per hour.

This procedure works well if you maintain a constant speed while driving the measured mile. However, you could also cover the mile in 1 minute by driving 100 mph for 20 seconds and 40 mph for 40 seconds. Again you would be traveling an average of 60 mph. If the police want to know if you were speeding, timing a measured mile may not tell them. The calculus and the police are interested in instantaneous rather than average velocity. To estimate instantaneous velocity, you have to make your measurements over a very short time, as the police do with lasers and radar guns. If we reduce the measured distance to 100 meters, this distance will be covered in about 3.7 seconds at 60 mph (sorry about mixing meters and miles). This still would not be short enough to be certain of getting an accurate instantaneous velocity since the driver could brake and change his velocity considerably during this time. However, if your reduced the measured distance to 1 meter, then the transit time would be about 37 ms (milliseconds) and the measurement will give a pretty good estimate of actual instantaneous velocity. If you go down to 1 mm, it will take about 37 μs (microsecond) and this will be a very good estimate.

In the abstract world of mathematics, analytic calculus uses the symbol "dt" to represent an *infinitesimal* duration in time. This allows the definition of a true instantaneous velocity. In the physical world every measurement takes some time. One always averages velocity over some period. Similarly, in the computer simulation world of numerical calculus, one uses a number for Δt. If you needed to simulate a car driving at high speed for a video game, 10 ms would be a reasonable choice for Δt.

Numerical solutions

In Chap. 11, Section 11.4, I showed a numerical solution for the membrane charging equation:

$$I_{in} = C \cdot \frac{\Delta V}{\Delta t} + g \cdot V$$

Using the explicit Euler method, the update rule for voltage worked out to be

$$V^+ = (1 - \Delta t \cdot \frac{g}{C}) \cdot V + \Delta t \cdot \frac{I_{in}}{C}$$

There are many different methods for solving differential equations numerically. Different methods will have different advantages and different drawbacks, both in terms of the quality of the solution and in terms of the computational demands. The major issues in solution quality have to do with accuracy and stability. In Chap. 11, I showed how the *explicit* Euler equation would become unstable if the time step was made too large. By contrast, the *implicit* Euler integration has the advantage of remaining stable for large time steps. The explicit Euler solution has the advantage of being much simpler to program for sets of linked differential equations.

While the explicit Euler solution bases its calculation of the present voltage on the past, the implicit Euler solution bases its calculation on the future. This is what makes it implicit. The concept of basing a calculation on a future value seems counterintuitive. It is possible because calculus is about prediction of future events in deterministic systems.

Just as we did with the explicit solution, we start by expanding out $\Delta V = \frac{V^+ - V}{\Delta t}$. However, now we use future voltage instead of present voltage elsewhere in the equation:

$$I_{in} = C \cdot \frac{V^+ - V}{\Delta t} + \boxed{g \cdot V^+}$$

The box highlights the difference from the derivation of the explicit equation: now we use the future conductance current $g \cdot V^+$ instead of the present conductance current $g \cdot V$. The rest of the algebra is pretty much as before:

$$\begin{aligned}
\Delta t \cdot I_{in} &= C \cdot V^+ - C \cdot V + \Delta t \cdot g \cdot V^+ \\
\Delta t \cdot I_{in} &= (C + \Delta t \cdot g) \cdot V^+ - C \cdot V \\
\Delta t \cdot I_{in} + C \cdot V &= (C + \Delta t \cdot g) \cdot V^+ \\
V^+ &= \frac{\Delta t \cdot I_{in} + C \cdot V}{C + \Delta t \cdot g} \\
V^+ &= \frac{C}{C + \Delta t \cdot g} \cdot V + \frac{\Delta t}{C + \Delta t \cdot g} \cdot I_{in}
\end{aligned}$$

Remembering that $V_{max} = I_{in} \cdot R = \frac{I_{in}}{g}$ and $\tau = \tau_{memb} = R \cdot C = \frac{g}{C}$, we can simplify the expression and compare it to the explicit version:

$$\text{Implicit}: \quad V^{+} = \frac{\tau}{\Delta t + \tau} \cdot V + \frac{\Delta t}{\Delta t + \tau} \cdot V_{max}$$
$$\text{Explicit}: \quad V^{+} = (1 - \frac{\Delta t}{\tau}) \cdot V + \frac{\Delta t}{\tau} \cdot V_{max}$$

In the example of Chap. 11, we used $\Delta t = 0.001$ ms, $\tau = 1$ ms, and $V_{max} = 1$ mV. We can plug the numbers into the implicit solution and compare the numerical update rules.

$$\text{Implicit}: \quad V = 0.999000999 \cdot V \quad + \; 0.000999000999$$
$$\text{Explicit}: \quad V = 0.999 \cdot V \quad + \; 0.001$$

These produce slightly different values over the course of five steps.

t (ms)	mV Implicit	mV Explicit
0.000	0	0
0.001	0.000999	0.001
0.002	0.001997	0.001999
0.003	0.002994	0.002997
0.004	0.003990	0.003994
0.005	0.004985	0.004990

A major advantage of the implicit equation has to do with stability. As previously noted, a large Δt of 2 ms will produce an unstable result when using the explicit solution. By contrast, the implicit solution will still give the correct result:

t (ms)	mV Implicit	mV Explicit
0	0	0
2	0.666667	2
4	0.888889	0
6	0.962963	2
8	0.987654	0
10	0.995885	2

Because the implicit solution is stable, it is possible to go straight to the solution of an integration in one step by using an even larger Δt. If we use $\Delta t = 100$ ms, then $V = 0.99$ after only one time step. This will not work with sets of nonlinear interlocking differential equations — we cannot generally do whole neural simulations in one time step.

Mathematical symbols

A barrier to understanding biology is the complex jargon that has developed to describe organic things — Latin words for creatures, long chemical names for compounds, whole geographies of cells and organ systems. In math, the problem is understanding the symbols.

We have introduced $\frac{\Delta x}{\Delta t}$ as the numerical representation of change with time. The simplest calculus representation just involves using a regular "d" instead of the Δ : $\frac{\mathrm{d}x}{\mathrm{d}t}$, this is "the first derivative of x with respect to t," or simply, by reading the letters, "d x d t." It is written as a fraction since it can be treated algebraically as a fraction. Since change with time is the most common application of calculus, there is a special shorthand for representing it, a dot over the variable: $\frac{\mathrm{d}x}{\mathrm{d}t} = \dot{x}$. Spoken, this is "$x$ dot." In the case of a function $f(x)$, a first derivative with respect to x can be represented as $f'(x)$.

The above is the first derivative. There are also higher derivatives: second derivative, third derivative, etc. The first derivative of distance with respect to time is velocity. The first derivative of velocity with respect to time is acceleration. Therefore, acceleration is the second derivative of distance with respect to time. This is represented as $\frac{\mathrm{d}^2x}{\mathrm{d}t^2}$ or $\ddot{x}$. Newton's first law, F = ma, is also written as F = m$\dot{v}$ or F = m$\ddot{x}$. The second derivative of a function of x with respect to x is written $f''(x)$.

Analytic solution to the charging curve

Above, we numerically solved the charging curve $I_{in} = C \cdot \frac{\Delta V}{\Delta t} + g \cdot V$. Here's the analytic solution for charging the membrane with a constant current. This probably won't make much sense if you've never studied calculus.

The charging of the membrane RC circuit is described by the equation $I_{in} = C \cdot \frac{\Delta V}{\Delta t} + g \cdot V$. The finite numerical $\frac{\Delta V}{\Delta t}$ is represented in calculus as $\frac{\mathrm{d}V}{\mathrm{d}t}$. Rearranging the terms gives the canonical form:

$$\frac{C}{g} \cdot \frac{\mathrm{d}V}{\mathrm{d}t} = -V + \frac{I_{in}}{g}$$

$\frac{\mathrm{d}V}{\mathrm{d}t}$ is not really a normal fraction. However, it is written as a fraction because the terms can be handled separately as if it were. This makes sense when you consider that the expression comes from taking a limit of $\frac{\Delta V}{\Delta t}$ as Δt approaches zero. We separate terms involving t from terms involving V, and then integrate both sides as follows. In the following expressions, K_1 and K_2 are constants that will need to be determined based on initial conditions.

$$\begin{aligned}
\int -dV/(-V + \frac{I_{in}}{g}) &= -\int \frac{g}{C} \cdot dt \\
\ln(-V + \frac{I_{in}}{g}) &= \frac{g}{C} \cdot t + \mathrm{K}_1 \\
V - \frac{I_{in}}{g} &= \mathrm{K}_2 e^{-g \cdot t/C} \\
V &= \mathrm{K}_2 \cdot e^{-g \cdot t/C} + \frac{I_{in}}{g}
\end{aligned}$$

We'll use $V(t) = V_0$ as the initial condition for setting K_2:

$$\begin{aligned}
V_0 &= \mathrm{K}_2 + \frac{I_{in}}{g} \\
\mathrm{K}_2 &= V_0 - \frac{I_{in}}{g}
\end{aligned}$$

So the solution is

$$V = (V_0 - \frac{I_{in}}{g}) \cdot e^{-\frac{g}{C} \cdot t} + \frac{I_{in}}{g}$$

Substituting in the terms for $\tau = \frac{g}{C}$ and $V_{max} = \frac{I_{in}}{g}$ we get

$$V = (V_0 - V_{max}) \cdot e^{-t/\tau} + V_{max}$$

We can confirm what we illustrated with the numerical solution. The curve is exponential. As time gets very large, V asymptotes at V_{max}. The time constant τ is rise time to $1 - e^{-1} \sim 63\%$ of total excursion.

16.6 Electrical engineering

Digital has now beat out analog in almost every realm: CDs instead of LPs for music, digital telephones, and now digital TV. Study of analog electronics has become almost quaint, with its foreign names from the heroic age of electrical discovery. Whether or not the neuron turns out to use some sort of digital code, neuronal signaling will still need to be explained by analogy with the batteries, resistors, and capacitors of folks like Volta, Ampère, Ohm, Galvani, Franklin (he's not foreign), Kirchhoff, and Faraday.

The three big laws: Ohm, Kirchhoff, and the other one

Ohm's law deals with resistors. Kirchhoff's law deals with currents. The law of capacitance has no name. In electronics, capacitance tends to slow things down and becomes a nuisance. Maybe that's why no one wanted to claim it. In neurons, the capacitor's ability to hang on to charge permits signals to add up.

Kirchhoff's law is conservation of current. Like the laws of conservation of mass and energy in their respective domains, this law says that all stuff comes from somewhere and goes somewhere. Nothing appears out of nowhere or disappears into nothingness. (These laws don't hold in the quantum realm.) Kirchhoff's law says that recycling of current is required as long as you stay within the circuit. However, there is an inexhaustible landfill, called ground, where all current eventually goes. Going to ground, current doesn't disappear — it just dissipates.

Fig. 16.2 shows how Kirchhoff's law is typically applied at nodes where wires are connected. From the positive voltage (or potential) at the bottom of the figure, positive current will flow "downhill" to ground. Ground always remains at 0 volts. If the side of the circuit away from ground is held at a negative voltage, positive current will flow out of the ground down to this more negative potential. Current (I) is measured in amps (A), which represents flow of charge (Q) per unit time (t). In Fig. 16.2 the two branches are the same, so the current will split equally at the point where the two wires separate. Half of the 3 amps goes down one branch and half goes down the other. When the wires come back together, the currents add back up. Charge is conserved; current is conserved. The charge then goes into the ground and might show up next in a lightning bolt or just creep into someone's radio as static.

Ohm's law

Current can flow through wires, through salt water, through anything that has mobile charge carriers. Anything that will carry current is a conductor of electricity. The quality of a conductor is called its conductance (g). No conductor is perfect, however (superconductors come close). Any time that current flows it will encounter some resistance to its flow. Thus any conductor can equally well be described in terms of resistance (R). A good conductor will have a low resistance. Conductance and resistance are just two ways of describing the same thing: $g = \frac{1}{R}$; $R = \frac{1}{g}$. Resistance is measured in ohms (Ω). The unit of conductance, inverse ohm, is sometimes called mho (ohm spelled backward). It is also called siemens (S, not s).

Ohm's law states that current times resistance equals voltage ($V = IR$). It therefore also states that conductances times voltage equals current ($I = gV$). $V = IR$ is more commonly used in electrical engineering; $I = gV$ is more commonly used in neuroscience. Given a choice, current will follow

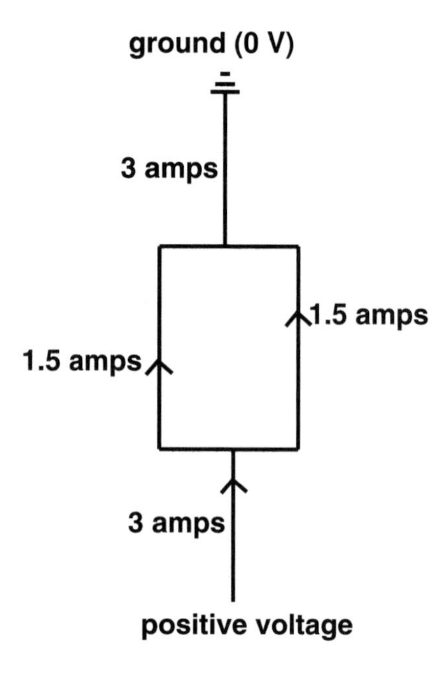

Fig. 16.2: Kirchhoff's law and ground.

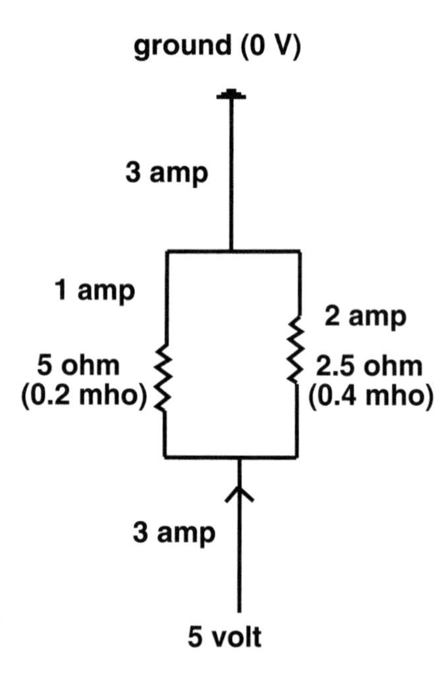

Fig. 16.3: Current follows the path of least resistance. The resistors are the sawtooth symbols.

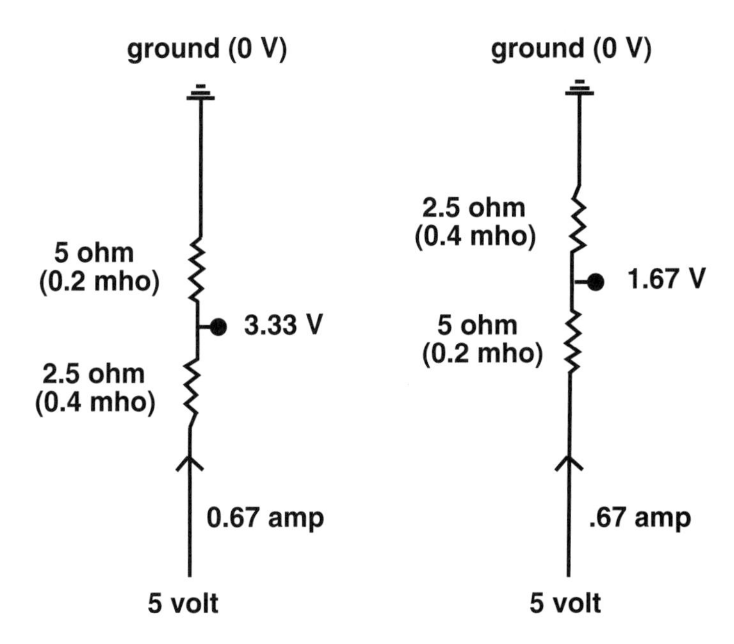

Fig. 16.4: Resistors in series.

the path of least resistance. If we modify Fig. 16.2 to include resistors in the two branches, most of the current will go into the better conductor (Fig. 16.3). If conductance is twice as good in one branch, that branch will get twice as much current. With one side of the circuit at 5 volts and the other at ground, the voltage drop across each resistor is 5 volts. The two resistors are parallel routes to ground so each resistor has this 5-volt drop across it. For the smaller resistor (larger conductor), the current will be 2 amps $= V/R = 5\ V/2.5\ \Omega = g \cdot V = 0.4\ S \cdot 5\ V$. For the larger resistor (smaller conductor), the current will be 1 amp $= V/R = 5\ V/5\ \Omega = g \cdot V = 0.2\ S \cdot 5\ V$.

Normally, we build a circuit out of known components and have the values for voltage (from a battery) and the values of the two resistors. We then calculate the current through each branch and add them up to get the total current. We can then use Ohm's law in the other direction and calculate the total resistance of the circuit. In this case, total resistance $R = V/I = 5$ Volts/3 amps $= 1.6667\ \Omega$. Equivalently, $g = I/V = 0.6\ S$. Using symbols for the resistance values and working through a bunch of algebra gives the rule for adding resistances in parallel branches: $\frac{1}{R_{parallel}} = \frac{1}{R_1} + \frac{1}{R_2}$. The equivalent rule for conductance is easier: $g_{parallel} = g_1 + g_2$. In this case, $0.6\ S = 0.2\ S + 0.4\ S$.

Resistors can also be arrayed in series. The serial arrangement is called a voltage divider. Perhaps it would be better to call it a voltage subtractor.

Each resistor drops the voltage a certain amount, providing a portion of the voltage in between. Fig. 16.4 shows two different serial resistance circuits. These two circuits are not connected to each other — each has its own separate battery. The easiest way to calculate the current is to know that resistance adds in series $R_{\text{series}} = R_1 + R_2$, just as conductance adds in parallel. Hence the total resistance in both circuits is $5\ \Omega + 2.5\ \Omega = 7.5\ \Omega$. The current in both cases is $I = V/R = 5\ V/7.5\ \Omega = \frac{2}{3}\ A$. In the left circuit, the small resistor is encountered first, providing a voltage drop of $V = IR = \frac{2}{3}\ A \cdot 2.5\ \Omega = 1\frac{2}{3}\ V$. This is the drop in voltage from the starting potential of 5 V so the voltage at this intermediate location is $5\ V - 1\frac{2}{3}\ V = 3\frac{1}{3}\ V$. The same current flowing through the next resistor drops the voltage down to zero at ground $V = IR = \frac{2}{3}\ A \cdot 5\ \Omega = 3\frac{1}{3}\ V; 3\frac{1}{3}\ V - 3\frac{1}{3}\ V = 0\ V$. This circuit drops off one-third of the voltage in the first resistor and the other two-thirds in the second resistor. Switching the resistors, as at right, drops off two-thirds of the voltage first and the other one-third second from the divide to ground.

Capacitance

Anthropomorphically speaking, electricity just wants to flow downhill to the ground. Anything that impedes this flow is called impedance. Resistance is one kind of impedance. The two other kinds of impedance are capacitance and inductance. Inductance isn't much of a factor in neurons. The three kinds of impedance differ in their response to time-varying signals. Resistors don't care about the speed of change: $V = IR$ whether voltage is changing quickly or slowly.

A capacitor will pass current easily in the presence of a quickly changing voltage but will not pass any current in the presence of constant voltage (Fig. 16.5). It is therefore a high impedance for slow-changing voltage signals and a low impedance for fast-changing voltage signals. Capacitance means the capacity to hold charge. The equation $Q = CV$ says that more charge Q can be held at a given voltage V if the capacitor has high capacitance C. In Chap. 11, Section 11.4 and in Section 16.5, I explained how to go from charge to current by calculating the change in charge with time (first derivative with respect to time): $\frac{\Delta Q}{\Delta t} = I$. By taking the first derivative of both sides we get the impedance relation: $I = C \cdot \frac{\Delta V}{\Delta t}$.

Current flow through capacitors is dependent on how fast voltage is changing. Capacitance tends to take current away from fast-changing voltages — shunting the current off to ground. Fig. 16.5 shows the passage of a fast-moving voltage signal through a wire that is connected through capacitors to the ground. This is the situation in transmission lines, which always have some capacitative coupling to grounded material. As long as the voltage is oscillating slowly, the signal will not leak out of the wire through the capacitors (top). If the voltage is oscillating at a high frequency, however,

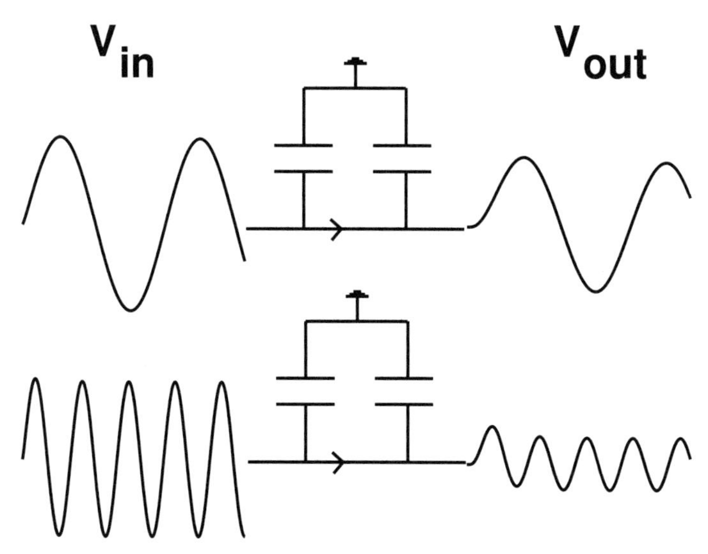

Fig. 16.5: Capacitative drop-off in an alternating current (AC) increases with increasing signal frequency. No drop-off would be seen with a constant direct current (DC).

there will be a substantial drop-off (bottom). Capacitance also produces a drag on the signal, slowing it down. For these reasons, capacitance is typically a nuisance in electronics — preventing equipment from running as fast as one might want. This is the problem that computer chip designers are always fighting. They want to run the chips faster so that they can do faster logic. But if they try to send signals through too fast, the signals get sucked away by parasitic capacitance and don't get to their destination.

In neurons, however, capacitance is a good thing. Since the brain is not strictly clocked like a computer is, one cannot depend on signals coming in at exactly the same time. Capacitance broadens and slows signals, allowing for the temporal summation of signals that come in at slightly different times.

References

Here's a list of some references that are interesting, fun, useful, or of historical interest (sometimes all four). Some of these are textbooks that are complementary to the present work. Others are general references. I also give some primary literature references that describe research that I presented or alluded to in the text.

I've divided the references into sections based on the chapters of the book, although in many cases a reference is pertinent to more than one section.

General

• MA Arbib. *The Handbook of Brain Theory and Neural Networks.* MIT Press, Cambridge, MA, 1998. This miniature encyclopedia is a good general references that covers most major topics in computational neuroscience.

• JA Anderson and E Rosenfeld. *Neurocomputing: Foundations of Research.* MIT Press, Cambridge, MA, 1988. This collection of papers contains several of the artificial neural network papers cited below.

Perspectives

• PS Churchland. *Neurophilosophy: Toward a Unified Science of the Mind-Brain.* MIT Press, Cambridge, MA, 1986. One of the first books that looked at what philosophy could learn from neuroscience. Considers the mind–body problem and the question of subjectivity (qualia).

• PS Churchland and TJ Sejnowski. *The Computational Brain.* MIT Press, Cambridge MA, 1994. A broad review of issues in computational neuroscience.
• A Damasio. *The Feeling of What Happens: Body and Emotion in the Making of Consciousness.* Harcourt Brace, New York, 1999. About as close as anyone's gotten to getting a bead on the mind-body problem.
• DO Hebb. *Organization of Behavior.* John Wiley & Sons, New York, 1949. The classic Hebbian book. Honestly, I've never actually read it, but the little parts I've looked at are good.
• D Marr. *Vision.* Freeman, San Francisco, 1982. Description of a modeling approach to vision that presents Marr's ideas of levels of investigation.

Basic neuroscience

• ER Kandel, JH Schwartz, and TM Jessell. *Principles of Neural Science.* McGraw-Hill, New York, 4th edition, 2000. The grandaddy of neuroscience textbooks. An excellent introduction. The only downside is that it has become very long – it was shorter back when I read it four editions ago.
• JG Nicholls, AR Martin, PA Fuchs, and BG Wallace. *From Neuron to Brain.* Sinauer Associates, Sunderland, MA, 2001. Another classic text, from which I stole the title.

Representations

• PE Ceruzzi. *A History of Modern Computing.* MIT Press, Cambridge MA, 2000. A computer science history that gives an introduction to the history of representations.
• D Knuth. *The Art of Computer Programming.* Addison-Wesley, New York, 1973. The classic multi-volume work on data structures and programming techniques. However, if you wish to learn to program, this is not the place to start. Instead, figure out what language you want to learn and then pick up a tutorial book with exercises.

Computers

• M Campbell-Kelly and W Aspray. *Computer: A History of the Information Machine.* Harper-Collins, New York, 1997. A well-written history of computer science.
• J Von Neumann, PM Churchland, PS Churchland, and K Von Neumann. *The Computer and the Brain.* Yale University Press, New Haven, CT, 2000. A quite readable book based on von Neumann's lectures on the concept of brain function that led him to help invent the computer.

Cybernetics

• N Wiener. *Cybernetics.* MIT Press, Cambridge, MA, 2nd edition, 1965. This is the book that introduced ideas and concepts of cybernetics, as well as coining the term. It's a hard read.

Concept neurons

• J Hertz, A Krogh, and RG Palmer. *Introduction to the Theory of Neural Computation.* Addison-Wesley, Redwood City, CA, 1991. This is a thorough and well-written treatment of artificial neural networks. It's rigorous mathematically and offers a good introduction for physicists and engineers.
• LV Fausett. *Fundamentals of Neural Networks.* Prentice Hall, New York, 1994. A more widely accessible introduction to artificial neural networks.
• W McCulloch and W Pitts. A logical calculus of ideas immanent in nervous activity. *Bulletin of Mathematical Biophysics* 5:115–133, 1943. This is the McCulloch and Pitts work that helped inspire von Neumann. It's quite mathematical.
• D Hubel and T Wiesel. Receptive fields, binocular interaction and functional architecture in the cat's visual cortex. *Journal of Physiology – London* 160:106–154, 1962. Classic paper identifying response patterns in visual cortex.
• CD Salzman and WT Newsome. Neural mechanisms for forming a perceptual decision. *Science* 264:231–237, 1994. Paper describing the thought-insertion experiment.
• F Rieke, D Warland, R de Ruyter, and V Steveninck. *Spikes.* MIT Press, Cambridge, MA, 1999. A mathematical treatment of neural coding by spike trains and spike timing.
• J Gleick. *Chaos: Making a New Science.* Penguin, New York, 1988. A wonderful history of the discovery of the mathematics and physics of chaos. There are no direct applications to computational neuroscience in this book, but it offers a very readable introduction to ideas in dynamical systems and complex systems.
• S Singh. *The Code Book: The Evolution of Secrecy from Mary, Queen of Scots to Quantum Cryptography.* Doubleday, New York, 1999. A fun, readable history of codes and decoding. Nothing directly about nervous systems but the description of the methods used to disentangle lost languages provides some parallels.
• JA Anderson, editor. *An Introduction to Neural Networks.* MIT Press, Cambridge, 1995. This textbook has a chapter on limulus that gives more details than I have given. It is also one of the few texts in the field that are readily accessible to undergraduates. Compared to the present work, this book has less emphasis on biology and more on cognitive neuroscience.

Supervised learning: the delta rule and back-propagation

• DE Rumelhart and JL McClelland. *Parallel Distributed Processing: Explorations in the Microstructure of Cognition. Vol. 1: Foundations.* MIT Press, Cambridge, MA, 1986. JL McClelland and DE Rumelhart. *Parallel Distributed Processing: Explorations in the Microstructure of Cognition. Vol. 2: Psychological and Biological Models.* MIT Press, Cambridge, MA, 1986. Highly readable companion volumes that formed the manifesto of the PDP movement. Still very worthwhile reading.
• R O'Reilly, Y Munakata, and JL McClelland. *Computational Explorations in Cognitive Neuroscience.* MIT Press, Cambridge, MA, 2000. An introduction to the use of artificial neural networks in a cognitive context.
• D Rumelhart, G Hinton, and R Williams. Learning representations by back-propagating errors. *Nature* 323:533–536, 1986. The paper that introduced back-propagation.
• TJ Anastasio and DA Robinson. Distributed parallel processing in the vestibulo-oculomotor system. *Neural Computation* 1:230–241, 1989. T Anastasio. Modeling vestibulo-ocular reflex dynamics: from classical analysis to neural networks. In *Neural Systems: Analysis and Modeling.* Kluwer, 1993. Two original papers about the application of back-propagation to the VOR.
• JA Anderson. A simple neural network generating an interactive memory. *Mathematical Biosciences* 14:197–220, 1972. T Kohonen. Correlation matrix memories. *IEEE Transactions on Computers* C-21:353–359, 1972. Two papers that independently introduced the concept of outer-product formation of associative memory networks.
• JJ Hopfield. Neural networks and physical systems with emergent collective computational abilities. *Proceedings of the National Academy of Sciences* 79:2554–2558, 1982. Hopfield's classic paper in which he showed why autoassociative networks would have point attractor dynamics.

From soap to volts

• C Koch and I Segev. *Methods in Neuronal Modeling: From Synapse to Networks.* MIT Press, Cambridge, MA, 2nd edition, 1998. C Koch. *Biophysics of Computation: Information Processing in Single Neurons.* MIT Press, Cambridge, MA, 1998. Two books by leading biophysicists who describe how cells might be able to do complex processing.
• D Johnston and SMS Wu. *Foundations of Cellular Neurophysiology.* MIT Press, Cambridge, MA, 1994. A neuroscience textbook with an emphasis on neurophysiology.
• CF Stevens. *Neurophysiology: A Primer.* J. Wiley, New York, 1966. A classic brief description of neurophysiology, which is now out of print. It presents the original treatment of slow potential theory.

Hodgkin-Huxley model

• D Junge. *Nerve and Muscle Excitation.* Sinauer, Sunderland MA, 1992. An introduction to basic neurophysiology.
• B Hille. *Ionic Channels of Excitable Membranes.* Sinauer Associates, Sunderland, MA, 3rd edition, 2001. A very thorough treatment of ion channels with an excellent chapter on the Hodgkin-Huxley model.
• AK Hodgkin and AF Huxley. A quantitative description of membrane current and its application to conduction and excitation in nerve. *Journal of Physiology – London* 117:500–544, 1952. The original modeling paper, now 50 years old. It is quite readable.

Compartment modeling

• I Segev, J Rinzel, and GM Shepherd, editors. *The Theoretical Foundation of Dendritic Function: Selected Papers of Wilfrid Rall with Commentaries.* MIT Press, Cambridge, MA, 1995. A collection of the classic papers of compartment modeling.
• JJB Jack, D Noble, and RW Tsien. *Electric Current Flow in Excitable Cells.* Oxford, New York, 1983. An analytic treatment of passive dendritic processing. Now out of print.
• DB Jaffe and NT Carnevale. Passive normalization of synaptic integration influenced by dendritic architecture. *Journal of Neurophysiology* 82:3268–3285, 1999. A nice simulation paper presenting passive normalization.
• K Neville and WW Lytton. Potentiation of Ca^{++} influx through NMDA channels by action potentials: a computer model. *Neuroreport* 10:3711–3716, 1999. A simulation paper of mine (with colleague) describing possible effects of active dendritic channels on Hebbian learning.

From artificial neural network to realistic neural network

• RD Traub, JGR Jefferys, and MA Whittington. *Fast Oscillations in Cortical Circuits.* MIT Press, Cambridge, MA, 1999. Oscillations in neural networks presented by a pioneer in the development of realistic neural modeling with physiology colleagues.
• ME Hasselmo and JM Bower. Cholinergic suppression specific to intrinsic not afferent fiber synapses in rat piriform (olfactory) cortex. *Journal of Neurophysiology* 67:1222–1229, 1992. Barkai E and Hasselmo ME. Modulation of the input/output function of rat piriform cortex pyramidal cells. *Journal of Neurophysiology* 72:644–658, 1994. Two papers explaining the need for synaptic suppression when porting the Hopfield model to realistic neural networks.
• J Chover, L Haberly, and WW Lytton. Alternating dominance of NMDA and AMPA for learning and recall: a computer model. *Neuroreport* 12:2503–

2507, 2001. A paper written by my colleagues and me that presents the dual matrix hypothesis.

Neural circuits

• G Shepherd, editor. *The Synaptic Organization of the Brain.* Oxford University Press, Oxford, 1997. A multiauthor work that describes connectivity in different regions of the brain.
• RJ Douglas, KAC Martin, and D Whitteridge. A canonical microcircuit for neocortex. *Neural Computation* 1:480–488, 1989. A suggestion for a basic cortical circuit.

Units

• *Units* program — *http://www.gnu.org/software/units/units.html* (free software)

Linear algebra

These two programs are specialized for performing linear algebra operations (and yes, much more):

• Matlab: *http://www.mathworks.com/*
• Octave: *http://www.gnu.org/software/octave/octave.html* (free software)

Numerical calculus

• Press WH, SA Teukolsky, BP Flannert, and WT Vetterling. *Numerical Recipes in C++: The Art of Scientific Computing.* Cambridge University Press, Cambridge, 2002. A wonderful introduction to all kinds of numerical computing. In addition to the algorithms and programs, individual chapters give good introductions to various math topics. Other books in the series give algorithms for C, Fortran, and Pascal.
• S Wolfram. *The Mathematica Book.* Cambridge University Press, New York, 4th edition, 1999. This book describes the use of Mathematica, a program for doing both numerical and symbolic mathematics. Mathematica is an excellent computer tool for learning math.

Electrical engineering

• P Horowitz and W Hill. *The Art of Electronics.* Cambridge University Press, Cambridge, 2nd edition, 1989. The introductory chapter is a great description of RC circuits and impedance relations. The rest of the book give lots of practical information for building circuits.

Glossary

ablation: In neurology, an area of destruction or damage in a region of brain. Verb: ablate.
ablative disease: Disease that involve loss of brain matter such as stroke or head trauma.
absence epilepsy: Seizure disorder of childhood characterized by brief (several seconds) periods of loss of consciousness with staring but without collapse or limb movements. This disorder can be confused with daydreaming. However, electroencephalogram shows characteristic abnormal three per second spike-and-wave discharges.
absolute refractory period: In neurophysiology, the time after firing of an action potential during which another action potential cannot be initiated.
accommodation: In neurophysiology, decrease in spike firing rate with time (also called adaptation). In neuropsychology, reduction in response to sensory stimulation with time (also called habituation). In ocular physiology, relaxation of the lens.
accumulator: In computer science, a single specialized register where arithmetic results are stored. Abbrev.: ACC. The term is no longer current; modern computers have many such registers.
acetylcholine: A neurotransmitter that serves as a neuromodulator in the central nervous system and is the principal transmitter at the neuromuscular junction. Adj.: cholinergic.
acquired disease: A disease that is picked up during life. Contrasted with congenital.

action potential: Spike in membrane voltage conducted down an axon to transmit information to other neurons. Action potentials can also occur in dendrites.
activation function: See squashing function.
active channels: Ion channels that change permeability (conductance) with changes in voltage. Contrasted with passive channels.
active membrane: Membranes that have voltage-sensitive channels. Also called excitable membrane.
adaline: An early version of the sum-and-squash unit used in artificial neural networks. It is an acronym for ADAptive LInear NEuron.
adaptation: In neurophysiology, reduction in firing rate over time with continuing stimulation.
adaptive filter: In signals and systems, a signal processing technique that uses a learning algorithm to change the filtering properties depending on aspects of the signal.
afferent: An axon, neuron, or tract that brings information in toward a neuron or a neural structure.
afterhyperpolarization: Negative voltage that overshoots resting membrane potential following an action potential. Abbrev.: AHP.
agonist: A drug or ligand that activates a receptor. Opposite of antagonist.
Algol: An early computer language that was precursor to Pascal.
algorithm: In computer science, a sequential series of discrete steps used to solve a problem. Comparable to the steps of a recipe used in cooking.
alien-hand syndrome: In neurology, a disconnection syndrome resulting from cutting of the corpus collosum. The two cerebral hemispheres then operate independently, resulting in two largely separate consciousnesses. The left typically remains dominant because of its control of speech. The "self" of the left hemisphere regards the left hand as alien since it is controlled by the right hemisphere.
all-or-none: A binary phenomenon without any intermediate state. Term used to describe the action potential.
allele: A particular form of a gene; e.g., different alleles of a single gene produce different eye colors.
allosteric: In biochemistry, alteration of the activity of an enzyme or ion channel produced by binding of a chemical onto a site separate from the main binding site.
alpha waves: Waves of 8 to 12 Hz in the electroencephalogram.
alternating current: Electrical current delivered as a sinusoidal wave of changing voltage amplitude. This is the form of current delivered at a wall outlet. Abbrev.: A.C.
amino acid: A chemical containing an amino group (nitrogen and hydrogen) and a carboxyl acid (carbon, oxygen, and hydrogen). Amino acids are the basic building blocks of proteins. Some amino acids serve as neurotransmitters.

AMPA receptor: A synaptic receptor for glutamate that mediates excitatory (depolarizing) signals. Abbrev. for α-amino-3-hydroxy-5-methyl-4-isoxazole propionic acid, an artificial agonist at the receptor.
ampere: Measure of current. 1 ampere=1 coulomb/second. Abbrev.: amp, A.
amplitude modulation: In signals and systems, transmission of a signal by change in the size of a carrier wave. Abbrev.: A.M.
amygdala: A deep-brain nucleus involved in emotional and appetitive behavior. Part of the limbic system.
analog: A signal characterized by a continuous quantity. Contrasted with discrete, digital, or binary; e.g., temperature is an analog property.
analog computer: A computer that organizes RC circuits so as to emulate physical phenomenon, using the fact that the differential equations of electronics are identical to those of mechanical dynamics.
analytic: Capable of being solved exactly using algebra, calculus, or other methods of mathematics. Analytic solutions are contrasted with numerical solutions, which are always approximations.
anatomy: The study of structures in biology. Contrasted with physiology.
AND: A Boolean operator. Symbol: $\wedge$. See Chap. 16, Section 16.3.
animalia: The animal kingdom.
anion: A negatively charged ion, e.g., Cl^-.
annealing: In physics, cooling so as to solidify in a lower energy state. In neural networks, using analogous mathematics to solve problems of minimization by reducing a high-energy network to an attractor state.
anomalous rectification: In electrophysiology, a channel that mediates inward current with membrane hyperpolarization.
antagonist: A drug or ligand that blocks a receptor. Opposite of agonist.
antidromic: In electrophysiology, signal flow against the normal signaling direction. Contrasted with orthodromic.
apamin: A potassium channel blocker that is a constituent of bee toxin.
aphasia: Loss of language following brain damage to the dominant hemisphere.
apical dendrite: A large dendrite ($\sim$ 1 mm) that typically protrudes from the apex of the soma in a pyramidal cell.
Archilochus: A dead white male who said something clever about small brown mammals (page 182).
architecture: In computer science, the details of the hardware design of a particular machine. In neural network theory, the connectivity of a network.
area MT: A cortical area that mediates visual perception of motion as well as other visual functions.
array: A data structure that places information in locations addressed by a consecutive numerical index. These are commonly used in numerical programming languages like FORTRAN and C. On a computer, a vector is stored as a one-dimensional array, and a matrix as a two-dimensional array.

artifact: An experimental finding that arises from the experimental design or experimental equipment and not from the thing being studied.
artificial intelligence: The development of computer programs to replicate complex intellectual skills. Abbrev.: AI.
artificial life: A science dedicated to building computer data structures to simulate evolving organisms.
artificial neural network: A system of interconnected computing units with simple properties loosely based on the behavior of neurons. Abbrev.: ANN.
Ascii: An international computer standard for binary encoding of letters, numbers, symbols, as well as certain screen and keyboard commands.
asphyxiation: Death due to lack of oxygen.
assembler language: Human-readable version of machine language. Assembler uses brief strings like MUL (for multiply) as a stand-in for the numerical representation of machine language. Verb: assemble — conversion from assembler to machine language. Also called assembly language.
assignment operator: In computer science, a symbol that represents the procedure whereby a variable (e.g., x) takes on a particular value (e.g., 5). In C: $x = 5$; in Algol $x := 5$; in Lisp (set 'x 5).
association cortex: An area of cortex that is not connected to primary afferents (sensation) or efferents (motor output). These areas are believed to coordinate or associate information from lower levels of cortex.
associative long-term potentiation: Increase in strength in a weakly stimulated synapse in the presence of strong activation in a nearby synapse.
associative memory: In neural networks, a network that associates one pattern with another. Sometimes used as a shorthand for autoassociative memory. Also associative network.
associative property: An arithmetic property that allows regrouping of terms: e.g., $(a+b)+c = a+(b+c)$; $(a \wedge b) \wedge c = a \wedge (b \wedge c)$ (Boolean AND).
astrocyte: A glial cell believed to be involved in maintaining the chemical composition of extracellular space.
asymptote: A value that is approached but never reached by a function.
asynchronous updating: A neural network technique where unit states are reset in random order.
ATP: Adenosine triphosphate. A compound used as energy storage for cells. ATP and its metabolites (ADP, AMP, cAMP) are also used as transmitters and second messengers.
attention: A neural mechanism to focus brain function on a particular sensory input.
attractor: A stable state or stable sequence of states in a dynamical system. A point attractor is a single point, e.g., the low point in a landscape where water will gather in puddles. A limit cycle is a repeated oscillation that remains stable over time — like the oscillation of a pendulum. A strange attractor is seen in chaotic dynamics.

attractor network: A neural network whose dynamics causes the state of the network to move to a single state (point attractor) or a sequence of states (limit cycle) representing the stored patterns.
autoassociative memory: A neural network memory that maps inputs onto themselves. Because of pattern completion, this form of memory can be used to identify a pattern despite an incomplete or degraded input.
autoimmune: Diseases produced by the body's own immune system attacking other body parts.
autonomic nervous system: The section of the nervous system that innervates the body's organs. Abbrev.: ANS.
Avogadro number: $\sim 6.02 \cdot 10^{23}$. Defined by the number of atoms in 1 gram of carbon-12.
axial resistance: In neurophysiology, the resistance along the length of a dendrite. Also called longitudinal resistance. Abbrev.: R_a, R_l, or R_i.
axoaxonic: Describing an axon that connects to another axon.
axon: Long, thin projections that typically carry action potential signals long distances in the nervous system.
axon collateral: A branch of an axon.
axon hillock: An initial expanded area of axon. Believed to be a spike initiation zone.
axon terminal: The end of an axon or axon collateral, usually presynaptic to another cell.
back-propagating spikes: In neurophysiology, action potentials that go backward from the soma up the dendrite.
back-propagation: An algorithm to permit reduction of error in multilayered networks by sending the error signal back through the layers. Also called back-prop.
band-pass filter: In signal and systems theory, a system that allows only a specific range of frequencies through, removing all frequencies below and above.
basal ganglia: A large complex of nuclei rostral (anterior) to the thalamus. Believed to be involved in motor activity, planning, and initiative.
base: In electrical engineering, the controlling input to a transistor. In mathematics, the value of the number system being used, e.g., base 8, base 10.
BASIC: An early interpreted computer language developed at Dartmouth College.
basin of attraction: In dynamical systems, area of state space from where system state will proceed to a particular attractor.
battery: A source of constant voltage.
bauble: A small pretty object that has no value.
belladonna: A drug that blocks the action of acetylcholine at the muscarinic receptor (a muscarinic antagonist).
beta waves: Waves of 13 to 26 Hz in the electroencephalogram.

bias: In electrical engineering, the voltage that must be applied to the base to get current flow from collector to emitter. In artificial neural networks, a false input that is always set in order to effectively shift activation function threshold during learning.
bilateral innervation: Projections to or from a brain area to both sides of the body. Contrasted with the more common contralateral innervation.
binary: The base two number system (Chap. 16, Section 16.3).
binary operator: In computer science, an operator that takes two arguments; e.g., '+' takes two arguments and produces the sum. Note that the binary in "binary operator" refers to *two* arguments and has nothing to do with binary numbers.
binding problem: Question of how the brain figures out which parts and attributes of a perception can be assigned to the same object. One aspect is how object attributes can be reconnected after being processed in different areas of the cortex. Must be solved by the brain in order to produce the perception of a single object rather than separate perceptions of color, motion, shape (see illusory conjunction).
bit: A binary digit — one place in the base 2 system.
bit-wise complement: Binary operation that changes every 0 to 1 and every 1 to 0.
bitmap: A data structure that stores an image using bits that represent single spots (pixels) of a picture.
black box: In engineering, term used to describe a device whose internal functioning is unknown.
blind spot: The area of visual space that projects onto the optic disk where the optic nerve exits. There are no photoreceptors at this location.
Boolean algebra: Mathematical system for calculation of truth values. Symbols for true (T) and false (F) are used with operators such as $\vee$ (OR) and $\wedge$ (AND).
bouton: Synonym for synaptic bouton — a form of presynaptic specialization.
brain: The part of the mammalian central nervous system within the skull.
brainstem: The lower part of the brain that connects to the spinal cord. Divides into three parts: mesencephalon, pons, and medulla.
Broca's area: A frontal area involved in producing language.
Brodmann area: A cytoarchitecturally defined area of cortex. These areas were originally mapped by Brodmann in the late 19th century.
bug: An error in a computer program.
bus: In computer science, the central wiring through which computer components communicate.
butterfly effect: A hypothesis in dynamical system theory: a butterfly flapping in China can change the weather a year later in South America. This is an evocative example of the sensitivity to initial conditions seen in chaotic systems. Weather is a chaotic dynamical system.

byte: In computer science, a unit of data or memory. Nowadays usually equal to 8 bits.
C: A popular programming language. C++ is a more modern variation on the language.
CA1, CA3: CA1 and CA3 are two areas of the hippocampus. CA stands for cornu ammonis.
cable equation: A partial differential equation describing the attenuation of voltage along an undersea cable or a passive dendrite. Attenuation is due to resistance along the cable as well as leakage through capacitance and resistance through the sides.
calculus: The area of mathematics dealing with change (differential calculus) and finding areas or volumes (integral calculus). See Chap. 16, Section 16.5.
capacitance: Ability of separated conductors to store electrical charge. See Chap. 16, Section 16.6.
capacitative coupling: Tendency of any transmitted oscillatory signal to induce charge buildup in neighboring conductors through field effects. For example, alternating current sent through a transmission line will induce charge in nearby people, creating temporary capacitors, with the line as one plate and the person as the other plate.
capacitor: In electronics, a device with parallel conducting places separated by an insulator.
carbon-based: Used to refer to living things, built primarily from carbon, hydrogen, oxygen, and nitrogen.
carrier frequency: In signals and systems, a constant or central frequency of a signal that does not itself convey information but is modulated in some way in order to transmit data.
catalyst: In chemistry, a material that increases the reaction rate of two chemicals by bringing them together. Enzymes serve as biological catalysts.
cation: A positively charged ion, e.g., Na^+.
cauda equina: The "horses tail" of nerves at the end of the spinal cord.
caudal: In anatomy, orientation toward the tail of the body. Due to the curvature of the neuraxis in the human, this direction is posterior for the brain and inferior for the spinal cord.
caudate: An input area of the basal ganglia. Part of the striatum.
cell assembly: Groups of simultaneously active neurons active together due to their mutual connections. Also called Hebb assembly.
cell body: See soma.
cell membrane: Lipid bilayer separating outside from inside of a cell.
central nervous system: The brain (including retina and brainstem) and spinal cord of higher organisms. Abbrev.: CNS.
central processing unit: The major data processing chip of the computer. In modern computers this can be a single silicon wafer with millions of transistors etched into it. Abbrev.: CPU.

cerebellum: A little cortex located behind the brainstem believed to be a movement or sensorimotor coordination center.
cerebrospinal fluid: The fluid that lies outside the brain and within the ventricles. Abbrev.: CSF.
cerebrum: The large outer lobes (cerebral hemispheres) of the mammalian brain. Adj.: cerebral.
chandelier cell: Inhibitory interneuron of cortex that synapses on pyramidal cell axons.
chaos: Unpredictable behavior in a deterministic dynamical systems with no random factors. In a chaotic system, small changes in initial conditions will completely alter the subsequent evolution of the system.
charge: Excess or deficiency in electrons in a substance. Measured in coulombs.
charge-couple device: Electronic array of photodiodes that captures the image in a digital camera. Abbrev.: CCD.
charging curve: In electronics or membrane physiology, the exponential increase in voltage with constant current injection.
charybdotoxin: A potassium channel blocker that is a constituent of scorpion toxin.
checksum: In data transmission, a data field that represents a sum of previously transmitted data used to check that no bits were lost.
chemical cascade: Signal amplification through a series of enzymatic transformations of second- and higher-order molecular messengers. At each stage, a single enzyme catalyzes production, resulting in amplification from a small number to a large number of signaling molecules.
chemical synapse: A synapse that uses neurotransmitters that diffuse across the synaptic cleft to bind receptors postsynaptically.
chromosome: DNA containing structure in the cell nucleus.
chunking: Organizing multipart information into a single datum for ease of remembering; e.g., remembering a telephone number as a three-digit followed by a four-digit number or as a single spelled word instead of as seven independent numerals.
circadian rhythm: In physiology, the daily cycles of body and brain.
CISC: Complex instruction set chip. In computer architecture, a central processing unit design that executes complicated instructions. With growing CPU complexity, the contrast with RISC is now historical.
clamp: In neurobiology, voltage or current may be clamped (see voltage clamp, current clamp). In neural networks, input units are clamped, meaning that their values are held fixed and not reset according to an update rule. In some cases, output units are also clamped — this is called teacher forcing.
clear bit: Binary bit value corresponding to 0 or False.
climbing fibers: A projection in the cerebellum. See Fig. 15.3.
clock: In computer science, a central oscillator that provides timing signals to coordinate signal transmission across the bus.

CNS: See central nervous system.
cocktail party effect: In psychophysics, the ability of the brain to selectively filter out large numbers of signals in order to attend to a single signal — as is used to carry on a conversation in a loud cocktail party.
code: In computer science, a program or the process of programming. In cryptology, a communication that has been altered in order to disguise its contents. In neuroscience, means by which information is passed in the nervous system.
coefficient: A number multiplying a variable in an equation. For example, 5 is the coefficient of x in the equation $5 \cdot x = 10$.
collaterals: Branches off the main trunk of an axon.
collector: In electrical engineering, one of the connections of a transistor.
column: In neuroanatomy, an internally connected volume of cortex often defined functionally by the presence of similarly responding cells.
command neuron: A neuron whose activity produces a stereotyped motor response. Found in some invertebrates.
communication theory: Mathematical theory for calculating the quantity of information that can be transmitted using particular types of signals. Closely related to information theory.
commutative property: An arithmetic property that allows swapping of terms: e.g., $a + b = b + a$; $a \wedge b = b \wedge a$ (Boolean AND).
compact: In electrophysiology, an adjective describing a neuron in which signals propagate throughout the dendritic tree with little decrement (long length constant).
compartment modeling: A standard biological modeling technique that divides an organism or part of an organism into homogeneous compartments that exchange some substance of interest. In electrophysiological modeling, the compartments are equipotential sections of dendrite that exchange current with neighboring compartments depending on differences in voltage.
compiler: Computer program that converts a high-level computer language such as C or Fortran into machine language. Verb: compile. Process: compilation.
complex number: A number that has both a real and imaginary part. The imaginary part is a multiple of i, the square root of -1. Complex numbers are used to describe oscillatory signals.
compound eye: Common eye design in invertebrates. A large eye is made up of many small eyes, each of which has its own cornea, lens, and photoreceptor.
computer: A term originally used to describe people who were employed to do calculations. Now used to describe machines that do calculations.
conditional statement: In computer science, a programming command testing truth to decide on the branching of control. In many computer languages, this takes the form: IF *true?* THEN *do A* ELSE *do B*.
conductance: See conductor.

conduction failure: In neurophysiology, failure of an action potential to transmit to the end of the axon.
conduction velocity: In neurophysiology, the speed of action potential propagation along an axon.
conductor: Something that conducts electricity — e.g., a wire or salt water. Symbol g, units siemens (S) or mhos. Conductance is the inverse of resistance: $g = 1/R$. See Chap. 16, Section 16.6.
cone: A type of retinal photoreceptor that detects color, primarily found in central retina.
confabulation: Pathological making up of stories to fill in inconsistencies in perception or memory, typically due to brain illness associated with dementia, delirium, or psychosis.
confectionery: Candy store.
congenital: Something present from birth.
connection matrix: A two-dimensional array of numbers give connection strengths between units of a neural network. Synonym: weight matrix.
connection strength: A parameter that determines the ability of a presynaptic neuron to drive a postsynaptic neuron. Also called weight.
consolidation: In neuropsychology, hypothetical process whereby memories are moved from temporary storage in the hippocampus to more permanent storage in cerebral cortex.
content-addressable memory: A memory system that allows retrieval of information based on a part of the information itself. Contrasted with the pointer-based addressing of a computer or a file cabinet.
continuous network: A network in which unit states change continuously in time instead of being updated at discrete time step increments. Of course, when a simulation is done on a computer, continuous time is approximated with discrete time steps anyway.
contralateral: Referring to the other side of the body. Opposite of ipsilateral. The cortex innervates contralateral body.
convergence: The number of inputs coming into a neuron. In mathematics, the ability of a numerical calculation or infinite series to reach a finite result.
core memory: In computer science, an old name for random-access memory (RAM). This term dates from an old technology that built RAM from magnetized iron cores.
cornu ammonis: Hippocampus, abbreviated CA in area names CA3, CA2, CA1.
corollary: In mathematics, a side proof that reveals another aspect or conclusion of a theorem.
coronal plane: A forward-facing vertical plane for sectioning the body (see Fig. 3.3).
corpus callosum: A large white matter tract that connects the two cerebral hemispheres.

correlation is not causality: Science buzz phrase indicating that just because two things have been observed to occur one after the other doesn't mean that the first one caused the second one to happen.
cortex: The outer gray matter of the cerebrum (cerebral cortex) or cerebellum (cerebellar cortex). Plural: cortices.
corticospinal tract: Motor pathway that runs from motor cortex to spinal cord. Also called pyramidal tract.
co-transmitters: Neurotransmitters released together at a single synapse.
coulomb: The measure of charge: 1 coulomb $\sim 6.2 \cdot 10^{18}$ electrons.
CPU: See central processing unit.
credit-assignment problem: In neural network memories, the difficulty of figuring out which unit or which connection was responsible for an error in a pattern. Neural network algorithms such as back-propagation have been developed to solve this.
critical period: Limited period of development of an organism when a particular brain area (e.g., vision, language) is plastic and can be molded by the environment.
cryptography: The art and science of coding and decoding.
CT scan: Computed tomography scan. An imaging method that re-creates two-dimensional cross sections by adding up x-ray conductances measured from different angles. Also called CAT scan.
current: Movement of charge: $\frac{\Delta Q}{\Delta t}$. Symbol I, measured in amperes (amps, A).
current clamp: Injection of continuous current to a neuron. Voltage is measured.
CVA: Cerebrovascular accident. See stroke.
cytoarchitectonics: Distinguishing brain areas by differences in neuron types and organization.
cytoplasm: The fluid inside a cell but outside of organelles.
Dale's principle: The hypothesis that a single neuron expresses only one transmitter. Although this has turned out to be wrong in detail (see co-transmitters), it's still used as a general rule to indicate that a single cell will not be both excitatory and inhibitory.
data structure: Representation format for organizing information in computer software.
database: In computer science, data structures for storing large amounts of information so as to make them compact and readily accessible.
De Morgan's theorem: A property of Boolean algebra that relates AND to OR and vice versa. See Chap. 16, Section 16.3.
DEC: Digital Equipment Corporation. A company that was an early innovator in producing small computers that could be run and used by an individual rather than a large team. See Parallel Data Processor.
decimal: Base 10.
declarative memory: Synonym for episodic memory.

degenerative disease: Diseases that involve progressive loss of neurons or neural functioning, e.g., Parkinson and Alzheimer disease.
degree: Unit of angular measure. There are 360 degrees in a circle. Also, unit of temperature measurement.
delayed rectifier: Synonym for the Hodgkin-Huxley potassium channel.
delta rule: In neural network theory, an algorithm for updating weights so as to correct errors and solve the credit-assignment problem.
delta waves: Slow waves (≤ 3 Hz) in the electroencephalogram. Mostly seen in sleep.
dendrite: Cellular extensions of the neuron with postsynaptic specializations. Distinguished from axons by being generally fatter, shorter, more tapered, and unmyelinated.
dendritic tree: The full set of dendrites.
dendrodendritic: Describing a dendrite that connects to another dendrite.
denominator: The bottom of a fraction, e.g., the 3 in $\frac{2}{3}$.
dentate gyrus: An area of hippocampus with cells that project to area CA3.
depersonalization: Mental or neurological condition producing loss of a person's sense of his own identity.
depolarization: Positive deviation from the negative resting membrane potential.
depolarization blockade: In electrophysiology, inactivation of sodium channels by prolonged depolarization causing failure of spiking.
derealization: Mental or neurological condition producing sense of disconnection from surrounding reality.
derivative: A basic operation of calculus. Ratio of amount of change in a function (y or $f(x)$) to the amount of change in an independent variable (x). This equals the slope of the tangent to the curve of y graphed against x. The derivative is represented by $\frac{\mathrm{d}y}{\mathrm{d}x}$. If x is position, then $\frac{\mathrm{d}x}{\mathrm{d}t}$, the change in position with respect to time, is velocity.
description length: In computer science, a technique for assessing the size of an algorithm by considering both the amount of memory devoted to the program and the memory devoted to data.
desensitization: Reduced responsiveness of a receptor to a ligand after prolonged exposure.
deterministic: In mathematics, a system whose behavior is entirely defined by its present state and its governing equations with no random inputs. Contrasted with stochastic.
dielectric constant: In electronics, describes the polarizability of a compound in an electric field. Capacitors use substances with a high dielectric constant between the plates in order to increase capacitance.
diencephalon: Part of the brain connecting the cerebral hemispheres with the brainstem. Includes thalamus and hypothalamus.

difference equation: A functional form that describes the differences between values in a sequence.
differential equation: An equation where derivatives appear on one or both sides of the equal sign.
differentiation: The taking of derivatives to determine the rate of change of one quantity with respect to another (e.g., time or space).
diffusion: The movement of particles in solution out of a high concentration toward lower concentration.
dimensional analysis: Determination as to whether the units being used in two aspects of an equation or conversion are compatible. Also called unit analysis.
dimensional reduction: In dynamical systems, a mathematical technique for analyzing a high-dimensional system by collapsing into a lower-dimensional system.
dimensionality: In dynamical system theory, the number of state variables in a system. The state of the system is a point and the full dynamics is a curve in state space with this number of dimensions.
diode: In electronics, a component that passes current in only one direction.
direct current: Electrical current delivered at a constant voltage. This is the form of current delivered by a battery. Abbrev.: D.C.
discrete: A measure or enumeration that only takes on a finite number of specified values. Opposed to continuous or analog.
discrete mathematics: Calculations that utilize specific numbers rather than symbols representing continuous values. Numerical analysis done by computers is discrete mathematics.
disinhibition: Removal or reduction of inhibitory effect in a neuron or circuit.
distributed representation: A representation coding information diffusely through multiple units. Opposed to local representation or grandmother cells. Also called distributed code.
divalent: Describing an ion with two excess electrons or protons (e.g., Ca^{++}).
divergence: The number of outputs coming from a neuron. Failure of a numerical calculation to converge to a finite result.
DNA: Deoxyribonucleic acid. The double helix that serves as primary genetic material.
domain: In mathematics, the set of values that can serve as inputs to a function. See range.
dominant hemisphere: The cerebral hemisphere that controls language, usually the left hemisphere.
dopamine: A neurotransmitter. Loss of dopamine cells causes Parkinson disease.
dorsal: Toward the back side of the neuraxis: posterior in the spinal cord and superior in the brain.

dot product: A linear algebra operation that takes two equal-length vectors and creates a scalar that is the sum of pair-wise products from corresponding positions in the two vectors. Also called inner product.
downregulate: In biology, reducing the amount or activity of an enzyme, receptor or other active protein due to negative feedback from the products or results of that enzyme or receptor.
driving force: In electrophysiology, the difference between the Nernst or reversal potential for a particular ion and the membrane potential: $V - E_{rev}$.
duty cycle: In signals and systems, the proportion of the period of a binary periodic signal that is spent at the high value.
dynamical system: Originally, a dynamical system was a set of any objects that moved and influenced each other through gravitation (e.g., the solar system). The term has generalized to include any system with multiple state variables that interact in a time-dependent fashion. In practice this means any system that can be described by a set of differential or difference equations.
dyslexia: Difficulty in reading due to brain disorder. Can be congenital or acquired.
dystopia: A vision of future society as an awful place. Opposite of utopia.
edge effects: Artifacts in neural network activity occurring in units at or near the boundaries of the network due to the reduced number of connections at the edge.
efferent: Leading or projecting away from neural structure.
electric eye: A photodiode paired with a constant beam of light that is used to detect the passage of a person that interrupts the beam. The electric eye has been partly supplanted by infrared, microwave, and ultrasonic door monitors.
electrical self-stimulation: A mechanism for allowing an animal to deliver small electric currents to activate a particular spot in its own brain.
electrical synapse: A gap junction between cells that allows current to flow directly from the interior of one cell to the interior of another cell.
electrochemical equilibrium: The balance point between diffusive force in one direction and membrane potential gradient in the other direction. The membrane potential where this occurs is the Nernst potential.
electrocorticogram: Brain waves recorded using electrodes placed directly on the brain.
electrode: A conductor used to probe an electrically active material. See microelectrode.
electroencephalogram: A recording of electrical signals from the brain made on the skin of the head. Abbrev.: EEG.
electrolyte: A substance that will carry electricity when dissolved in water.
electron microscopy: An imaging technique that uses electrons to get far better resolution than is possible with light microscopy.

electrophysiology: Set of techniques for measuring the electrical properties (voltages and currents) of neurons, of other electrically excitable cells, and of cellular connections.
electrotonic: In electrophysiology, concerning the response of a cell to a constant current injection.
emergent properties: Properties resulting from the combined effects of many elements in complex systems that cannot be explained by knowing the properties of these elements.
emitter: In electrical engineering, one of the connections of a transistor.
emulation: Use of a physical, as opposed to a virtual, model system. Contrasted with simulation.
emulator: In computer science, a program written to reproduce the behavior of a different computer system or component (like a CPU).
en passant synapse: A synapse at a site where an axon passes by a dendrite.
endogenous: Arising from or belonging to an organism or other system. Contrasted with exogenous.
endorphin: An endogenous peptide neurotransmitter that is active at the same location where opium or morphine acts.
energy function: A mathematical equation that assigns a potential energy value to every point in a multidimensional space.
engineering notation: A standard for representation of numbers that uses prefixes (e.g., kilo, mega, giga) for powers of 1000. See Chap. 16, Section 16.2.
enteric nervous system: The nervous system that runs the gut. It can function on its own even after being completely disconnected from other parts of the nervous system.
enzyme: A protein that catalyzes a chemical reaction, causing it to occur at a rate far faster than would occur in the absence of the enzyme.
ephaptic: Extrasynaptic interneuronal communication mediated by the effects of electric field generated by one neuron on another.
epilepsy: A brain disorder in which repeated seizures occur.
epiphenomenon: A measurable side effect of a phenomenon that has no functional relevance.
episodic memory: Human memory of events in life.
epsilon: In mathematics, a very small number.
equilibrium potential: The Nernst potential that produces electrochemical equilibrium for a particular ion.
equivalent: In chemistry, a unit corresponding to the quantity of ions needed to carry 1 faraday.
equivalent cylinder model: A reduction of a branched dendrite model into a single cylinder that preserves the electrical properties of the original.
error-correcting code: A redundant representation of information that includes an internal check of accuracy that allows slight transmission errors to be corrected by the receiver.

error minimization: In artificial neural network supervised learning theory, the goal of reducing the difference between network output and desired target to a minimum.
eukaryote: In biology, organisms that have cells with a nucleus. This includes animals, plants, and yeast.
excitatory: Serving to excite or turn on a neuron — often said about synapses, weights, or connections.
excitatory postsynaptic potential: Depolarization of the postsynaptic membrane potential due to synaptic action. Abbrev.: EPSP.
exclusive OR: See XOR.
existence proof: In mathematics, a proof that demonstrates that a certain thing (e.g., a type of function) must exist, even if we cannot yet, or can never, determine what it actually is. It is said that the brain provides the existence proof for the ability of a machine to do language, vision, etc.
exocytosis: Release of material from a cell due to fusion of a vesicle with the cell membrane. This is how neurotransmitters are released into the synaptic cleft.
exogenous: Compound or material introduced to a system from outside. Contrasted with endogenous.
expert system: Computer program that attempts to replicate human reasoning in a particular domain such as medicine or law.
explicit Euler integration: A numerical integration technique that uses the past value of state variables in the approximation. See Chap. 11, Section 11.4.
exponent: The power a in x^a.
extracellular: The volume outside of cells in the body.
facilitation: In neuroscience, transient increase in excitability. Generally, used for effects that are short in duration as compared to potentiation.
false unit: Unit of a neural network that does not do any processing but simply serves as an access point for presenting inputs or bias.
farad: Measure of capacitance. 1 farad=1 coulomb/volt. Abbrev.: F.
faraday: A mole of electric charge. Equals approximately 96,485 coulombs.
feature: In computer science, desirable functionality in software, whether deliberately programmed in or not. Contrasted with bug. In image processing, an elemental attribute to be extracted from an image during initial processing.
feedback: Information or signal that is communicated back to a component from a later processing component.
feedforward: In artificial neural networks, a network architecture that sends information in one direction only. Contrasted with recurrent. In neuroscience, refers to projections (particularly inhibition) that arrives before or together with a major projection.
field potential: In electrophysiology, an electrical measurement representing the summed activity of many neurons.

filter: In signal and systems theory, a program or physical device that transforms a signal. There are a variety of named filter types that handle signals differently: low-pass, high-pass, band-pass, etc.
fitness function: Function used in genetic algorithms to describe how well a result fits a target.
fixed point: In the context of a dynamical system, a point in state space where the system stops and stays, the state vector remaining the same thereafter.
fluid mosaic model: The standard conceptual model of the cell membrane: the lipid bilayer is in a fluid state somewhere between a liquid oil and a solid fat. Proteins and other uncharged molecules can float around in the membrane.
flux: Flow. In neurophysiology, used to describe the flow of an ion across the membrane.
FORTRAN: FORmula TRANslation language, the earliest compiled computer language.
fovea: The central part of the retina that has the most acute vision.
frequency: The number of oscillation in a fixed period of time. Measured in hertz = oscillations/second.
frequency coding: The notion that increases in spike frequency are an important part of the code of the nervous system. Also called rate coding.
frequency modulation: In signals and systems, transmission of a signal by changes in the frequency of a carrier wave. Abbrev.: F.M.
frontal lobe: The most anterior lobe of the brain. Believed to play a role in initiative and planning.
functional magnetic resonance imaging: A technique for imaging physiological activity using magnets and radio waves. Abbrev.: fMRI.
GABA: γ-aminobutyric acid. A major inhibitory neurotransmitter in the central nervous system.
$GABA_A$: A GABA receptor that mediates rapid inhibitory postsynaptic potentials.
$GABA_B$: A second-messenger linked GABA receptor that mediates slow inhibitory postsynaptic potentials.
gain: In signals and systems, the amplification factor provided by a system.
gamma waves: Waves of 20 Hz and above in the electroencephalogram.
ganglion: A collections of neurons. Usually refers to neurons lying outside the central nervous system in vertebrates (except in case of the *basal ganglia*). In invertebrates, ganglia are the central processing structures (the brains). Plural: ganglia.
gap junction: An ion channel that connects two cells together.
gating: In neurophysiology, the ability of channels to be opened or closed either by membrane potential (voltage gated) or by the arrival of a neurotransmitter or other chemical (ligand gated).
Gaussian function: The bell curve used in statistics; the basic form is $f(x) = e^{-x^2}$.

gene: The unit of information storage in the coding nucleic acid (usually DNA) of a cell. Classically a single gene coded for a single protein; this is an oversimplification.
generalization: Flexibility in learning that allows a system to correctly classify an object not previously learned.
genetic algorithm: A computer data-fitting algorithm modeled after the process of evolution through changes in DNA. Many sets of parameters are chosen randomly and tested against a fitness function. Those that are most fit are then combined with each other through crossing-over and mutated to produce a next generation to be tested.
genetic code: Representations of amino acids as well as stop and start signals by triplet sequences of four nucleotides (A,T,G,C) in DNA and RNA.
genome: All of the genes of an organism.
glial cell: The most common type of cell in the nervous system. These cells play roles in support, hormone release, and balancing extracellular concentrations. Two important types are astrocytes and oligodendroglia.
globus pallidus: An area of the basal ganglia. It has two parts: pars externa and pars interna.
glucose: The sugar used as food by cells.
glutamate: An amino acid that is the major excitatory neurotransmitter in the central nervous system. Also called glutamic acid.
glycine: An amino acid that is also an inhibitory neurotransmitter.
Goldman-Hodgkin-Katz equation: Equation describing the contribution of multiple ions to membrane potential based on their independent permeabilities and concentrations.
graceful degradation: In neural networks, the ability of a network to show graded reduction in function with damage. Contrasted with the catastrophic breakdown property of computers.
gradient: A slope.
gradient descent algorithm: In artificial neural network theory, process for reducing the energy level of network state by heading gradually "downhill" from higher to lower potential energy.
grandmother cell: An individual neuron that represents a specific thing, place, or person (such as one's grandmother). An evocative term for local representation.
granule cell: Name for small, densely packed neurons found in dentate gyrus and cerebellum.
grapheme: The written representation of a sound in a language. See phoneme.
gray matter: Brain volumes made up of cell bodies. Cerebral cortex, basal ganglia, and thalamus are gray matter.
ground: In electronics and electrophysiology, a connection to a standard zero electrical potential, provided by attaching to a large conductor (typ-

ically the ground under your feet) that can readily source or sink charge. When used as a verb: attach to ground or to a grounding wire.
gyrus: The cerebral hemispheres are folded; a gyrus is one of the ridges.
H.M.: Initials of a man who had bilateral hippocampectomy (removal of the hippocampus on both sides) to cure his epilepsy. Over the past several decades, he has been relentlessly studied by neuropsychologists trying to infer the function of the hippocampus from his many memory dysfunctions.
habituation: Reduced response to a stimulus after repeated stimulation. A model of simple learning. See accommodation.
hack: In computer science, a programming trick.
Hamiltonian: The total energy (kinetic plus potential) of a system. Extended from use in statistical mechanics to information theory.
hardware: In computer science, the underlying machine that does data processing.
Heaviside function: A discontinuous step function with discontinuity at 0. If $x < 0$, $h(x) = 0$ else $h(x) = 1$.
Hebb rule: The hypothesis that synapses will be strengthened when the pre- and postsynaptic neurons are active at the same time. This is referred to in multiple expressions: Hebb's law, Hebb's postulate, Hebb synapse, Hebbian synapse, Hebbian learning, etc.
hertz: The unit of cycles per second for a sinusoidal oscillation. Abbrev.: Hz.
heteroassociative: Referring to a memory system that maps one thing onto another, e.g., names onto faces.
hexadecimal: Base 16; also called hex.
hidden layer: A layer of units in a feedforward neural network that are not exposed to either the input or the output.
hidden unit: A unit in a feedforward network that does not receive inputs or produce outputs.
high-pass filter: In signal and systems theory, a system that removes the low frequencies from a signal and only lets the high frequencies get through to the output.
hippocampus: A cortical area in the temporal lobe involved in episodic memory (memory for events).
histology: The study of the organization of living cells into tissues.
Hodgkin-Huxley equations: Differential equations describing the origin of the action potential. Each of three particles is described by differential equation of the form $\tau_x \cdot \frac{\mathrm{d}x}{\mathrm{d}t} = x_\infty - x$, where x is m, h, or n. The infinity and tau curves (Fig. 12.5) are defined in terms of $\alpha(V)$ and $\beta(V)$ as follows:

$$x_\infty(V) = \frac{\alpha_x(V)}{\alpha_x(V) + \beta_x(V)} \qquad \tau_x(V) = \frac{1}{\alpha_x(V) + \beta_x(V)}$$

The α and β curves are defined differently for each of the three particles:

$$\alpha_m(V) = \frac{0.1\,(V+40)}{1-e^{-0.1(V+40)}} \quad \beta_m(V) \; = 4 \cdot e^{-0.0556(V+65)}$$

$$\alpha_h(V) = 0.07 \cdot e^{-0.05(V+65)} \quad \beta_h(V) \; = \frac{1}{1+e^{-0.1(V+35)}}$$

$$\alpha_n(V) = \frac{-0.01\,(V+55)}{e^{-0.1(V+55)}-1} \quad \beta_n(V) \; = 0.125e^{-0.0125V+65}$$

Voltage is given by the parallel conductance model (Fig. 12.2):

$$C \cdot \dot{V} = \overline{g}_{Na} m^3 h\,(V - E_{Na}) + \overline{g}_K n^4\,(V - E_K) + \overline{g}_{leak}\,(V - E_{leak})$$

where each $\overline{g}$ is an individual maximal conductance, each E the potential for a battery, and C the membrane capacitance.

holding potential: In electrophysiology, a voltage being maintained at the beginning of an experiment. Voltages other than resting membrane potential are maintained by injecting positive or negative current.

hologram: An imaging method that records frequency and phase rather than intensity of light.

homologous evolution: In biology, the production of similar structures in different creatures through Darwinian evolution. The existence of similar wings in birds and bats is an example.

Hopfield network: A fully connected Hebbian recurrent network that shows attractor dynamics and works as a content-addressable memory.

horizontal plane: The horizontal plane for sectioning the body (see Fig. 3.3).

horseshoe crab: Common name for Limulus.

humunculus: In neuroscience, distorted representation of the body mapped from receptive fields of neurons on cortical areas. See Fig. 7.1. Also used to refer to the putative little guy in the brain who watches everything that goes on.

Huntington disease: An inherited degenerative brain disease involving the basal ganglia. It is a movement disorder characterized by dance-like movements.

hydrophilic: In chemistry, a charged or polar compound that will exist comfortably in water but not in fat. Opposite of hydrophobic.

hydrophobic: In chemistry, a compound that doesn't mix with water but will mix (partition) into fat. Same as lipophilic, opposite of hydrophilic.

hyperpolarization: Negative deviation from resting membrane potential.

hypoglycemia: Low blood sugar.

hypothalamus: A collection of brain nuclei responsible for central control of various hormones and of appetite and thirst.

hypothesis-driven: Ideal that scientific research should generally (perhaps always) be motivated by the search for confirmation or refutation of a specific theory.

I-beam: In civil engineering, a steel structural unit shaped like a capital I in cross section. Commonly used to build skyscrapers and bridges.
I-f curve: In electrophysiology, the current-frequency curve. The firing rate of a neuron graphed as a function of amount of injected current.
I/O: In engineering, an abbreviation for input/output. Describes the output that will be produced by a system in response to a particular input.
IBM: International Business Machines, one of the first major manufacturers of computers. It dominated the industry for several decades.
if and only if: Mathematics phrase for sufficient and necessary. Symbolized $\leftrightarrow$ or "iff."
illusory conjunction: A perceptual error that can occur with brief visual presentations of multiple objects. For example, many different colored shapes are presented briefly, and the visual system mistakenly binds some shapes with the wrong color.
imaginary number: In mathematics, a number that is a multiple of i, the square root of -1.
impalement: In electrophysiology, a technique for making measurements inside of a cell by sticking an electrode through the membrane.
impedance: Opposition to current flow. See Chap. 16, Section 16.6.
impedance matching: In electronics, providing similar resistance at a junction between circuits so as to optimize flow by preventing reflection of current back into the source circuit.
implicit Euler integration: A numerical integration technique that uses the future value of state variables in the approximation. See Chap. 16, Section 16.5.
***in vitro*:** Study of biological tissue after removal from the organism; means "in glass."
***in vivo*:** Study of biological tissue in the intact organism; means "in life."
inactivation: Closing of ion channels with time.
inertial guidance system: A navigation device that detects motion in order to calculate position.
inferior olive: A brainstem nucleus involved in movement.
infinite loop: In computer science, a programming error (bug) that results in the computer endlessly executing the same commands.
infinitesimal: In mathematics, an infinitely small duration or size.
information theory: A mathematical framework for describing the information content of messages.
inhibitory: Serving to turn off a neuron.
inhibitory postsynaptic potential: A change in membrane voltage due to synaptic activation that tends to reduce the activity of a postsynaptic neuron. Usually hyperpolarizing. Abbrev.: IPSP.
initial condition: A starting point for a dynamical system.
inner product: See dot product.
innervate: Form neural connections onto some structure.

input impedance: In neurophysiology, a measure of the effective resistance (V/I) encountered by a current injected into a cell. Similarly in electronics for any circuit.
instantaneous frequency: A minimal frequency measure, made between two spikes in a spike train or by taking the inverse of a single period of any oscillatory signal.
instantiation: A physical implementation of a hypothesis or concept.
instruction register: In computer science, a location on the central processing unit where a command is stored immediately before execution.
instruction set: In computer science, the set of commands built into a central processing unit.
integral: A basic operation of calculus. Represented by $\int$, a symbolic "S," it gives the sum of all the values of a curve — the area under the curve. The reverse operation of derivative.
integrate-and-fire model: A simple neural model that sums inputs and produces a spike when the model reaches a predetermined threshold.
integration: In mathematics, a calculus operation that sums up the area under a curve. In neuroscience, the tendency of neurons to add up synaptic signals.
interneuron: A neuron with axons that project locally. In the cortex many of these appear to be inhibitory.
interpreter: A computer program that reads a program line by line and performs the requested operation. Contrasted with compiler.
interspike interval: The period of time between two action potentials.
intracellular: The inside of biological cells.
intrinsic disease: Diseases that strike directly at the functioning of neural systems.
invertebrate: An animal without a spine. Can be divided into squishies (octopus, squid, leech) and crunchies (insect, limulus, lobster).
inward current: In electrophysiology, current that flows through a conductor from extracellular fluid to cytoplasm. Inward conductance current produces outward capacitative current, which will depolarize the cell.
inward rectifier: In electrophysiology, a channel that mediates inward current with membrane depolarization.
ion: A charged atom. The most important ions in the nervous system are Na^+,K^+,Cl^-,Ca^{++}. The superscript identifies the charge.
ion channel: In neuroscience, an integral membrane protein that allows the passage of ions down their chemical gradient. Ion channels are often selective and may be voltage-sensitive.
ipsilateral: Referring to the same side of the body. Opposite of contralateral.
Ising model: A classical physics model describing the submicroscopic organization of magnets. The way that the individual magnetic domains line up with each other has been taken as an analogue of cooperative organization in neural networks.

isolable: Something that can be isolated.
isopotential: Having the same electrical potential everywhere. Same as equipotential.
JAVA: A modern computer language optimized for portability.
joule: Measure of energy.
Kelvin: A temperature scale with its zero at absolute zero. Degree size is the same as in the Celsius (centigrade) scale.
kinesthesia: The sensation of one's own movements. This will include joint position sense and tactile clues.
kinetic energy: The energy associated with motion.
Kirchhoff's law: Conservation of charge, conservation of current.
kit-and-kaboodle: A silly old phrase meaning "the whole thing bundled together."
kludge: In computer science, a programming trick that is very hard for others to understand, making a program unreliable or impossible to maintain.
knock-out: In genetics, an animal born without certain proteins due to the removal of certain genes before fetal development.
Krebs cycle: The series of enzymatic reactions used to extract energy by oxidizing (burning) the 3-carbon products of glucose breakdown.
labeled-line coding: Coding that uses the identity of a specific wire or active neuron to identify a message. For example, activity in one set of neurons indicates pain, while activity in another set signals vibration.
lateral: In anatomy, toward the side. Opposed to medial.
lateral geniculate nucleus: An area of the thalamus that relays information from the retina. Abbrev.: LGN
lateral inhibition: In neurophysiology and neural networks, inhibition of flanking units.
learning: Storage of new information in a biological or artificial system.
learning rate: In neural networks, a coefficient that can be adjusted to speed up or slow down learning.
least mean squares rule: See delta rule. Abbrev.: LMS rule.
length constant: The distance in a passive dendrite over which a steady-state voltage drops to about 37% (e^{-1}) of its initial value. Also called space constant.
ligand: A chemical that binds (ligates) to some receptor.
ligand-gated channel: An ion channel that opens and closes depending on the arrival of some chemical — usually either a neurotransmitter or second messenger.
limbic system: A set of interconnected brain structures that may be involved in emotion (see Papez circuit).
limit cycle: See attractor.
Limulus: The short Latin name for the horseshoe crab.
limulus equation: An update rule for modeling lateral inhibition in the limulus eye: $\vec{s} = \sigma(\vec{p} + W \cdot \vec{s})$.

linear: In signals and systems, having a response or output that is directly proportional to the input. In mathematics, functions without powers higher than 1.
linear algebra: A form of mathematics that uses vectors and matrices (see Chap. 16, Section 16.4).
linearly dependent: In linear algebra, denotes a vector can be represented by a sum of multiples of other vectors.
linearly independent: In linear algebra, denotes a set of vectors for which no vector can be represented as a sum of multiples of any or all of the other vectors.
lipid bilayer: The architecture of membranes in living things. See Chap. 11, Section 11.2.
LISP: A list processing programming language popular in artificial intelligence.
lissencephaly: Smooth cortex lacking normal gyri and sulci found in some primates and in some children with abnormal brain development.
local code: See local representation.
local minimum: In attractor dynamics, a low-energy point that is not the lowest-energy point in the field.
local representation: A representation where each individual unit represents a specific thing. Opposed to distributed representation.
logarithm: Exponent needed to raise a base to produce a given number; e.g., $log_{10}(100) = 2$ because $10^2 = 100$.
logistic function: A function of the form $1/(1+e^{-x})$. This function ranges from 0 to 1 and is sometimes used as a squashing function.
long-term depression: A persistent weakening of synaptic strength. Abbrev.: LTD.
long-term potentiation: Persistent strengthening of synaptic strength after paired pre- and postsynaptic activity. Considered a neural analogue of the Hebb synapse. Abbrev.: LTP.
longitudinal resistance: See axial resistance.
low-pass filter: In signal and systems theory, a system that removes the high frequencies from a signal and lets only the low frequencies get through to the output.
LTP: See long-term potentiation.
lumen: In physics, unit of light intensity. In biology, the space at the center of a tube, e.g., the intestine.
Mach bands: A visual illusion seen at the edge of a contrast boundary. The bright side of the edge looks extra bright and the dark side of the edge looks extra dark.
machine language: In computer science, the set of binary operations that are understood and executed by the central processing unit (CPU).
macroscopic: Big enough to be seen with the naked eye.
mantissa: The significant digits in scientific notation, e.g., 6.022 in $6.022 \cdot 10^{23}$.

map: In neuroscience, an area of the nervous system that represents some aspect of sensory or motor space in a topographic manner. In mathematics, transformation of one set of variables to another.
mapping: Mathematically the transformation of one representation into another. For example, a matrix multiplication can be used to map vectors in one space to vectors in another.
Markov model: A modeling technique used in describing active proteins, such as ion channels or enzymes. A finite number of conformational states of the protein are defined. Transition maps and rates among the states describe how the model passes from one configuration to another. In the case of an ion channel, some configurations will be open and others closed to ion conductance.
matrix: Array of numbers arranged in rows and columns. See Chap. 16, Section 16.4.
Maxwell equations: The four differential equations that describe the interaction between electricity and magnetism.
McCulloch-Pitts model: An early artificial neural network model using summation of excitation with veto inhibition.
medial: In anatomy, toward the middle. Opposed to lateral.
medulla: The lower part of the brainstem connecting to the spinal cord.
membrane potential: Voltage measured at the inside of a membrane compared to ground outside.
memory: In computer science, usually refers to random-access memory (main memory). In general, any form of information or signal storage.
mesencephalon: The upper part of the brainstem, connecting to the diencephalon.
metabolism: Biological process of extracting and utilizing energy.
metabolite: A substrate or product of metabolism.
mhos: Siemens.
microelectrode: Tiny electrodes used to measure electrical activity in or around neurons or other cells.
microglia: A type of glial cell involved in protecting the central nervous system from infection.
micron: A micrometer: 10^{-6} meters.
micropipette: A small glass tube filled with conducting solution that can be used as a microelectrode.
microtubule: A cell support protein found in cells.
mirror attractors: In memory networks, additional attractors that form during learning that are the negatives of the learned patterns.
mitochondria: In cell biology, the organelles responsible for energy production. Mitochondria have their own genome, suggesting that they were originally free-living organisms before banding together with others to form the eukaryotic cell.
molar: Measure of concentration in solution equal to 1 mole of solute per liter.

mole: Standard unit of material quantity equal to approximately $6.02 \cdot 10^{23}$ atoms or molecules (the Avogadro number).
monotonic: In mathematics, property of a function that continues ever upward (monotonically increasing) or ever downward (monotonically decreasing).
monovalent: Describing an ion with a single excess electron (e.g., Cl^-) or proton (e.g., Na^+).
mossy fibers: A descriptive term for an axonal pathway. There are pathways by this name in the hippocampus (from dentate gyrus to CA3) and cerebellum (from spinal cord to granule cells).
motor: In neuroscience, descriptor for systems involved in producing movement of the organism.
motor cortex: Cortical areas that control movement.
MRI: Magnetic resonance imaging. A computed tomography technique that produces an image using nuclear magnetic resonance (NMR), the interaction of atoms with radio waves and magnets. This was called NMR until a savvy marketer realized that it would be difficult to sell a medical procedure that used the word "nuclear."
multiple sclerosis: A neurological disease involving autoimmune attack on myelin in the central nervous system, which produces axonal conduction failures and ephaptic transmission in tracts.
multiplex: In data transmission, sending more than one signal or type of signal on a transmission line at the same time.
multiunit recording: In electrophysiology, extracellular recording of action potentials from multiple neurons simultaneously. Contrasted with field potential or single-unit recording.
muscarinic: Adjective referring to a set of acetylcholine receptors in the central nervous system.
myelin: An insulating material wrapped around axons to permit faster conduction.
nadir: A low point. Opposite of zenith.
natural log: Logarithm using $e \sim 2.18281828\ldots$ as the base. Symbol: ln; e.g., $\ln(10) \sim 2.303$ because $e^{2.303} \sim 10$.
negative feedback: A signal communicated back to a prior system component that decreases the output or activity of that component.
neglect syndrome: A disorder of perception seen after damage to the nondominant hemisphere that involves a failure to perceive contralateral space, sometimes including the person's own body.
neocortex: Major area of cortex in higher mammals. Contrasted with more primitive archicortex.
Nernst potential: The membrane potential associated with an ion at electrochemical equilibrium accross the membrane. It can be calculated using the Nernst equation, which relates the membrane potential (battery) associated with an ion to its concentrations inside and outside. $E = \frac{RT}{zF} \ln(\frac{[A]_{out}}{[A]_{in}})$

where R is the gas constant, T the temperature, F the faraday, and z is the valence of ion A.
nerve: A bundle of axons in the peripheral nervous system. Nerves run to muscles and from sensory organs in the periphery.
neuraxis: In anatomy, the curve defined by the extent of the central nervous system.
neuroanatomy: Study of brain and other nervous tissue using observation of fixed specimens.
neuromodulator: A neurotransmitter with a relatively prolonged effect.
neuromuscular junction: The synaptic connection between nerve terminal and muscle.
neuron: The principal information processing cells of the nervous system.
neuron doctrine: Cajal's theory that neurons are separate cells. Contrasted with Golgi's reticular hypothesis suggesting that the neural cells formed one large syncytium.
neuropeptide: Short amino acid sequence that serves as a neurotransmitter.
neuropharmacology: Study of how drugs alter neural function.
neurophysiology: Study of brain and other nervous tissue by observing change in some measurable attribute such as voltage or the concentration of a chemical.
neuropil: Areas of the nervous system with few cell bodies, dominated by dendrites, axons, and synapses.
neurotransmitter: A chemical released across a chemical synapse to communicate from one neuron to another.
nicotinic: Adjective referring to a set of acetylcholine receptors in the central nervous system and at the neuromuscular junction.
NMDA receptor: An excitatory synaptic receptor that binds glutamate, provides a long time-course excitatory postsynaptic potential. Activation of NMDA underlies some forms of long-term potentiation (LTP). Abbrev. for N-methyl-D-aspartate, an artificial agonist at the receptor.
NO: Nitric oxide. A gaseous neurotransmitter.
node of Ranvier: A gap in the myelin sheath where action potentials can be generated.
noise: In signals and systems theory, random perturbations that interfere with transmission of a signal.
nondominant hemisphere: The cerebral hemisphere that doesn't control language. The right hemisphere in most people, it is believed to play a role in spatial tasks.
nonlinear: Having a response that is not directly proportional to the input.
normalization: In signal processing, reducing variation among signals of differing amplitudes in order to handle them similarly.
NOT: A Boolean operator. Symbol: $\sim$. See Chap. 16, Section 16.3.
nucleus: In neuroanatomy, a collection of cell bodies in the central nervous system. In cell biology, an organelle that holds the genome.

nullcline: In dynamical systems, the set of points on the phase plane where trajectories will either go straight up-down or straight left-right.
numerator: The top of a fraction, e.g., the 2 in $\frac{2}{3}$.
numerical integration: Approximating the solution of an integral equation using numbers instead of infinitesimals.
object-oriented language: In computer science, a computer language that defines types of objects to be manipulated in consistent ways. For example, drawing programs use object-oriented commands such as rotate, which will operate on many different graphical objects.
occipital cortex: The area of cortex at the back of the brain that is involved in visual processing.
occipital lobe: Posterior area of cortex housing visual cortex.
octal: Base 8.
Ohm's law: Voltage equals current times resistance ($V = IR$). Written in terms of conductance: current equals conductance times voltage ($I = gV$).
olfaction: Sense of smell.
oligodendroglia: A type of glial cell that provides axonal insulation by wrapping myelin-filled processes around an axon. Also called oligodendrocytes.
ommatidium: A single unit-eye of a compound eye.
one's complement: In computer science, a method of doing subtraction, see Chap. 4, Section 4.6.
operating system: The lowest-level software on a computer, used to integrate control of disk drives, keyboard, and monitor with the CPU and peripherals.
operator: In mathematics, a function. In computer science, a symbol that takes one or more arguments and produces an output, e.g., + is an arithmetic operator that takes two arguments; rotate is a common operator for a graphical language that takes one argument.
optic chiasm: In neuroanatomy, the location where the optic nerves come together and partially cross, forming the optic tracts.
optic nerve: Bundle of axons running from the retina to the optic chiasm. The optic nerve is not really a nerve because the retina is actually part of the central nervous system and not a peripheral receptor.
optic tract: Bundle of axons running from the optic chiasm to the lateral geniculate nucleus.
OR: A Boolean operator. Symbol: $\vee$. See Chap. 16, Section 16.3.
orbit: In dynamical systems, a trajectory whereby one object (or state variable) follows a closed path back to its original location, generally around another object in the system.
ordinary differential equation: A differential equation with all derivatives with respect to a single variable (e.g., time). Contrasted with partial differential equation. Abbrev.: ODE.
organelle: In cell biology, cell organs including nucleus, Golgi apparatus, mitochondria, etc.

orthodromic: In electrophysiology, signal flow in the normal direction of signaling. Contrasted with antidromic.
orthogonal: Lying or intersecting at right angles. In linear algebra, two vectors whose dot product is zero.
oscillation: A signal or process that shows repeating activity.
outer product: A vector operation that produces a matrix out of a row vector times a column vector. See Chap. 16, Section 16.4.
outward current: In electrophysiology, current that flows through a conductor from cytoplasm to extracellular fluid. Outward conductance current produces inward capacitative current, which will hyperpolarize.
outward rectifier: In electrophysiology, a channel that mediates outward current with membrane depolarization.
overflow: In computer science, loss of bits that exceed word size after performing an arithmetic operation.
overgeneralize: In learning theory, a tendency to apply learned patterns too broadly and thereby make errors based on the use of a rule in a place where there is an exception, e.g., using "haved" instead of "had."
pair-wise: Combinations of n objects taken two at a time. Usually refers to performing some operation on each possible couple of a list of objects; e.g., pair-wise sums of 8,7,2,1 are 8+7, 8+2, 8+1, 7+2, 7+1, 2+1.
Papez circuit: The set of interconnected structures of the limbic system believed (by Papez) to contribute to emotional responses. Recent data suggest that this circuitry is more important for memory than for emotion.
parabola: A curve with a square of the independent variable: $f(x) = ax^2 + b$.
paradigm shift: Philosophy of science concept of Thomas Kuhn that suggests that science does not progress by small incremental steps but instead enjoys periodic revolutions where old theories are overthrown and new theories are erected in their place.
Parallel Data Processor: Digital Equipment Corporation's name for a line of its computers. Abbrev.: PDP. Includes the PDP-8 and PDP-11.
parallel distributed processing: In artificial neural network theory, the use of multiple simple units working at the same time (in parallel) with information that is spread out (distributed) over the units. Abbrev.: PDP.
parallel fibers: Axon projection from granule cells to Purkinje cells in cerebellum. See Fig. 15.3.
parallel processing: Calculating by using multiple data-flow paths simultaneously. Opposed to serial processing.
parameter: In dynamical systems, a fixed value that defines the system. Contrasted with state variables.
parameter variation: Altering the parameters that define a dynamical systems. Typically done in order to make sure that the system will behave similarly despite minor changes.
parameterization: Fitting data with functions and specific values.

parasitic capacitance: Effect that reduces current that is being transmitted using an oscillatory signal through capacitative coupling to ground.
parietal lobe: Large middle region of cortex between occipital and frontal lobes.
parity: In data transmission, descriptor as to whether previous set of bits had even or odd number of ones. A parity bit is used to check that no bits were reversed during transmission.
Parkinson disease: A degenerative brain disease involving loss of dopamine-producing cells in the substantia nigra. Produces a movement disorder causing tremor and reduced movement.
pars: Latin word for "part"; e.g., *substantia nigra, pars compacta* means "the compact [meaning tightly packed cells] part of the black stuff."
parse: In computer science, the task of interpreting lines of a program to separate out arguments, operators, comments, subroutine names, etc. In general, any effort to interpret a signal or language by breaking it up into parts.
partial differential equation: A differential equation where different derivatives are taken with respect to different variables (e.g., one derivative with respect to time and another with respect to distance). Abbrev.: PDE.
particles: In Hodgkin-Huxley theory, the name given to the state variables determining ion channel conductance in the Hodgkin-Huxley equations. These were hypothesized to be little particles that blocked ion flow.
Pascal: A computer language.
passive channels: Ion channels that maintain same permeability (conductance) at all times. Contrasted with active channels.
passive membrane: Membrane lacking voltage-sensitive conductances.
patch clamp: In electrophysiology, a technique for making direct measurements of membrane by attaching an electrode tightly to the membrane and pulling a small circle of membrane off of the cell on the electrode.
pattern completion: Ability of artificial neural network or natural systems to fill in missing pieces of an incomplete input or memory.
PDP: Abbrev. for both parallel distributed processing (concept in neural networks) and Parallel Data Processor (a model of computer)
peptide: A short sequence of amino acids.
perceptron: A unit of the single-layer linear neural networks developed in the early 1960s.
perforant path: A pathway that projects from entorhinal cortex to hippocampus.
perikaryon: Synonym for cell soma.
period: The duration of a single cycle of a repeating wave.
peripheral nervous system: The nervous system that lies outside of the brain and spine. Abbrev.: PNS.
peripherals: In computer science, additional pieces stuck on to the computer such as the monitor and keyboard.

Perl: A computer utility language.
PET: See positron emission tomography.
phase: In signals and systems theory, the delay of an oscillatory signal compared to a fixed signal of the same frequency. Measured in degrees or radians.
phase plane: A two-dimensional graph illustrating the dynamics of a system by mapping one state variable against another as they change in time. This plane is a two-dimensional section through state space.
phase space: n-dimensional space where each axis represents one of the n state variables of an n-dimensional dynamical system. The dynamics of the system could be fully mapped as a curve in this space.
phase transition: In chemistry and physics, change from one state of matter to another: e.g., liquid to gas. In neural network, analogous state transition involving major changes in activity.
phasic: In signals and systems theory, a transient response to a signal.
phoneme: The minimum length sound unit that carries information in a natural language. See grapheme.
phospholipid: Biological soap that is the main constituent of lipid bilayers. They form with a charged phosphate (phosphorus and oxygen) head and a long-chain hydrocarbon (the fatty part).
photoreceptor: An electronic device or biological cell that can detect light or other electromagnetic radiation (photons).
phrenology: A discredited field of study that used skull shape to determine mental abilities.
physiology: The study of functional interactions in biology. Contrasted with anatomy.
pi: π: ratio of circumference to diameter of a circle.
piece-wise linear function: A function made of joined line segments.
pixel: Picture element. A single spot in a bit-mapped picture.
plasticity: Changeability. Used to refer to synapses or neural assemblies that are altered by activity or environmental events.
point attractor: In dynamical system theory, a final steady state that is a common target for some set of initial conditions.
point neuron: A model neuron without physical extent — represented as a single isopotential compartment.
pointer: In computer science, a numerical address that indicates the location of a data structure.
Poisson process: A probability model for events that occur rarely without reference to past activity (memoryless). Can be used to describe neural spiking as well as emission of radioactive particles and arrival of customers at a service counter.
pons: The middle part of the brainstem.
pore: Another name for ion channel.

port: To move or translate a program or algorithm from one type of machine to another. For example, a new word processor will typically be ported from one machine architecture or operating system to another.
positive feedback: A signal communicated back to a prior system component that increases the output or activity of that component.
positive true: In computer science, the use of high voltage on the transistor to represent binary 1 or true.
positron emission tomography: A physiological imaging technique that localizes radioactive substances that release positrons. Abbrev.: PET.
posterior: Toward the rear. Depending on the position in the central nervous system, this corresponds to either caudal or dorsal.
postsynaptic: The receiving side of a synapse, more generally used to denote the follower side of any weight connection in a neural network.
postsynaptic potential: Change in voltage in the postsynaptic membrane caused by synaptic activation.
potential: Electrical potential, voltage.
potential energy: Energy stored in a form from which it can later be released, e.g., height in a gravitation field, a charge in an electric field.
presynaptic: The transmitting side of a synapse, more generally used to denote the side of a weight connection that provides the signal.
primary sensory cortex: Area of cortex that receives sensory information directly from thalamus or other noncortical area.
primary visual cortex: The primary sensory area for vision.
primates: Monkeys, apes, and people.
procedural memory: Human memory for how to perform a procedure. This includes motor tasks and memories for rule-based activities such as games.
program: In computer science, a sequence of steps written in a computer language that is self-contained and produces some desired data or other output.
program counter: In computer architecture, central processing unit storage that used to provide a pointer to a word in memory to be executed.
projection cell: A neuron with axons that leave the area. Pyramidal cells are the major projection cells of cortex.
projective field: Locations or movements that are produced by activity of a particular motor neuron.
prokaryote: In biology, an organism that has a cells with no nucleus, such as bacteria and some other single-celled organisms.
proprioception: The sensation of movement in muscles and joints.
prosopagnosia: Neurological condition in which faces of individual people can't be recognized
protein: Chain of amino acids that provides the building blocks of living creatures. Proteins can be structural (e.g., hair) or may be enzymes that make things (e.g., beer). Protein channels provide routes for ions and chemicals to enter or leave a cell.

proximal: Closer to the center, opposite of distal.
pseudo-code: In computer science, a style of writing out an algorithm in the general form of a program without worrying about conforming to the precise syntax required by a specific programming language.
psychophysics: Research techniques that use precisely defined stimuli to quantify responses or behavior.
punch cards: An old technique of data storage that used holes in cardboard cards as input to a computer.
Purkinje cell: One of the large projecting inhibitory cells of the cerebellum.
putamen: An input area of the basal ganglia. A part of the striatum.
pyramidal cell: A characteristic neuron of the cortex that has a cell body shaped like a pyramid.
pyramidal tract: Corticospinal tract.
Q_{10}: A measure of the increase in rate of an active protein (e.g., an enzyme, voltage-sensitive channel or pump) to temperature change. The rate increase factor is equal to $Q_{10}^{\Delta \text{temperature}/10}$.
qualia: Term used by philosophers to connote the internal, subjective sense of a sensation or experience.
quantum mechanics: Description of the motion of subatomic particles. Distinct from the classical mechanics of dynamical systems.
queue: In computer science, a common data structure that stores and retrieves multiple items in a *first-in, first-out* (FIFO) sequence
quod erat demonstrandum: In mathematics, a Latin phrase used to denote the end of a proof. It means "that which was to have been proved." Abbrev.: QED.
radian: Unit of circular measure equal to $\sim$ 57.2958 degrees. There are $2 \cdot \pi$ radians in a circle.
radix point: The period "." used in number systems to indicate where the whole part of a number ends and the fractional part begins. This is called the decimal point in the base 10 system.
random-access memory: The main (primary) memory of a modern computer. Abbrev.: RAM. Can be subdivided into sRAM (static RAM), which cannot be altered by the computer and dRAM (dynamic RAM), which can be altered. Software and data are stored in dRAM before execution.
range: In mathematics, the set of possible outputs of a function. For example, the typical neural network squashing function has a range of 0 to 1. See domain.
rate coding: Frequency coding.
RC circuit: Resistor-capacitor circuit. Such circuits can be used to provide different high-pass and low-pass signal filtering. The RC circuit is also a model of the electrical properties of a passive cell membrane.
realistic neural network: Network that attempts to accurately portray neurons and neural dynamics.

receptive field: Locations or stimulation range that activates a particular neuron. For touch this could be an area of skin; for vision this would a part of the visual field or for hearing a range of frequencies.
receptor: In neuroscience, a molecule to which a chemical (ligand) can bind.
reciprocal: Giving and taking. In neuroscience, describes to-and-fro connectivity between areas or individual neurons: area A (or cell a) projects to area B (or cell b), which then projects back to A (a).
rectify: In electronics, passing current in one direction only. The diode is the main rectifying component.
recurrent collaterals: Axon collaterals that come back and synapse either on the same neuron or on the same type of neuron.
reflex: A stereotyped closed stimulus-response loop.
register: In computer science, A specialized memory location residing on the central processing unit.
reinforcement learning: An artificial neural network training technique that utilizes a general reward signal rather than a specific target for instruction.
relative refractory period: A period during which a second neuron stimulation must be increased in order to generate another action potential.
relaxation: In physics, term referring to gradual reduction in energy of a system. Used in artificial neural network theory to refer to a learning process that reduces an energy function. See annealing.
relay cell: See thalamocortical cell.
relay nucleus: One of the sensory nuclei of the thalamus that transmits information from the sensory periphery to sensory cortex.
repellor: In dynamical system theory, a point in state space that the system moves away from. Also called an unstable equilibrium.
repolarization: In neurophysiology, reversal of depolarization.
resistor: A substance that impedes the flow of electricity linearly with voltage (see Ohm's law). Symbol R, units Ohms (Ω). Resistance is the inverse of conductance: $R = 1/g$.
resting membrane potential: Voltage of neuron membrane in the absence of stimulation. Usually somewhere near -70 mV. Abbrev.: RMP.
reticular activating system: Area of brainstem that maintains arousal. Damage to this area causes coma.
reticular cell: Thin spindly cells found in various areas of the brain. The reticular cells of thalamus are particularly well studied.
retina: The network of neurons at the back of the eye that detects light and does early visual processing. The retina is part of the central nervous system.
retinal slip: Lag of the eye during pursuit.
retinotopic: In neuroscience, a mapping onto an area of brain that preserves an orderly representation of the retinal surface. Two points close

to each other in retina are also represented by nearby cells in the brain structure in question.
retrograde transmission: In neurophysiology, transmission of a chemical signal backward across a synapse.
reversal potential: The electrical potential associated with an ion channel, based on the type of ion or mix of ions passed by the channel. So called because current through such a channel will reverse direction when the membrane holding potential passes the reversal potential.
RISC: Reduced instruction set chip. In computer architecture, a central processing unit design that executes only simple instructions, requiring complex instructions to be coded in software. With growing CPU complexity, the contrast with CISC is now historical.
RNA: Ribonucleic acid. The molecule that transmits information from the gene (DNA) to produce proteins.
rod: A type of retinal photoreceptor that detects low levels of light.
ROM: Read-only memory. A form of computer memory that is written on manufacture and can only be read thereafter.
Rosetta stone: A stone fragment containing three different scripts (Greek, demotic, and hieroglyphics) that was the breakthrough document for deciphering ancient Egyptian hieroglyphics.
rostral: The head end of the body. Opposite of caudal.
saccade: A rapid eye movement used to bring the eyes around to the object of attention.
sagittal plane: A side-facing vertical plane for sectioning the body (see Fig. 3.3).
salt: Ionically bonded crystalline compound produced by combining an acid and a base. Table salt is NaCl.
saltatory conduction: The jumping of an action potential from one node of Ranvier to the next in a myelinated axon.
sampling rate: In signals and systems theory, the frequency or density of signal measurement. Interactions between frequency of sampling and the frequency of the signal can lead to illusions of recurring patterns.
scalar: A single number. Contrasted with vector or matrix.
scale model: A reduced-size model that preserves relative dimensions.
scantron form: Form used for data entry that can be scanned by a computer. Typically these have little circles that must be blackened with a pencil to indicate a choice on an exam or questionnaire.
Schaeffer collaterals: Pathway in hippocampus from CA3 to CA1.
schizophrenia: A psychiatric disease characterized by disconnection from reality, frequently with delusions and auditory hallucinations.
scientific notation: Standard for representation of large and small numbers in science by using a mantissa and a base-10 exponent, e.g., 10,000 would be represented as $1 \cdot 10^4$. See Chap. 16, Section 16.2.
second derivative: The derivative of a derivative. For example, velocity is the first derivative of position with respect to time. The first derivative

of velocity with respect to time is acceleration. Acceleration is the second derivative of position with respect to time.
second messenger: An intracellular chemical that transmits a message from a postsynaptic release location to other locations in the cell.
segmentation error: In computer science, an error due to the attempt to read or write to computer memory that is not currently accessible.
semantic memory: Human memory for general facts.
semicircular canal: An organ of inertial sensation in the ear. There are three semicircular canals placed orthogonally in order to detect rotations in any of the three dimensions of space.
sensitivity to initial conditions: In dynamical systems, an attribute of chaotic systems. If a chaotic system is started at a slightly different point (the initial condition), its subsequent behavior will be completely different.
sensorimotor: Denotes the coordinating of incoming sensory signals with movement. This includes the modification of movement based on kinesthetic feedback.
sensory channel: Psychophysics term for neural pathways communicating a particular type of sensation.
serotonin: A neurotransmitter. Prozac and similar drugs work by blocking serotonin reuptake, thereby increasing amount of transmitter at synapses.
set bit: Binary bit value corresponding to 1 or True.
shunting inhibition: Inhibition caused by increase in conductance without hyperpolarization.
SI units: Abbrev. for the Système International d'Unités (International System of Units). This is the standard system of scientific units.
siemens: The unit of conductance. Equal to inverse ohms. Abbrev.: S. Also called mho.
sigmoid function: An equation giving an S-shaped curve (e.g., the logistic function). Utilized as a squashing function in neural networks.
sign magnitude: In computer science, a method of representing negative numbers by using a single bit as a code to indicate negative. See Chap. 4, Section 4.6.
signal integration: In signals and systems theory, the adding up of signals by a system.
signal transduction: Conversion of a physical signal in the environment into a chemical or electrical form for neural processing. The first step in a sensory system.
signals and systems: A field that studies generation, transmission, and processing of time-dependent signals.
silent synapses: Anatomically defined synapses without physiological effect.
silicon: A chemical element used in electronic circuits and devices. Used as a short-hand to refer to computers and breast implants. Abbrev.: Si.
simulation: A numerical imitation of the behavior of a physical system on a computer.

single-unit recording: In electrophysiology, extracellular recording of action potentials from one neuron. Contrasted with field potential or multiunit recording.
sink: In electronics, a destination for current. Opposed to source. Ground is the ultimate sink.
slow potential theory: Transduction of firing rate through temporal summation of long-lasting postsynaptic potentials. See Chap. 11, Section 11.7.
soap: A compound that has a fatty part that sticks to fat and a polar part that sticks to water.
software: In computer science, programs. Contrasted with hardware.
solute: The dissolved material in a solution.
solution: In mathematics, an answer. In biology, a liquid with something dissolved in it.
solvent: The liquid that a solute is dissolved in. Biologically, usually water.
soma: Large central area of a neuron from which extend the dendrites and axon.
somatosensory: Relating to the sense of touch.
source: In electronics, a location from which current arises. Opposed to sink.
space constant: The distance along a passive dendrite for a fixed voltage to decline by $\frac{1}{e}$ (37%) Symbol: λ.
space-time trade-off: In computer science, the need to balance a program's requirement for computer memory (space) with the amount of time the program will take to execute. In general, you can write a program that will run faster if you devote more space to it.
spatial summation: Adding up of synaptic potentials arriving at different places on a neuron.
spike: An action potential.
spike frequency: The frequency at which spikes occur. Usually between 5 and 100 Hz in most cell types.
spike initiation zone: A specialized area of soma or axon where a particularly high concentrations of sodium channels allows initial generation of an action potential.
spinal cord: The part of the central nervous system that lies in the spine.
spine: In neuroanatomy, a small thorn-like extension off a dendrite that is often the location of an excitatory synapse. Also, the backbone of vertebrate animals.
spontaneous activity: Resting firing of a neuron when not being stimulated.
spurious attractors: In memory networks, additional attractors that form during learning but do not correspond to any learned pattern.
squashing function: In neural networks, function that converts a broad domain of input values into a limited range of output values. Used to limit the range of state values for an artificial neural network unit.

stack: In computer science, a common data structure that stores and retrieves multiple items in a first-in, last-out (FILO) sequence.
state: The condition or value of a system or system component at a particular time.
state space: Phase space representing progression of unit state vector in a network.
state variable: In a dynamical system, an independent variable that changes with time.
state vector: A vector giving the current value of all states of all units in a neural network.
step function: A sharp thresholding function with discontinuity at 0. Same as Heaviside function.
step size: In numerical integration, the Δt used to advance the calculation.
stimulus: Something that can activate a sensory response.
stochastic: In mathematics, a system whose behavior is governed by random events. Contrasted with deterministic.
stomatogastric ganglion: A small ($\sim$ 30 neuron) ganglion that runs parts of the digestive and circulatory system in the lobster.
strange attractor: The type of attractor that is seen in chaotic dynamics. A chaotic system never revisits the same spot in state space. A strange attractor describes a region of state space that the trajectory will repeatedly pass through.
stria of Gennari: White stripe in primary visual cortex due to the many axons projecting from thalamus.
striate cortex: Another name for primary visual or occipital cortex, so called because it has a white-matter stripe (stria of Gennari).
striatum: The caudate and putamen, the input area of the basal ganglia.
stroke: Death of brain tissue due to loss of blood flow. Also called cerebrovascular accident (CVA).
subroutine: In computer science, a brief sequence of steps written in a computer language that is utilized by other routines to perform a data-processing task.
substantia nigra: An area of brainstem with connections to basal ganglia. Loss of dopamine cells in this area is the cause of Parkinson disease. The area is black due to accumulation of the melatonin formed from dopamine breakdown.
subthreshold: In electrophysiology, not reaching the threshold for firing an action potential.
sufficient and necessary: Science buzz phrase indicating that in a theory showing that cause A is important for producing phenomenon B, A is all you need (sufficient) to get B and that without A you won't get B (necessary). Synonymous with "if and only if."
sulcus: The cerebral hemispheres are folded; a sulcus is one of the valleys.
summation: In mathematics, adding up, symbolized by Σ. In neuroscience, the ability of neurons to combine synaptic inputs.

summed-square error: A common error function used for matching a single network output to a target: error $= \frac{1}{2}\Sigma_i(out_i - t_i)^2$. Using an error vector $\vec{e} = \vec{out} - \vec{t}$, we can express this as $\frac{1}{2}(\vec{e} \cdot \vec{e})$.
superior olive: A brainstem nucleus involved in hearing.
supervised learning: In artificial neural network learning theory, algorithms to teach a neural network by presenting targets that are to be learned.
suprachiasmatic nucleus: A nucleus of hypothalamus that contains cells mediating the "master clock" of the circadian rhythm.
sylvian fissure: The large front-pointing indentation separating the temporal from frontal and parietal lobes.
symbol processing: An artificial intelligence technique that manipulates discrete descriptors of information in a particular knowledge domain.
symbol table: In computer science, a table that lists the symbols in a computer program (e.g., names of variables, subroutines, and data structures) alongside the physical address in memory where these will be stored.
synapse: Point at which information is transmitted from one neuron to another. A connection between units in a neural network.
synaptic cleft: The region separating the presynaptic and postsynaptic membranes at a chemical synapse.
synaptic transmission: The sending of information across the synapse, usually by means of release and reception of a neurotransmitter.
synaptic triad: A synaptic complex in the thalamus involving co-localized excitatory inputs from an afferent onto dendrites of both a thalamocortical cell and a thalamic interneuron. A neighboring inhibitory dendrodendritic synapse from the interneuron to the thalamocortical cell is the third part of the triad.
synaptic vesicle: Organelles found in the presynaptic cell that store and release neurotransmitter.
synchronous updating: In artificial neural network theory, the setting of states for all units to their new values at the same time.
syncytium: A tissue of continuously connected cells that share a common cytoplasm.
synesthesia: A benign neuropsychological condition that results in crosstalk between sensory modalities, e.g., a synesthetic might experience a particular smell when seeing a certain shape or experience a color when hearing a particular melody.
syntax: In computer science, the detailed requirements for the order of commands and arguments in a computer language.
systems neuroscience: Physiological investigation pertaining to perception or behavior of an animal or person.
target: In artificial neural network theory, an explicitly provided value used as a learning goal for a neural network.

taxonomy: A set of descriptions for different manifestations of a central phenomenon. In biology, the term is used for the descriptions of species.
teacher forcing: Clamping a dynamical neural network to sequences of states in order to help it learn specific state progressions.
teleology: A point of view that assumes that natural objects are designed for a purpose. Although most scientists believe that random evolution and natural selection explain biological features, it is often easier to describe an organ or organism as if it had been designed. For example, the lens is placed where it can focus light on the retina.
temporal integration: In signals and systems theory, the adding up of signals over time. In electrophysiology, the adding up of postsynaptic potentials arriving over time at a synapse.
temporal lobe: A lobe of the brain lying below the sylvian fissure.
temporal summation: Adding together of postsynaptic potentials arriving one after another at a synapse.
terminal arbor: A bunch of axon branches at the end of an axon.
***terra incognita*:** Latin for "unknown country."
tetrodotoxin: A poison from the puffer fish that blocks sodium channels. Allegedly the poison used by the voodoos for creating zombies.
thalamic reticular nucleus: Collections of broad flat cells that wrap around the principal thalamic nuclei. Also referenced by the Latin name: *nucleus reticularis thalami*. Abbrevs.: TRN and nRT.
thalamocortical cell: The principal cells of thalamus that project to cortex. Also called relay cells.
thalamus: Central gray matter nucleus that forms entry point for sensory systems to the brain. It is also involved in sleep.
theta waves: Slow waves (4 to 7 Hz) in the electroencephalogram.
thought insertion: Placing thoughts into the head of a human or animal. This is an occasional delusional complaint in schizophrenia.
three-body problem: In dynamical systems, the dynamics of three heavenly bodies with mutual gravitational attraction (e.g., sun, earth, moon). This problem cannot generally be solved analytically. Now that these systems can be solved numerically, it has been shown that some parameter choices can lead to chaotic dynamics.
threshold: In neuroscience, the voltage needed to generate an action potential. In general, an activation level that causes a unit to go from a quiescent to an excited state.
threshold logic unit: McCulloch-Pitts neuron. Abbrev.: TLU.
time constant: In a first-order differential equation, the time it takes for a state variable to change by $1 - \frac{1}{e} \sim 63\%$.
tissue: In biology, a collection of connected cells in an organism.
tonic: A sustained response to a continuing stimulus, opposite of phasic.
torus: A geometrical shape that looks like a donut. From a point on the surface of a torus you can follow a circular path all the way around in either of two directions.

tract: A bundle of axons that run together in the central nervous system. Comparable to a nerve in peripheral nervous system.
tractable: Practicable, possible, doable.
trajectory: In mechanics, the movement of an object in a gravitational field. More generally, the progress of a state variable in a dynamical system.
transcendental functions: Nonalgebraic functions, including trigonometric functions such as sine and cosine.
transfer function: See squashing function.
transistor: An active electronic device that functions as a controllable switch.
transistor-transistor logic: In computer science, the original technique used to organize transistors to perform logical operations. This has since been superseded by the organization of many transistors using VLSI. Abbrev.: TTL.
transmembrane protein: Protein that goes through the lipid bilayer to face both extracellular and intracellular space. Such proteins can be pumps, ion channels, receptors, etc.
transpose: In linear algebra, an operation that flips a matrix so as to make the columns rows and rows columns.
tricarboxylic acid cycle: Krebs cycle.
trisynaptic pathway: Pathway in hippocampus from entorhinal cortex to dentate gyrus to CA3 to CA1.
truth table: A list of outputs for a Boolean operator.
Turing test: Test of the ability of an artificial intelligence program to be so damn intelligent that it could fool people into believing they're talking to a person.
two's complement: In computer science, a method of doing subtraction, see Chap. 4, Section 4.6.
two-point discrimination: The ability to distinguish two pins placed close together on the skin.
ultrastructural: In anatomy, refers to structures that cannot be seen with light microscopy but can be seen with electron microscopy.
unary operator: In computer science, an operator that takes one argument, For example, "–" is a unary operator that turns a number into its negative.
unit: A single element in a neural network.
unit analysis: See dimensional analysis.
Unix: An operating system popular for engineering and scientific applications.
unsupervised learning: In artificial neural network learning theory, algorithms that allow a neural network to learn patterns by exposure to data without the defining of specific tasks or targets.
update rule: In artificial neural network theory, an equation for assigning the state of a unit based on its inputs. In numerical integration, the assignments for state variables after a time step based on their prior values.

upregulate: In biology, increasing the amount or activity of an enzyme, receptor or other active protein. This is seen in the case of receptors that are deprived of their ligand by drugs that block the receptor or prevent transmitter release.
V1: Primary visual cortex.
vector: A one-dimensional array of numbers.
ventral: The belly side of the body. For the central nervous system this would be anterior in the spinal cord and inferior (down) in the brain.
ventricles: Large cavities in the brain containing cerebrospinal fluid. Also the large chambers of the heart.
vesicle: In cell biology, an organelle for chemical storage. See synaptic vesicle.
vestibulo-ocular reflex: A reflex that stabilizes the eyes in the head when the head moves to maintain vision. Abbrev.: VOR.
veto inhibition: Inhibition that completely turns off the neuron regardless of amount of excitation.
virtual experiment: Process of exploring a simulation as if it were a natural object. The complexity of many simulations makes it impossible to understand them without running such experiments.
virus: In biology, cell parasites that contain genetic material (either DNA and RNA) and proteins to get around but not to reproduce.
visual pursuit: Slow eye movements triggered by the passage of an object through the visual field. Pursuit cannot be triggered consciously — there must be something to look at.
VLSI: Very-large-scale integration. A technology that allows manufacturer of silicon chips containing millions of transistors.
voltage: The work needed to move a unit positive charge from a reference (typically ground) to a particular location. Symbol: V, units: volts. Also called potential or electrical potential.
voltage clamp: Maintenance of a constant voltage in a neuron by injecting or withdrawing current to compensate for membrane currents.
voltage-gated channel: Voltage-sensitive channel.
voltage-sensitive channel: An ion channel that opens and closes in response to changes in membrane potential.
volume transmission: In neurophysiology, transmitters that spread broadly and thereby signal to multiple neurons rather than point to point at a synapse.
VOR: See vestibulo-ocular reflex.
voxel: Volume element. A single spot in an object being imaged with a multidimensional imaging technique such as CT or MRI.
Weber-Fechner law: Psychophysical finding that perception of a stimulus is typically proportionate to the logarithm of stimulus intensity in most sensory systems: Perception = constant · log(stimulus).
weight: Numerical value representing strength of a connection between units in a neural network.

weight matrix: See connection matrix.
weight space: Phase space representing progression of weight matrix in a network during learning.
weight vector: A row of a network weight matrix representing convergence onto a single unit.
Wernicke's area: An area of temporal lobe involved in processing of language.
white matter: In neuroscience, brain areas that appear white due to preponderance of myelinated axons.
white noise: In signals and systems theory, random perturbations that occur equally at all frequencies.
whole-cell patch: In electrophysiology, a technique for making measurements inside of a cell by attaching an electrode tightly to the outside of the membrane and then blowing out a hole.
Widrow-Hoff rule: See delta rule.
winner-take-all: In neural networks, dynamics that permit the activity of a single unit to emerge and dominate the network.
word: In computer science, a unit of memory. Differs in size in different computer models.
wraparound: Organizing a neural network so that the units at one end or edge connect with units at the other end or edge.
wulst: An area of avian brain that is involved in learning and memory.
XOR: A Boolean operator. See Chap. 16, Section 16.3.
Y2K problem: In computer science, Y2K means Year 2000, using Y for year and K for kilo. Refers to the problem that arose due to the use of a two-digit representation for calendar year in programs: "00" could be either 2000 or 1900.
zoology: The study of similarities and differences among animal species.

Index